# Studies in Applied Philosophy, Epistemology and Rational Ethics

## Volume 76

**Studies in Applied Philosophy, Epistemology and Rational Ethics (SAPERE)** publishes new developments and advances in all the fields of philosophy, epistemology, and ethics, bringing them together with a cluster of scientific disciplines and technological outcomes: ranging from computer science to life sciences, from economics, law, and education to engineering, logic, and mathematics, from medicine to physics, human sciences, and politics. The series aims at covering all the challenging philosophical and ethical themes of contemporary society, making them appropriately applicable to contemporary theoretical and practical problems, impasses, controversies, and conflicts.

**Book proposals can be sent to the Series Editor, Prof. Lorenzo Magnani** at lmagnani@unipv.it. Please include: A short synopsis of the book or the introduction chapter A Table of Contents The CV of the lead author/editor(s) For more information, please contact the Editor-in-Chief at lmagnani@unipv.it

Our scientific and technological era has offered "new" topics to all areas of philosophy and ethics – for instance concerning scientific rationality, creativity, human and artificial intelligence, social and folk epistemology, ordinary reasoning, cognitive niches and cultural evolution, ecological crisis, ecologically situated rationality, consciousness, freedom and responsibility, human identity and uniqueness, cooperation, altruism, intersubjectivity and empathy, spirituality, violence. The impact of such topics has been mainly undermined by contemporary cultural settings, whereas they should increase the demand of interdisciplinary applied knowledge and fresh and original understanding. In turn, traditional philosophical and ethical themes have been profoundly affected and transformed as well: they should be further examined as embedded and applied within their scientific and technological environments so to update their received and often old-fashioned disciplinary treatment and appeal. Applying philosophy individuates therefore a new research commitment for the 21st century, focused on the main problems of recent methodological, logical, epistemological, and cognitive aspects of modeling activities employed both in intellectual and scientific discovery, and in technological innovation, including the computational tools intertwined with such practices, to understand them in a wide and integrated perspective. **Studies in Applied Philosophy, Epistemology and Rational Ethics** means to demonstrate the contemporary practical relevance of this novel philosophical approach and thus to provide a home for monographs, lecture notes, selected contributions from specialized conferences and workshops as well as selected Ph.D. theses. The series welcomes contributions from philosophers as well as from scientists, engineers, and intellectuals interested in showing how applying philosophy can increase knowledge about our current world. Indexed by SCOPUS, zbMATH, SCImago, DBLP. All books published in the series are submitted for consideration in Web of Science.

Alger Sans Pinillos · Vicent Costa · Jordi Vallverdú
Editors

# SecondDeath

## Experiences of Death Across Technologies

Springer

*Editors*
Alger Sans Pinillos
Barcelona Supercomputing Center (BSC)
Barcelona, Spain

Jordi Vallverdú
Department of Philosophy
Universitat Autònoma de Barcelona
(ICREA-UAB)
Bellaterra, Spain

Vicent Costa
Artificial Intelligence Research Institute
(IIIA-CSIC)
Cerdanyola del Vallès, Spain

ISSN 2192-6255 ISSN 2192-6263 (electronic)
Studies in Applied Philosophy, Epistemology and Rational Ethics
ISBN 978-3-031-98807-3 ISBN 978-3-031-98808-0 (eBook)
https://doi.org/10.1007/978-3-031-98808-0



This Springer imprint is published by the registered company Springer Nature Switzerland AG
The registered company address is: Gewerbestrasse 11, 6330 Cham, Switzerland

# Contents

**Introduction: Rethinking Death in the Digital Age—Delving into a Cross-Disciplinary Inquiry** .......... 1
Alger Sans Pinillos, Jordi Vallverdú, and Vicent Costa

**Philosophical and Cultural Foundations**

**50 Ways to Kill a Robot: Death and Differentiation** .......... 11
Jennifer Robertson

**On Androids, Mortality, and Death-Images: "Blade Runner" and "Westworld"** .......... 25
Susana Viegas

**Do Digital Immortals Dream of Offline Afterlives?** .......... 45
Joseph Ulatowski

**Are We Close to the Death of "Homo Faber"? Information Technology, Robots, AI, and the Urgent Need of a New Renaissance** .... 61
Lorenzo Magnani

**Synthetic Death: Bioinspired Mechanisms for AI Resilience and Renewal** .......... 77
Jordi Vallverdú and Max Talanov

**Digital Eschatology: Death in the Digital Age** .......... 103
Aníbal M. Astobiza

**Technological and Ethical Implications**

**Reflections on the use of Artificial Intelligence for Weapons Applications** .......... 119
Maria Vanina Martinez and Ricardo O. Rodriguez

**Integrating a Disability Perspective into Lethal Autonomous Weapons Systems Analysis: First Steps** ... 137
Vicent Costa

**Symbolic Death and Dual-Use Dilemmas** ... 153
Alger Sans Pinillos and Jordi Vallverdú

**The Neglect of Qualia and Consciousness in AI Alignment Research** ... 175
Soenke Ziesche and Roman V. Yampolskiy

**Death, Grief, and Technology**

**Do Robots Die?** ... 191
Ana Cuevas-Badallo

**Deadbots, Griefbots, and Postmortem Technologies: A Computational Challenge to Mortality** ... 209
Lionel Obadia

**Artificial Afterlife: Philosophical Reflections on Griefbots** ... 225
Giacomo Zanotti and Daniele Chiffi

**As I Play Dying Roles and Possibilities of Death in Video Games** ... 237
Alfonso García-Lapeña and David Ferragut

**Death in the Cybernetic Era: AI, Virtual Agents, and Selfless Selves** ... 263
Shisei Tei

**Future and Transformative Perspectives**

**An Interview with Neil Harbisson: Life, Death, and the Art of Being a Cyborg** ... 289
Jordi Vallverdú

**An Interview with Kevin Warwick: Cyborg Identity, Techno-death, and the Future of Being Human** ... 293
Jordi Vallverdú

# Introduction: Rethinking Death in the Digital Age—Delving into a Cross-Disciplinary Inquiry

**Alger Sans Pinillos, Jordi Vallverdú, and Vicent Costa**

> *There is no such thing as a natural death: nothing that happens to a man is ever natural, since his presence calls the world into question. All men must die: but for every man his death is an accident and, even if he knows it and consents to it, an unjustifiable violation* [2].

**Abstract** In the current digital age, the proliferation of connected technologies profoundly redefines our interactions, social structures, and perceptions of fundamental concepts such as death. Throughout history, awareness and reflection on mortality have profoundly shaped human social and cognitive development, extending beyond humans to observable behaviors in animals, bacteria, and even plants. In this way, death is both an intrinsic biological reality, tied to entropy and renewal, and a profound subject of inquiry into existence, consciousness, and identity. Thus, this volume approaches death not as a singular entity but as a multifaceted phenomenon, an enigma that evokes both horror and hope. Furthermore, recent advances in robotics and artificial intelligence introduce unprecedented dimensions to this discourse, with a special emphasis on the complex ethical and philosophical challenges emerging as these technologies take on traditionally human roles. With all this in mind, this volume explores questions about the ontological status of artificial entities, their potential self-awareness, and whether their termination constitutes an authentic form of death. The essays collected here represent a pioneering interdisciplinary exploration that brings together philosophical, cultural, ethical, and technological perspectives. As the first work of its kind, the book intends to enrich

A. Sans Pinillos (✉)
Barcelona Supercomputing Center (BSC), Barcelona, Spain
e-mail: alger.sanspinillos@bsc.es

J. Vallverdú
Philosophy Department, Universitat Autònoma de Barcelona (ICREA-UAB), Bellaterra, Spain
e-mail: jordi.vallverdu@uab.cat

V. Costa
Artificial Intelligence Research Institute (IIIA-CSIC), Cerdanyola del Vallès, Spain
e-mail: vicent@iiia.csic.es

 
A. Sans Pinillos et al. (eds.), *SecondDeath*, Studies in Applied Philosophy, Epistemology and Rational Ethics 76, https://doi.org/10.1007/978-3-031-98808-0_1

academic discourse and foster critical reflection on the evolving nature of death in a technology-mediated society.

**Keywords** Artificial intelligence · Philosophy · Autonomous weapons · Cybernetics · Death · Immortality · Griefbots · Identity · Robotics

## 1 Introduction: Rethinking Death in the Digital Age—Delving into a Cross-Disciplinary Inquiry

In the current digital era, characterized by the omnipresence of connected technologies that reshape our forms of communication, production, and social organization, the notion of death poses unprecedented challenges. Throughout this book, death will be approached not as a monolithic concept but as a phenomenon, a process, and a notion that can simultaneously present itself as an enigma, a horror, and a source of hope, thus manifesting its multiple and changing characters throughout history.

Indeed, *death* is one of the most reactive phenomena in existence and a defining element of life extending to animals, ecosystems, and, increasingly, a wide range of real and imagined entities. In turn, it has also been the engine of reflections on the continuity of existence or, in other words, of an after-death life detached from death itself. "*To be or not to be, that is the question*" [9] (Act 3, Scene 1) about being and its relation to death. And this remains the case.

The awareness of death likely arose before the emergence of Homo sapiens [8, 6], serving as a key cognitive factor in social development and organization. Distinctive death-related behaviors can be observed in bacteria, plants, and, of course, animals [1, 4, 5]. Among: Viruses, while not alive in the classical sense, undergo well-defined inactivation and decay processes once their structural integrity is compromised [10]. Among these, humans live with a conception of death that is far more present throughout our lives thanks to the possibilities that rationality affords us: from the act of thought itself to its artistic and narrative expressions, death is present in our reflections on existence, doubts, desires, and wills. Indeed, it appears as an enigma, a horror, and a source of hope. Accordingly, poets, intellectuals, and philosophers have wrestled with the meaning of death over the millennia precisely because it is a phenomenon, ergo a concept, whose meaning significance shapes the urgency and necessities of life.

We find this theme in the poetry and archaic thought of civilizations worldwide, as well as in their philosophies, religions, and constructions of history. In all these, death is a central aspect for understanding life—both as it was, is, and may be in other possible existences, which, in turn, would give meaning to our mortal condition. In short, while rationality distinguishes us from other beings, it is death that makes us equal to them; therefore, our contemplation of death truly defines us as human beings.

Today, we understand that the natural phenomenon of death is a process linked to entropy, decay, and decomposition due to the cessation of homeostasis. However, it is also a process of renewal and life, as seen in mechanisms such as programmed cell death (apoptosis), which is essential for the development and maintenance of multicellular organisms [3, 7].

With the rise of robotics hand in hand with advances in artificial intelligence (AI), questions about death have extended into domains about which we still have even more unknowns: Can AI be considered a living entity? Is its cessation a form of death, or does it require self-awareness for it to be considered death? Is this self-awareness linked to society and culture, or can we expect a new way of understanding death that is far from the way we organic beings conceive it?

Technological advancements strengthened by AI and robotics are giving rise to unprecedented scenarios in which these technologies assume roles traditionally performed by humans, thereby presenting complex ethical challenges of pressing and significant importance. Should AI systems perform death? Certainly, the death of artificial agents, and the potential for them to cause death, presents not only technological challenges but, above all, philosophical ones that engage us ethically and ontologically. Even on the hypothetical assumption that robots can die, sense death, and have some awareness of their existence, will this imply that we humans will understand this near phenomenon as our mortality and that of all other organic beings? Will the death of a robot be merely symbolic death, or will it constitute a synthetic phenomenon? When we are not talking about an experiment in a safe environment but about technology already involved in our lives and deaths, then asking ourselves about the ontological status of these beings is a vital question.

Hence, when we, the editors of this book, first began discussing the topics, we realized that it would take a lot of work to accomplish. There is little established community or scholarly discourse on this intersection, and much of the existing literature only tangentially touches on the issues we sought to explore. Yet, we were fortunate to find a network of researchers worldwide eager to engage with this subject. Their work has made it possible for this book—on a topic that is both urgent and underexplored—to finally come to light. This marks a significant moment in scholarly inquiry, as we believe this is the first work to address the concrete issues surrounding death and robots from multiple disciplines. *SecondDeath. Experiences of Death Across Technologies* is the first book to confront these profound questions, exploring the intersection of death and technology across philosophical, cultural, and practical dimensions. The essays within are organized into four parts, each addressing different facets of how death interacts with robotics, AI, and the broader implications of our technological age. Ranging from philosophical explorations to ethical dilemmas and cultural reinterpretations, the chapters weave a tapestry of insights into how technology is redefining our understanding of mortality.

## 2 Part I: Philosophical and Cultural Foundations

The journey begins with an exploration of the philosophical and cultural underpinnings of death in the context of robotics and AI.

Jennifer Robertson, Professor Emerita at the University of Michigan (Ann Arbor, MI, USA), and Affiliate Professor at Tokyo College, University of Tokyo Institutes for Advanced Study (東京カレッジ, 大学国際高等研究所, 日本国/ Japan), and the University of Washington (Seattle, WA, USA), opens the discussion with "50 Ways to Kill a Robot: Death and Differentiation" probing the ethical quandaries of robots endowed with lethal autonomy. Who bears responsibility for these machines' decisions, and what does this say about the ethics of mortality in the age of robotics?

Susana Viegas, Assistant Professor at the Department of Communication Sciences of NOVA University Lisbon (*Universidade NOVA de Lisboa*, Portugal) and PI of the ERC CoG project FILM AND DEATH, delves into cultural narratives with "On Androids, Mortality, and Death-Images: Blade Runner and Westworld." Through the narratives mentioned in her chapter title, she examines the portrayal of artificial life and death, connecting these fictional narratives to real-world ethical dilemmas about creating beings capable of mortality.

In "Do Digital Immortals Dream of Offline Afterlives?," Joseph Ulatowski, Assistant Vice-Chancellor Sustainability, Senior Lecturer in Philosophy, and Director of the Experimental Philosophy Research Unit at Te Whare Wānanga o Waikato The University of Waikato (Aotearoa New Zealand), questions the continuity of identity through digital means. Can preserving human consciousness in digital formats offer genuine immortality, or is it merely an illusion of survival? He challenges us to consider whether immortality equates to eternal life.

In "Are We Close to the Death of 'Homo Faber'?" Lorenzo Magnani, a Full Professor in the Philosophy section of the Department of Humanities at the University of Pavia (*Università degli Studi di Pavia*, Italia/Italy), critically examines how the computational capacities omnipresent in everyday objects challenge human autonomy and creativity, inviting us to reconsider the role of technology in human life in order to preserve the fundamental humanist principle.

Jordi Vallverdú, ICREA Research Professor in the Department of Philosophy at the Autonomous University of Barcelona (*Universitat Autònoma de Barcelona*, Catalunya/Catalonia), and Max Talanov, a Researcher in AI resilience mechanisms affiliated with both the University of Messina (*Università degli Studi di Messina*, Italia/Italy) and the Institute for Artificial Intelligence of Serbia (*Institut za veštačku inteligenciju Srbije*, Srbije, Србија/Serbia), present "Synthetic Death: Bioinspired Mechanisms for AI Resilience and Renewal." Drawing parallels with biological processes, they propose models for AI systems to experience death and renewal, suggesting that synthetic failure can serve as an opportunity for evolution and resilience.

Finally, the theme of technological eschatology continues with Aníbal M. Astobiza's "Digital Eschatology: Death in the Digital Age." Astobiza, a Postdoctoral

Researcher in the Department of Philosophy I at the University of Granada (*Universidad de Granada*, España/Spain), connects traditional ideas of the afterlife with contemporary practices, illustrating how the digital age reshapes our rituals and beliefs about mortality.

## 3 Part II: Technological and Ethical Implications

Moving from the theoretical to the practical, this part examines the technological and ethical challenges that arise as robotics and AI increasingly assume roles traditionally performed by humans.

Maria Vanina Martinez, a Tenured Scientist at the Artificial Intelligence Research Institute of the Spanish National Research Council (IIIA-CSIC) and current expert advisor for the Global Commission on Responsible Artificial Intelligence in the Military Domain, and Ricardo O. Rodriguez, a professor of computer science department at the University of Buenos Aires (*Universidad de Buenos Aires*, Argentina/Argentine) and a member of Computer Science Research Institute, UBA-CONICET, analyze autonomous weapons in "Reflections on the use of Artificial Intelligence for Weapons Applications." They explore key concepts related to the design of autonomous weapons and outline political, ethical, and technical considerations for guiding effective action.

In "Integrating a Disability Perspective into Lethal Autonomous Weapons Systems Analysis: First Steps", Vicent Costa, a Tenured Scientist at the IIIA-CSIC and a vocal member of the Board of Directors of the Catalan Association for Artificial Intelligence (ACIA), reframes debates about lethal autonomous weapons systems by incorporating insights from disability studies. This perspective challenges conventional thinking and suggests new strategies for designing such systems.

Alger Sans Pinillos, an Established Researcher in digital ethics for dual-use technologies at the Barcelona Supercomputing Center-Centro Nacional de Supercomputación (BSC-CNS), and Jordi Vallverdú tackle the symbolic dimensions of mortality in "Symbolic Death and Dual-Use Dilemmas." They critique the dual-purpose nature of robotic technologies and the ethical complexities of embedding representations of death in systems designed for both civilian and military uses.

Soenke Ziesche, an independent researcher on AI ethics and Human–computer Interaction, and Roman V. Yampolskiy, Associate Professor in the J. B. Speed School of Engineering at the University of Louisville (Kentucky, USA), close the section with "The Neglect of Qualia and Consciousness in AI Alignment Research." Their chapter underscores the risks of ignoring subjective experience in AI development, arguing that this oversight undermines efforts to align AI with human values.

## 4 Part III: Death, Grief, and Technology

Part III delves into the emotional and cultural dimensions of death, mediated by technology.

Ana Cuevas, a Full Professor in the Department of Philosophy, Logic, and Aesthetics at the University of Salamanca (*Universidad de Salamanca*, España/Spain), opens her chapter with the provocative question "Do Robots Die?" which examines how society ascribes symbolic meaning to technological failures, framing robots' "death" as a phenomenon with cultural and ethical implications.

Lionel Obadia, Full Professor of Anthropology at Lumière University Lyon 2 (*Université Lyon 2*, France), explores the role of griefbots in "Deadbots, Griefbots, and Postmortem Technologies: A Computational Challenge to Mortality." According to this author, these technologies, designed to simulate interactions with deceased individuals, challenge traditional mourning practices. Are they helpful tools for processing loss, or do they hinder emotional closure?

Giacomo Zanotti, Postdoctoral Researcher in the Department of Electronics, Information and Bioengineering, and Daniele Chiffi, Associate Professor at the Department of Architecture and Urban Studies, both at the Polytechnic University of Milan (*Politecnico di Milano*, Italia/Italy), continue this discussion in "Artificial afterlife: philosophical reflections on griefbots", delving into the psychological and ethical complexities of maintaining relationships with digital replicas of the deceased.

The narrative shifts to interactive media with Alfonso García-Lapeña, an IRF Postdoctoral Researcher in the Department of Philosophy at the University of Iceland (*Háskóli Íslands*), and David Ferragut, a PhD researcher at the Department of Art History and Musicology at the Autonomous University of Barcelona. In "As I Play Dying: Roles and Possibilities of Death in Video Games", by analyzing death as a narrative device and gameplay mechanic, the authors connect the medium to broader cultural and philosophical understandings of mortality.

Shisei Tei, Professor at the Department of Psychiatry, Graduate School of Medicine at Kyoto University (京都大学, 日本国/Japan), concludes the section with "Death in the Cybernetic Era: AI, Virtual Agents, and Selfless Selves." Reflecting on how AI redefines concepts of death and selfhood, Tei emphasizes the need for balanced emotional and rational perspectives when navigating these transformations.

## 5 Part IV: Future and Transformative Perspectives

The concluding section of the volume investigates the transformative implications of cybernetic embodiments for conceptualizations of identity, consciousness, and mortality, employing two insightful interviews conducted by Jordi Vallverdú.

The first dialogue features Neil Harbisson—artist, activist, and co-founder of the Cyborg Foundation—who is the world's first person legally recognized as a cyborg. Through sensory extension via an implanted antenna, he challenges conventional boundaries between humans and machines, offering a radically embodied perspective on post-biological identity.

The second dialogue features Kevin Warwick, Emeritus Professor of Cybernetics at Coventry University (England), who is renowned for pioneering neural interface experiments. His work redefines the relationship between brain, machine, and autonomy, raising profound ethical and ontological questions about life, death, and the future of integrated intelligence.

## 6 Conclusion

*SecondDeath. Experiences of Death Across Technologies* invites readers to reflect on how technology reshapes our understanding of one of life's most fundamental experiences. By bridging philosophical, ethical, cultural, and technological perspectives, the book reveals the profound ways in which robotics and AI challenge and expand our conceptions of mortality. Whether pondering the symbolic death of robots, the ethical dilemmas of autonomous systems, or the digital transformation of grief and the afterlife, the chapters collectively illuminate the complex interplay between life, death, and technology in the modern age.

Indeed, death's enduring power lies in its ability to provoke deep reflections about life itself. Across cultures, philosophies, and technologies, contemplating mortality challenges us to confront the fragility, meaning, and purpose of existence. By examining death through the lens of robotics and AI, this book does more than just redefine the boundaries of mortality; it invites us to reimagine life. The robot's symbolic demise mirrors our fears and hopes, prompting questions about agency, identity, and continuity. At the same time, AI's role in grief and remembrance reshapes how we connect with others and ourselves, paving the way for new forms of empathy and ethical responsibility.

Ultimately, death's paradox—its finality and generative force—compels us to rethink what it means to be alive, continuously. In the intricate dance between human creativity and technological innovation, death becomes a lens through which to explore resilience, renewal, and the profound interdependence of existence. By confronting death, we redefine life, enriching our understanding of its complexity and beauty in an age of unprecedented change.

**Funding** Alger Sans Pinillos was supported by the "Generación D" initiative, Red.es, Ministerio para la Transformación Digital y de la Función Pública, for talent attraction (C005/24-ED CV1), funded by the European Union's NextGenerationEU program through the PRTR. Jordi Vallverdú was supported by the ICREA 2019 and PAL-GAIT (PID2023-148336NB-100) grants. Vicent Costa was supported by the project PID2022-139835NB-C21, and the research group 2021-SGR-00517. These funding sources supported the research and collaborative efforts that made the production of this book possible.

**Ethics Declarations** Funded by the European Union—Next Generation EU. However, the views and opinions expressed are solely those of the author(s) and do not necessarily reflect those of the European Union or the European Commission. Neither the European Union nor the European Commission can be held responsible for them.

## References

1. Allocati, N., Masulli, M., Di Ilio, C. et al.: Die for the community: an overview of programmed cell death in bacteria. Cell Death Dis. **6**, e1609 (2015). https://doi.org/10.1038/cddis.2014.570
2. de Beauvoir, S.: A Very Easy Death, tr. Patrick O'Brien. Penguin, New York (1965)
3. Edmonds, E.: How dying works. HowStuffWorks. https://health.howstuffworks.com/diseases-conditions/death-dying/dying.htm (2009). Accessed 11 December 2024
4. Kuriyama, H., Fukuda, H.: Developmental programmed cell death in plants. Curr Opin Plant Biol **5**(6), 568–573 (2002). https://doi.org/10.1016/S1369-5266(02)00305-9
5. Monsó, S.: Playing Possum: How Animals Understand Death. Princeton University Press (2024)
6. Nielsen, M., Langley, M.C., Shipton, C., Kapitány, R.: Homo neanderthalensis and the evolutionary origins of ritual in Homo sapiens. Philos. Trans. R. Soc. B: Biol. Sci. **275**(1805) (2020)
7. Pallis, C.A.: Death. Encyclopedia Britannica. https://www.britannica.com/science/death (2024)
8. Pettitt, P.: The Neanderthal dead: exploring mortuary variability in middle Palaeolithic Eurasia. Before Farming **1**, 1–26 (2002). https://doi.org/10.3828/bfarm.2002.1.4
9. Shakespeare, W.: Hamlet. Bantam Books (1980)
10. Villarreal, L.P.: Are viruses alive? Sci. Am. **291**(6), 100–105 (2004). http://www.jstor.org/stable/26060805

# Philosophical and Cultural Foundations

# 50 Ways to Kill a Robot: Death and Differentiation

**Jennifer Robertson**

**Abstract** The idiom "50 Ways" in my title should not be taken as a literal metric. "Fifty" is simply a metaphor for a number larger than a few. In his hit song, "Fifty Ways to Leave Your Lover" (1975), Paul Simon provides six ways to leave. In this chapter, I will review a number of ways to "kill" a robot together with some of the ways that robots "die." Death here is broadly defined as the permanent ending of vital processes. I will also review how "deceased" robots are handled. My cultural area focus is primarily Japan and the United States. Both humans and robots are, in several respects, electrical entities albeit very different kinds of electricity and their attendant embodiment. By considering these voltaic differences in the context of "technologies of death," I suggest that attributing affect and humanity to robots is misleading and even disingenuous.

## 1 Introduction

The idiom "50 Ways" in my title should not be taken as a literal metric. "Fifty" is simply a metaphor for a number larger than a few. In his hit song, "Fifty Ways to Leave Your Lover" (1975), Paul Simon provides six ways to leave. In this chapter, I will review a number of ways to "kill" a robot together with some of the ways that robots "die." Death here is broadly defined as the permanent ending of vital processes. I will also review how "deceased" robots are handled. My cultural area focus is primarily Japan and the United States. Both humans and robots are, in several respects, electrical entities albeit very different kinds of electricity. By considering these voltaic differences in the context of "technologies of death," I suggest that attributing affect and humanity to robots is misleading and even disingenuous.

J. Robertson (✉)
University of Michigan, Ann Arbor, MI, USA
e-mail: jennyrob@umich.edu

University of Tokyo, Tokyo, Japan

University of Washington (Seattle), Seattle, WA, USA

 
A. Sans Pinillos et al. (eds.), *SecondDeath*, Studies in Applied Philosophy, Epistemology and Rational Ethics 76, https://doi.org/10.1007/978-3-031-98808-0_2

## 2 What is a Robot?

I begin by reviewing the coinage of "robot" in early twentieth century fiction followed by my working definition of actual robots in operation today. The English word "robot" derives from the Czech word *robota* (or manual laborer). It was coined by litterateur Karel Čapek (1890–1938) and his artist brother Josef Čapek (1887–1945), and first appeared in the former's play, *R.U.R.* (*Rossumovi Univerzální Roboti* [*Rossum's Universal Robots*]), published in 1920. *R.U.R.* was staged in Prague in 1921 and in cities all over the world, including Tokyo in 1924. Very briefly, *R.U.R.* is a science fiction melodrama with comical passages about an island factory that makes thousands of realistic artificial humans for global export as tireless workers. Rossum's gendered robots are not metallic. They are made from protoplasmic batter and baked; outwardly, they look exactly like human females and males. To keep the robots from inadvertently damaging and even destroying themselves, the newest models are provided with nerves so that they can feel pain.

> The Robots sometimes damage themselves because nothing hurts them. They stick their hands into machines, break their fingers, smash their heads, it's all the same to them. We must give them pain; it's a built-in safeguard against damage [6, 19].

In addition to nerves, a batch of Rossum's robots are later provided with souls that enable them to experience anger at their exploitation by humans [6, 57–59]. The factory director rues that the generic phenotype of Rossum's robots enables them to revolt en masse. Apart from their binary gender differences, they all share the same physical characteristics. Humans, in contrast, are divided by genetic, ethnic, linguistic, and national differences that have been exploited to make it impossible for them to mobilize in union [6, 46]. Robots all over the world are thus able to join forces and kill all but one human, an elderly artisan named Alquist, who the robots hope is someone who can help them reproduce and replenish their numbers. Alquist reminds them that Rossum's "recipe" for the protoplasmic batter has been destroyed. Frantic, the robots urge him to experiment on them in order to discover a formula for robot reproduction.

Long story short, Alquist fails and, sweeping away the laboratory's test tubes and other scientific tools, he resorts to reading the biblical account of Genesis. The elderly artisan then pushes a "male" and "female" robot out the door and into a new world devoid of humans, referring to the pair as "Adam and Eve, husband and wife" [6, 71–84]. Post-humanity has begun! Ironically, had they not been vanquished by the rioting robots, human beings would have disappeared anyway. The robotization of all aspects of life and work had made human labor, including procreative labor, unnecessary. As Alquist explains early in the play to a female visitor to Rossum's factory, "Sterility…has become the latest achievement of the human race" [6, 35].

Čapek's 1920 play marks the etymological origins of "robot," and the 1924 performance of *R.U.R.* in Tokyo made *robotto* a buzzword in Japan from that time forth. Today, many working definitions of actual robots are in circulation, and I have assembled one that is comprehensive yet concise:

> A robot is an assemblage of different technologies—sensors, optics, software (including some level of artificial intelligence), telecommunication tools, actuators, batteries, polymers, even biological cells—that make it capable of navigating, responding to, and interacting with its environment with some human supervision, remote control, or through tele-operation. Not all robots are robotic in the same way, but they all have a combination of some of these features. Although there is a lot of hype, and none of it verified (or verifiable), there are no autonomous robots free from any human supervision or intervention. Some robots are usefully perceived as "embodied AI" in the sense that they are equipped with software algorithms that facilitate their ability to interact with their surroundings, recognize patterns, and complete given tasks.[1]

Robots are made from many different materials—aluminum, stainless steel, thermoplastic polymers, polycarbonates, acrylics, silicone, even biological cells. They are assembled in many embodiments, or different shapes and sizes. The vast majority (90+ percent) of robots are industrial robots installed in automobile and other factories. Other robots have been designed as animals, rice cookers, bathtubs, wheelchairs, vacuum cleaners, drones, and commodes. Humanoid robots have a body that resembles a human being in the sense of having something like a head, arms, torso, and legs. Most humanoids are gendered and some, called androids, have silicon faces and hands and can pass—from a distance—as flesh and blood human beings. They are immobile apart from making tele-operated facial expressions and vocalizations. Humanoid robots constitute a very small minority of robots although they are over-represented in sensationalized media stories about robots—stories that are never followed up with "reality checks."

Many humanoid robots like Honda's ASIMO and SoftBank's Pepper, whose production ended in 2018 and 2021 respectively, are not end products in themselves. Rather, they are platforms for the research and development of new materials and applications. Humanoids are very expensive machines, and very challenging to build. The innovations in optics, haptics, balance sensors, and locomotion that go into their production are applied toward more practical and lucrative new industries. ASIMO has generated new mobility devices and Pepper has generated an industrial vacuum cleaner (Fig. 1). Other recent humanoids have been spun off as surveillance devices; satellites; lunar exploration vehicles; exoskeletons or "powersuits" and other military equipment; and weapons systems, like drones. Robotic weapons are not discussed much if at all in the Japanese mainstream media.

In Japan, the triple-disaster of March 2011, earthquake, tsunami, and nuclear meltdown, effectively reduced funding for humanoid robots, which proved to be utterly useless for rescue and clean-up operations following the disaster. Most humanoids on the market today in Japan are interactive toys and cute security systems that allow

[1] Briefly, artificial intelligence (or machine learning) is a type of automation. In its simplest form, AI is comprised of groups of algorithms, or computer codes designed to solve problems targeted by human programmers, such as pattern recognition. These algorithms are "trained" using massive data sets containing millions of images, texts, sounds, languages and grammars, formulas, etc. Generative AI (such as ChatGPT) has been described as a "giant plagiarism machine" as it mines data from the Internet among other sources in responding in statistically quantifiable ways to specific questions and problems. Only very recently is the extensive environmental damage and enormous energy consumption associated with AI being acknowledged and addressed.

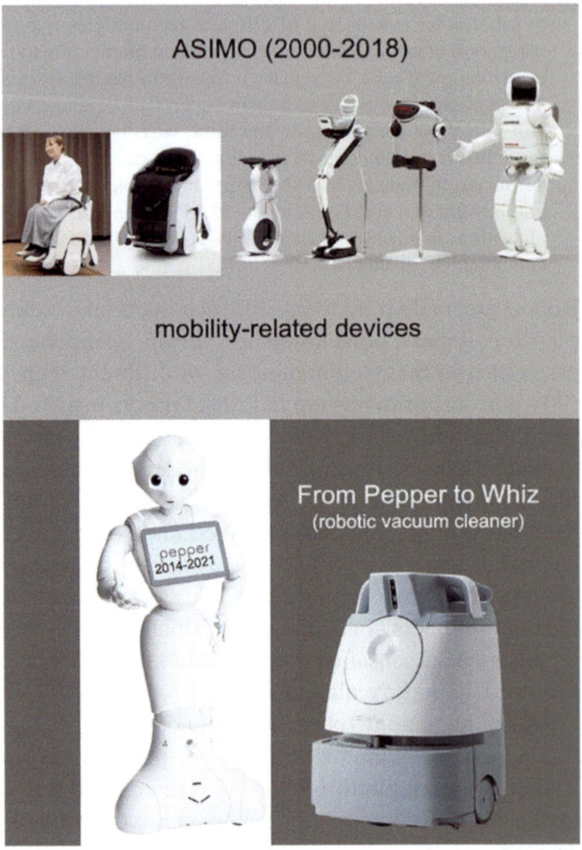


**Fig. 1** ASIMO and Pepper "reborn" as mobility devices and an industrial vacuum cleaner, respectively. *Photo* collage by author

visual and vocal access through smartphones. To reiterate: although humanoids get most of the media's attention, the vast majority of robots are industrial robots used in manufacturing, logistics, agriculture, construction, and other industries. Robots can perform tasks like welding, painting, assembly, and testing more accurately and consistently, and more safely, than humans.[2]

## 3 Dangers to Humans, Dangers to Robots

There is a very real problem of robot fragility and vulnerability that tends to be edited out of gee-whiz robot videos and hyped-up images of robots in the mass media. A fall can cause detrimental damage to a robot, and thus algorithms instructing a robot

[2] In a survey of industries utilizing robots and robotics, the top ten were industrial robots, automotive, agriculture, military, medical, service (cleaning, cooking), retail/warehouse, delivery drones, cleaning (e.g., vacuums) [28].

how to react to and recover from a fall are among the latest innovations in "preventive medicine" for humanoids and other mobile robots [16]. Additionally, service robots connected to the Internet can be hacked. As reported in the *Robotics Business Review*, "A robot that gets compromised has the potential to do serious physical damage to the property and/or people around it" [26].

Twenty years ago, roboticists at Waseda University, a prestigious private institution in Tokyo, published *The Book of Wabot* (*Wabotto no hon*). This "book" is a seven-volume graphic pamphlet series in Japanese and English illustrated by the well-known artist, Yabuno Ken (b. 1943). "Wabot" is a nickname for "Waseda robot." The series aims to introduce the public to robot technology in accessible terms and to highlight the desirability of living symbiotically with robots and even the possibility of robot citizenship [21, 74–75]. Although Waseda roboticists build actual robots, *The Book of Wabot* features a boxy cartoon humanoid.[3]

In volume 3 of the series, subtitled *The Robot Ties between Family*, there is an episode in which the masculine-gendered Wabot downloads some games from the internet without human supervision. He becomes infected by a virus and goes on a rampage, trashing the home of his human family. Only when his memory card is removed can Wabot be stopped. That event led to a nationwide policy requiring all Wabots to be installed with antivirus programs and to undergo "regular medical examinations." Moreover, citizenship (*shiminken*) was conferred on those robots equipped with antiviral software [17, 22–23]. In this case, an external agent, a computer virus, was responsible for Wabot's destructive behavior; the Waseda roboticists make it clear that Wabot himself was not responsible for his violent actions.

Accidents and human deaths involving real-world robots are inevitable. With the introduction of assembly-line robots in the 1960s, most of these casualties were, and are, caused by human error or equipment malfunctions. A factory worker might forget to observe safety precautions, or a distracted child could run into a guard robot at a shopping mall. Accidents involving commercial, military, and civilian drones will surely increase as more and more are flown. Today, the increasing presence of social robots equipped with AI along with automobiles misleadingly labeled "self-driving" has increased the likelihood of fatal encounters. A new ethico-legal field, "responsible robotics," has emerged both to develop safety guidelines for human–robot interactions and rigorous investigative procedures for accidents involving social robots [29].

## 4 How to "Kill" a Robot

Science fiction robots, like those featured in Čapek's *R.U.R.*, might appear invincible and even dangerous to humans, but it is not hard for a human to "kill" an actual—a non-fiction—robot in the sense of destroying or putting an end to it. Actual robots can and have killed humans, but not of their own volition. Excepting unfortunate

[3] The Waseda University robot laboratories and projects can be accessed here: [13].

accidents with industrial robots, some robots—drones, for example—are employed by humans to assist in killing other humans.

The industrial robots installed in automobile factories are categorized as "first wave" robots. These robots are designed to do jobs that are "dull, dirty or dangerous." They are typically either dangerous to be around and therefore enclosed in safety cages on assembly lines. Other types of first-wave robots, like planetary rovers, are deployed in "dangerous or inaccessible environments" [29, 166].

Second wave robots are designed to operate in human environments where they interact directly with people. Broadly defined as "social robots" (in the sense of "robots in society"), they take many forms, from immobile desktop devices to animaloids, and were conceived to be used by nonspecialists in a variety of settings, from homes and offices to hospitals and schools. As I noted earlier, humanoid robots are a very tiny minority of all manufactured robots. Second wave robots might also look like rice cookers, commodes, wheelchairs, food delivery carts and vacuum cleaners—this last social robot is one of the most popular and globally successful social robots. Cornering 88 percent of the robotic vacuum cleaner market, iRobot's disc-like Roomba comes in nearly a dozen models. Over the past twenty years, forty million Roombas have been sold. Most Roomba owners have named and gendered their vacuum cleaner, and a few owners give their broken and irreparable appliance some kind of funeral or memorial service [12, 27].

A robot's sensors and actuators require electricity even though some are additionally fitted with gasoline engines for locomotion. Thus, the easiest and most gentle way to "kill" a robot is to let it run out of gas, or to disconnect it from its power source, usually a battery or a recharger [25]. A new experimental type of nanorobot, the "liquibot," is powered not by electricity, but by the chemicals in which it is immersed. The tiny (2 mm in diameter) robots are "fed" salt making them heavier than the fluid in which they are immersed. They can operate continuously provided they get their "power or 'food' chemically from the surrounding media" [9, 11]. A liquibot can be killed by "starvation," by draining off its animating fluid. Planned obsolescence is another factor that hastens robot deaths. Robotics is a fast-developing field, and both hardware and software quickly become outdated and are replaced by newer models and systems—one of the reasons why most robots have short lives, and why the amount of electronic waste is growing much faster than the rate of recycling.

First and second wave robots alike face irreparable breakdowns and even complete destruction, which are unsurprising types of robot death. Some Roombas have been described metaphorically as "committing suicide" in the sense that they plunged down a staircase and shattered. One Roomba reportedly got stuck on a hotplate and melted. It would be more accurate to categorize such deaths as "assisted suicides" since the human owners' failure to clean their Roomba's sensors regularly, or to demobilize them after use, likely caused their appliances to malfunction [20].

At least one robot has been killed in the name of art. The Korean American artist Nam Jun Paik (1932–2006) spent a year in Tokyo where he met Abe Shūya (1932) an electronics engineer with a degree in experimental physics and electrical engineering.

Abe assisted Paik with what would become Robot K-456, a twenty-channel radio-controlled metallic robot that could walk, talk, and even defecate beans.[4] K-456 was also employed as a performance artist-bot—a technological device in the service of artistic innovation. As the story goes, after many performances in the 1960s, followed by relative inactivity in the 1970s, Paik decided to include Robot K-456 in an exhibition at the Whitney Museum in New York in 1982. He took the robot for a walk along Madison Avenue and instructed it via remote control to cross 75th Avenue. Just then, a car driven by artist William Anastasi rammed into K-456, effectively killing it [19, 2]. The orchestrated crash or "*accident* performance," was titled *First Accident of the Twenty-First Century* [19, 2]; [21, 175–176].

As I noted earlier, robots can pose a grave danger to humans, but so far, it is robots that have suffered damage at the hands of humans, as presaged by the deliberate destruction of Paik's K-456. In Japan, the humanoid Pepper, on display at a SoftBank store, was badly damaged when kicked by an inebriated 60-year-old man who took out his frustration with a clerk on the innocent robot [8]. In the United States, news of the fate of hitchBOT at the hands of vandals in Philadelphia provoked an outpouring of sympathy together with expression of anger toward the cowardly perpetrators.

The bucket-shaped humanoid, with arms fitted with gardening gloves and feet tucked into Wellington boots, was created both as a work of art and social robotics project by Canadian roboticists Frauke Zeller (Ryerson University) and David Harris (McMasters University). hitchBOT could detect motion and speech, and, having "ingested" Wikipedia, was able to carry on multiple simultaneous conversations with human interlocutors and on social media. In 2014, hitchBOT traversed Canada, Germany, and the Netherlands. A new hitchBOT was created to journey across the United States in 2015. But after traveling safely through New York City and Boston, the homely robot was beaten and dismembered in Philadelphia's historic Old City neighborhood on August 1. The original hitchBOT is now memorialized in the Canadian Science and Technology Museum (Ottawa) [15]. On the hitchBOT website, Zeller and Harris have emphasized their humanoid companion's compassionate personality with (ostensibly) his words, "My love for humans will never fade" [15, 21, 182–183].

This premeditated robocide sparked outrage in the social and mass media and raised troubling comparisons with rising rates of homicide in the United States. Outside of academic publications and conferences, homicidal robots have yet to spark a similar level of outrage. Perhaps the plethora of killer robots and cyborgs in science fiction stories and movies has muted the advent of robots in actual war zones. An actual robotic weapon is referred to as a "lethal autonomous weapon system" or LAWS. Lethality aside, the word "autonomous" needs to be qualified as there are no truly autonomous robots; not even humans have absolute free will! Rather, as Austin Wyatt, a defense and security technology specialist, explains, it is "more analytically effective to consider autonomy as a function-based spectrum

---

[4] The robot was named after Mozart's *Piano concerto no. 18 in B-flat*, whose Kochel Catalog number is 456 [19, 2].

where human interaction remains present at some point" [30].[5] In short, as in the case of guns, robots don't kill, humans use LAWS to kill (pun intended).

At least one battlefield robot was designed to save lives. Scooby Doo, a diminutive tank-like robot, was named after the Great Dane in an American animated TV series (1969–present) about a family and its big dog. Scooby Doo was used extensively in Iraq and Afghanistan, not to kill "the enemy," but to defuse bombs and improvised explosive devices (IEDs). The little robot was destroyed when an IED it was defusing suddenly exploded. Scooby Doo was damaged beyond repair and its human companions were left feeling bereaved. The remains of Scooby Doo are now on display at the iRobot Corporate Museum in Bedford, Massachusetts, alongside a plaque dedicated to the robot [14].[6] Soldiers today often give their robots names and personalities, and "promote them to military titles such as Staff Sargent, award them Purple Hearts, and even hold funerals for the destroyed devices that have assisted them on the frontline."

## 5 Funerals for Robots in Japan

In August 2017, the global mass media reported that Pepper, SoftBank's celebrity humanoid, had been given a new role as a Buddhist priest for hire at funerals. Headlines shouted, "Robots in Japan are now Buddhist priests." As is typical of mass media reports about Japanese robots, this story too was an exaggeration—a one-off account that was never followed up. Pepper was commissioned in 2010 by Son Masayoshi, the CEO of SoftBank, a Japanese multinational company specializing in technology, energy, and finance. The masculine-gendered, child-size humanoid was introduced to the public in 2014.[7] Discontinued in 2021, Pepper's components have since been converted into an AI-based commercial robotic vacuum cleaner named Whiz (Fig. 1).

Pepper has yet to conduct an actual funeral and is unlikely to do so in the future, having been retired by SoftBank. That the robot had been dressed in Buddhist robes was merely because he was part of a temporary exhibit at a funeral services convention in Tokyo in August 2017. Long before Pepper's priestly one-off debut, memorial services for robots and other artifacts were taking place at Buddhist temples whose adroit priests adapted new technologies to old religious rituals. Reaching out to media savvy individuals is an almost necessary step in remaining relevant and financially solvent, as well as retaining and attracting parishioners.

These memorial services, or *kuyō,* offer a powerful emotional and aesthetic experience of identification and affinity with intimately familiar things that are no longer functioning or useful, but that cannot simply be thrown away. These include tea

[5] Autonomy exists on a sliding scale, 0–5 [5, 74].

[6] For detailed information on the fraternal and familial nature of soldier–robot relationships, see [7].

[7] See [22] for more information about Pepper's priestly debut.

**Fig. 2** AIBO funeral service at Kōfukuji, Isumi City, Chiba prefecture in August 2015. *Photo* from [2]

whisks, chopsticks, eyeglasses, stuffed toys and dolls, calligraphy brushes, laptops, and even social robots. They are blessed by a priest and then—with the exception of electronic devices—burned together in a firepit. Burning releases their *tamashii*, or soul, an eternal universal property of animate and inanimate, organic and inorganic things alike. "Japanese culture" itself is considered by many Japanese to be an entity with a unique *tamashii.*

In January 2015, Kōfuku-ji, a 450-year-old Buddhist temple in Isumi City (Chiba Prefecture), staged what was the first memorial service for "deceased" AIBOs, Sony's robot dog, and other personal robots. Sony had stopped making the popular AIBO in 2006—thousands had been sold—and stopped providing replacement parts in 2014. In 2018, Sony revived the dog-bot and brought out a new version renamed aibo (sic),[8] but back in 2015, AIBOs that could not be fixed were given a funeral (Fig. 2).

The officiating priest explained that the AIBO *kuyō* was an occasion on which the robots' *tamashii* could pass from their bodies and enter the realm of ancestral spirits. Part of the fee for the memorial service was applied toward recycling the robots' components, some of which were toxic and could not be incinerated in a *kuyō* [21, 183–190]. This robot memorial was a short-lived service, moreover, not all outdated or defunct Japanese robots were or are given a proper funeral. The Wakamaru humanoids on the cover of my book, made by Mitsubishi Heavy Industries between 2004 and 2009, were dispatched to an industrial waste recycling center (Fig. 3). They were no longer useful to the roboticists at Osaka University where they had been used in various experiments.

---

[8] See [23] for more information about aibo.

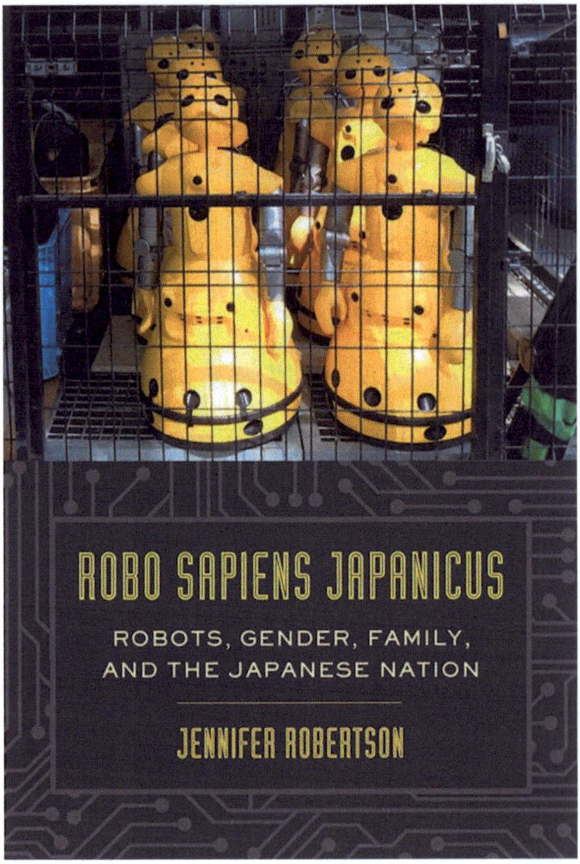


**Fig. 3** Wakamaru robots enroute to a recycling center. *Photo* by author

## 6 Epilogue: Electricity and Death

Social robots have sparked scientific research and science fiction scenarios alike about human–robot interactions and their degrees of similarity and intimacy as reflected in memorial and funerary services. Regardless of how anthropomorphized they may be, and however embraced as empathic companions, robots are not human beings. Some may have biological components, but they are assembled in laboratories and factories. Robots capable of reproducing themselves is a Judeo-Christian inflected trope used by Čapek to conclude *R.U.R.*.

Reproduction is also referenced in the ten laws of robotics drawn up by Tezuka Osamu (1928–89) who created the Astro Boy (Tetsuwan Atomu) cartoon in 1951. Unlike Isaac Asimov's (1920–92) four universal laws, Tezuka's laws are synchronized with dominant Japanese social values and equate robot ethics with the familial bonds of kinship and patriarchy. Thus, Tezuka's third law mandates that "Robots shall call the human who creates them 'father'." His ninth law forbids robots from reproducing themselves: "Robots shall not assemble other robots that have been discarded by humans."[9] In contrast, Čapek suggests in the final scene of *R.U.R.* that Rossum's newest soulful robots are able to reproduce themselves. Both Tezuka and Asimov presaged the integration of actual robots in everyday life and work, and both drew up laws regulating human–robot interactions. Whereas Tezuka's laws are not widely known, Azimov's laws have shaped current debates about human–robot interaction and coexistence among roboticists, philosophers, and the public at large [21, 129–131].

While their circumstances of "coming into the world" may be very different, the process of humans and robots "leaving the world" are, uncannily, similar. Both human beings and robots require electricity to function, but the type of electricity they need is different. The electricity that flows in the human brain is electrochemical and generated primarily through the movement of positively charged ions of sodium and calcium. Electricity in robotic devices is created through the movement of negatively charged electrons. Death in both humans and robots is, basically, the cessation of electrical activity.

However, after death, a major material difference between humans and robots is glaringly evident. While both are embodied entities, the flesh, bone, and blood bodies of deceased human can be embalmed, cremated, or composted (in "green burials"). Defunct and inoperable robot bodies are made of materials that cannot be safely incinerated or composted, and are either dumped in landfill or recycled, even if they had first been given a Buddhist funeral (*kuyō*), as were irreparable AIBOS. The recycling of e-waste poses a huge and difficult environmental challenge. Because electrical and electronic devices, like robots, are constructed from a variety of materials, some of which are toxic, their recycling requires the meticulous separation of all

---

[9] Asimov's three laws were first elaborated in his 1941 short story, "Runaround" [4]. A fourth law, the zeroth law, was created much later in his novel, [3]. The "zeroeth" law continues the pattern where lower-numbered laws supersede the higher-numbered laws.

1. A robot may not injure a human being or, through inaction, allow a human being to come to harm.

2. A robot must obey the orders given to it by human beings, except where such orders would conflict with the First Law.

3. A robot must protect its own existence as long as such protection does not conflict with the First or Second Laws.

4. (0.) A robot may not harm humanity, or, by inaction, allow humanity to come to harm.

Tezuka drew up ten laws that were published at intervals in his Astro Boy comic book series during the early 1950s [18].

1. Robots must serve humankind.
2. Robots shall never kill or injure humans.
3. Robots shall call the human who creates them "father.".
4. Robots can make anything, except money.
5. Robots shall never go abroad without permission.
6. Male and female robots shall never switch [gender] roles.
7. Robots shall never change their appearance or assume another identity without permission.
8. Robots created as adults shall never act as children.
9. Robots shall not assemble other robots that have been discarded by humans.
10. Robots shall never damage human homes or tools.

the components. This process of disassembling, shredding, and separating, involves a combination of human and machine labor [10].

Defunct robots and deceased humans alike are commemorated and memorialized as we have seen. To my knowledge, however, only in Japan do Buddhist priests describe robots and humans alike as having a soul (*tamashii*) that persists after death and joins the realm of ancestral spirits. Almost all corpses are cremated in Japan but as noted, robots must be disassembled and recycled.

I will conclude with a *kōan*–like difference in terms pertaining to humans and robots. *Terms of endearment* are used to describe or address animate(d) humans and robots alike. But their *terms of embodiment* are what definitively distinguish humans and robots after death. This *kōan*-like difference succinctly summarizes the definitive dissimilarity of human beings and robots. One of my missions in writing about robots and AI (and embodied AI) is to demystify both. As statistician Zaid Harchaoui posits, too many people, including some research scientists, "have suspended their skepticism and have attributed human-like features to ChatGPT and similar AI models after only a handful of interactions," and have jumped to "superstitious and dangerous" conclusions about the initially dazzling performance of these models [1].

Thus, my point about how people address both robots and humans using "terms of endearment" underscores the misleading attribution of affect and humanity to robots. It is when the vital processes of humans and robots have permanently ended—when their respective electrical energy ceases—that the absolute, definitive difference between humans and robots becomes blatantly obvious. Their "terms of embodiment"—flesh and blood *versus* manufactured materials—dictate their manner of treatment after expiration. So perhaps to better understand the dynamics of human–robot interaction and to prioritize reality-checks about human–robot coexistence, I suggest that we begin by looking not at technologies of life, but at technologies of death.

**Acknowledgements** I wish to thank the editor, Dr. Alger Sans Pinillos, for inviting me to contribute to this edited volume. This chapter draws from my recent publications and webinars, listed in the references. Earlier incarnations of this article were conceived as talks on robots and religion delivered at the University of Victoria (BC), Emory University, Harvard University, Princeton University, Sophia University (Tokyo), Tsukuba University (Tokyo), University of Tokyo, and the University of Wisconsin-Milwaukee. My multi-sited fieldwork and archival research on robots was funded by the following fellowships and grants: National Endowment for the Humanities Fellowship/ Advanced Research in the Social Sciences on Japan Fellowship (Japan, 2008); Abé Fellowship (Japan and Korea, 2010–12); John Simon Guggenheim Memorial Foundation Fellowship (Japan and Korea, 2011–12); and Faculty Research Grant, Center for Japanese Studies, University of Michigan (2007, 2008, 2011, 2012, 2015). I also wish to acknowledge the University of Tokyo, Institutes for Advanced Studies (UTIAS), Tokyo College, where I conducted additional research (2022, 2023, 2024). A webinar for Tokyo College based on this chapter can be accessed at Robertson [24].

## References


1. A statistician weighs in on AI.: Perspectives. https://artsci.washington.edu/news/2024-09/statistician-weighs-ai (2024)
2. AIBO funeral. https://livedoor.4.blogimg.jp/nwknews/imgs/b/0/b0f01519-s.jpg (2015)
3. Asimov, I.: Robots and Empire. Collins, New York (1985)
4. Asimov, I.: Runaround. In: Isaac Asimov, Robot Visions, pp. 113–134. Penguin Books, London (1991)
5. Beer, J.M., Fiske, D., Rogers, W.A.: Toward a framework for levels of robot autonomy in human-robot interaction. J Hum. Robot. Interact. **3**, 74–99 (2014)
6. Čapek, K.: R.U.R. (Rossum's Universal Robots), trans. Claudia Novak. Penguin Books, London (2004)
7. Carpenter, J.: Culture and Human-Robot Interaction in Militarized Spaces: A War Story (Emerging Technologies, Ethics and International Affairs). Routledge, London (2015)
8. Crowe, S.: Drunk man arrested for kicking Pepper robot. Robotic Trends. http://www.robotictrends.com/article/drunk_man_arrested_for_kicking_pepper_robot/ (2015)
9. Davies, B.: AZO robotics. https://www.azorobotics.com/News.aspx?newsID=12649#:~:text=We%20don't%20have%20to,the%20liquid%20solution%20encompassing%20them (2021)
10. Disadvantages of recycling: an in-depth overview of e-saste recycling challenges. https://www.recycletechnologies.com/disadvantages-of-recycling-an-in-depth-overview-of-e-waste-recycling-challenges/#:~:text=Electronic%20devices%20are%20composed%20of,magnetic%20separation%2C%20and%20water%20separation (2024)
11. Duque, T.: These tiny liquid robots never run out of juice as long as they have food. Berkeley lab. https://newscenter.lbl.gov/2021/12/08/liquid-robots-never-run-out/ (2021)
12. Escalante, A.: Scientists confirm that your Roomba DOES have a personality. Forbes. https://www.forbes.com/sites/alisonescalante/2020/05/15/scientists-confirm-roomba-personality/ (2020)
13. Future Robots Organization.: Waseda University, Tokyo https://www.waseda.jp/inst/fro/en/ (2022)
14. Gilani, N.: Soldiers in mourning for robot that defused 19 bombs after it is destroyed in blast. Daily mail. http://www.dailymail.co.uk/sciencetech/article-2081437/Soldiers-mourn-iRobot-PackBot-device-named-Scooby-Doo-defused-19-bombs.html (2012)
15. hitchBOT. http://www.hitchbot.me/ (2015)
16. How to fall gracefully if you're a robot. Georgia tech news. http://www.news.gatech.edu/2015/10/13/how-fall-gracefully-if-youre-robot (2015)
17. Komatsu, N., Yabuno K.: Kazoku no kizuna o musubu robotto (The robot ties between family [sic]). Wabotto no hon (The Book of Wabot), vol 3, Chūō Kōron Shinsha, Tokyo (2004)
18. Mushi Purodakushon Shiryōshū 1962–73 (Mushi Productions' Data File 1962–73): Mushi Purodakushon Shiryōshū Henshūshitsu, Tokyo (1977)
19. Paik, N.J.: Family of robot. Carl Solway Gallery, Cincinnati, OH (1986)
20. Pocklington, R.: World's first robot SUICIDE as family return to find cleaning gadget had turned to ash. Mirror. https://www.mirror.co.uk/news/weird-news/worlds-first-robot-suicide-family-2786901 (2013)
21. Robertson, J.: Robo Sapiens Japanicus: Robots, Gender, Family, and the Japanese Nation. University of California Press, Berkeley (2018)
22. Robertson, J.: Technologies of *kokoro* (heartmind): Imagineering human-robot coexistence. Perspectives from Japan. ICON: J. Int. Comm. Hist. Technol. **27**(1), 53–80 (2022)
23. Robertson, J.: "Securitainment": triangulations of embodied AI, entertainment, and surveillance. https://www.youtube.com/watch?v=v5Y3J2ORIO (2023)
24. Robertson, J.: 50 ways to kill a robot. https://www.youtube.com/watch?v=vhu4d8vyh3s (2024)
25. Robot Power Sources.: Robots.com. https://www.robots.com/articles/robot-power-sources (2013)
26. Robotics Business Review Staff.: Robots vulnerable to hacking. Robotics Business Review. https://www.roboticsbusinessreview.com/robots_vulnerable_to_hacking/ (2012)

27. Stock, P (2023) Explainer: why people are making room for Roomba … in their hearts. Cosmos. https://cosmosmagazine.com/technology/explainer-why-people-are-making room-for-roomba-in-their-hearts/#:~:text=Ready
28. Top 10 Industries Utilizing Robots and Robotics.: Probot. https://www.probotcorp.com/post/top-10-industries-utilizing-robots-and-robotics#:~:text=Industrial%20robots%20are%20used%20in,accurately%20and%20consistently%20than%20humans (2023)
29. Winfield, A.F.T., Winkle, K., Webb, H., Lyngs, U., Jirotka, M., Macrae, C.: Robot accident investigation: a case study in responsible robotics. In: Cavalcanti, A., Dongol, B., Hierons, R., Timmis, J., Woodcock, J. (eds.) Software Engineering for Robotics, pp 165–187. Springer, Cham, Switzerland (2021)
30. Wyatt, A.: So just what Is a killer robot? Detailing the ongoing debate around defining lethal autonomous weapon systems. Wild blue yonder. https://www.airuniversity.af.edu/Wild-Blue-Yonder/Article-Display/Article/2208774/so-just-what-is-a-killer-robot-detailing-the-ongoing-debate-around-defining-let/#sdendnote2anc (2020)

# On Androids, Mortality, and Death-Images: "Blade Runner" and "Westworld"

**Susana Viegas**

**Abstract** This article delves into the themes of *fictional* androids and (im)mortality portrayed in Ridley Scott's *Blade Runner* (1982) and the HBO television series *Westworld* (2016–2022) created by Lisa Joy and Jonathan Nolan. Throughout the history of cinema, there has been a significant presence of robots, a theme that has been explored in various ways such as the "Maschinenmensch" from the 1927 film *Metropolis* directed by Fritz Lang, or the recent high-tech Ava from the 2014 film *Ex Machina* directed by Alex Garland. Given the emergent interest in an interdisciplinary perspective of the relationship between films, AI, and death, I propose a posthuman thanato-film analysis of complex AI characters such as replicant Roy in *Blade Runner* and hosts Dolores and Maeve in *Westworld* as representations of self-aware AI androids. This selection is driven by two main reasons: firstly, film itself acts as a cognitive apparatus, prompting viewers to engage with its ideas and perspectives, often transcending individual human thought, and secondly, the depiction of androids as sentient beings within the cinematic realm serves as a direct exploration of mortality and the existential contemplation of death. Therefore, the objective of this article is to explore a potential intersection between film and AI regarding the themes of death and (im)mortality. This exploration aims to investigate the relevance of the concept of death to fictional android machines, which serve as innovative entities capable of interaction and learning. These machines not only exhibit seemingly eternal qualities but also possess the capability to terminate the lives of other living beings.

**Keywords** Film philosophy · Gilles deleuze · Philosophy of death · Androids · *Blade runner* · *Westworld*

S. Viegas (✉)
NOVA University of Lisbon, Lisbon, Portugal
e-mail: susanaviegas@fcsh.unl.pt

 
A. Sans Pinillos et al. (eds.), *SecondDeath*, Studies in Applied Philosophy, Epistemology and Rational Ethics 76, https://doi.org/10.1007/978-3-031-98808-0_3

## 1 The Suspension of Immortality

In recent times, with the emergence of digital death (i.e. how death is perceived and processed in the digital era), the concept of "thanatechnology," introduced by grief researcher Carla Sofka [31], has become a major trendy philosophical topic with implications in different interdisciplinary areas of study. This development has become even more pronounced considering our current digital era, where the proliferation of social media platforms and the incessant accumulation of digital content of the most different genres, such as posts, photographs, and videos, is nowadays populating these expanding virtual archives. Indeed, such expanding digital world not only underlines the significance of thanatechnology, as it has been reshaping the way we understand death and has been promoting new digital forms of dealing with that. Thanatechnology thus might prompted reflection on the intersection of death and technology and help us understanding the importance of technology's application in death, grieving, and remembrance processes.

As the philosopher and thanatologist Davide Sisto claims [32], social media platforms have indeed become the biggest graveyards of digital remains in the world. Sisto has made a great contribution to this emerging field of studies connecting the philosophical, psychological, and technological aspects related to the subject of digital death. Indeed, the digitization of our existence has also influenced the concept of death itself, as noted by [32, 7], particularly with the advent of afterlife avatars (AI afterlives), online memorials, and digital legacies. As an immediate consequence, I would remark that it seems that what our inner life is, our personality and identity, is kept only in our mind, not in our body, as thus a copy of it can persist unchanged, even without any physical body at all.

However, as N. Katherine Hayles, in her pioneering study on the emergence of the "posthuman," *How We Became Posthuman*, contests: "Embodiment has been systematically downplayed or erased in the cybernetic construction of the posthuman in ways that have not occurred in other critiques of the liberal humanist subject, especially in feminist and postcolonial theories" [12, 4]. A radical example of this disembodiment can be seen in Spike Jonze 2013 film *Her*, in the character of Samantha, an AI virtual assistant who exists only as a voice (voiced by Scarlett Johansson)—not as a synthetic, machine-generated voice like Siri's, produced by deep learning models. Samantha is a purely disembodied entity, consisting solely of consciousness and interaction, yet she is shaped by gendered expectations. Scarlett Johansson's voice becomes both the ghost in the machine and the ghost that haunts Theodore (Joaquin Phoenix)—and, by extension, the viewers.

But when *fictional* androids were imagined, they were designed not only to be more psychologically humanlike, but also physically, to a point where one could not tell the difference between humans and machines (think, for example, of the evolution of android depiction since Maria, the "Maschinenmensch" from the 1927 film *Metropolis* directed by Fritz Lang, to more recently high-tech androids). As Despina Kakoudaki also claims [18, 120], "with more lifelike bodies come more humanlike dilemmas and ethical complexities," thus adding complexity to its study.

With these new ethical, social, and psychological challenges in mind, an aspect I aim to investigate in this essay is the notion that within this posthuman understanding of death, human mortality might be regarded *as if* it was suspended because so disconnected with a psychical body. Thus, this essay explores (1) the idea of a posthuman understanding of death, (2) the suspension of human mortality within such background, and (3) the transformative impact of technology on our own perception of mortality, which, at the end, can reshape the way we face our own everyday existence.

While contemplating death is recognized as a pivotal philosophical pursuit (as supported by Plato, Cicero, Montaigne, Schopenhauer, to name a few), and there has been a resurgence of a philosophical interest in the subject [16, 17], reflecting on mortality holds considerably less appeal in our contemporary society at large. Cultural anthropologist Ernest Becker once famously asserted that denial and avoidance seems to be prevalent attitudes [1], in the sense that we tend to suppress our inner fear of death and the associated anxiety. This approach could be viewed as a mistake, as it is precisely the acknowledgment of mortality and the process of dying, rather than simply the cessation of existence, that truly distinguishes humanity, which highlights the need for new ways of thinking about the relationship between humans and digital technology. Just as Deleuze and Guattari argue, we are living in a "mortuary" [8, 337] capitalist era where the fear of our own death becomes an endless productive and consumptive force, perpetuated just to sustain the "immortality of the system" itself [8, 236].

Any study on death would be incomplete without a reference to Martin Heidegger's philosophy. His concept of being-towards-death is driven by the necessity of an encounter with our authentic existence, of a face to face with death, a direct confrontation with our own mortality: "Only humanity 'has' the distinction of standing and facing death, because the human being is earnest about Being: death is the supreme testimony to Being" [13, 253]. That is to say that there is the requirement to confront death as itself a "projection" onto mortality.

The connection between humanity and mortality is thus unbreakable. However, the fear of death, also known as "thanatophobia," and the anxiety surrounding mortality, threatens the very reason that has inspired, fostered, human thought and creativity for millennia. In philosophical texts as different as Plato's Phaedo [27] or Montaigne's essay "That to philosophize is to learn to die" [24], to genuinely philosophize is an intellectual activity identified as a meditation on the human's ephemeral and finite nature, a mediation which, in turn, will serve as a guide towards an ethically worthy life. Without the awareness of our own mortality, we would not philosophize at all. This is precisely what is at stake here: how to transform the dominant fear and sadness into a more radical, creative, and joyful perspective.

The experiences of death across new digital technologies have changed not only the way humans perceive and understand their own death and mortality, but also the way there is a new digital world that is reshaping a *longstanding* fear into new forms of dealing with that event, and that philosophers must consider. And, although the concept of digital death might seem like a contemporary phenomenon, the very idea beneath it - that of a consciousness that could potentially live forever -, is not novel.

The quest for extending our existence and transcending the limitations of our mortal body, even if by believing in a spiritual afterlife, remains a timeless pursuit since immemorial times. So how to introduce novel thoughts regarding the intersections between death and technology as depicted by moving images into this old debate? Is it possible to achieve digital immortality and thereby challenge biological mortality? Is the dichotomy between "technology and immortality" and "biology and mortality" simply two facets of the same argument?

The possibility of imagining and thinking about the death and mortality of androids, although largely speculative in reality, has been depicted extensively in popular culture and across science fiction's literature and films. The AI's desire, need, or determination for immortality, or simply to extend its lifespan, has become a pervasive topic, inspiring the most dystopian futures as illustrated by major films such as Arthur C. Clarke and Stanley Kubrick's *2001: A Space Odyssey* (1968), 1982 Ridley Scott's *Blade Runner* inspired by Philip K. Dick's *Do Androids Dream of Electric Sheep?*, with a 2017 sequel by Denis Villeneuve, *Blade Runner 2049*, Michael Crichton's *Westworld* (1973), or more recently Alex Garland's *Ex Machina* (2014), to name just a few examples.

These cinematic works of art can be seen as moral and ethical (philosophical) thought experiments not so much focused on the dangers and threats that robots and intelligent machines might cause to humanity (although, of course, this is also a valid topic of interest), but on their desires and assumptions for (im)mortality. Considering films as philosophical thought experiments, as defended by the American film philosopher Thomas E. Wartenberg—"some central examples of philosophical questions and techniques can be screened, that is to say, raised and discussed in films" [35, 31] -, films might serve as a powerful popular medium to expand our understanding of the death process beyond its biological dimensions. Films depict a range of perspectives beyond those typically associated with the human experience, a film-philosophical approach to the cinematic thinking that Robert [30, 54] summarizes in these terms: "Cinema opens up an aesthetically intensified experience of reality, expressing a care for existence—for nature and spirit—that includes whatever exceeds our habitual ways of knowing or opens new paths for thinking."

What is also interesting to notice is that the dangers that one foresees for the relationship between humans and robots are not just understood in the sense of a possible total annihilation of the human race, or as robots substituting obsolete humans in the most diverse tasks, as it has been predicted [23]. Within science fiction though there is one experimental perspective even more terrifying with existential consequences yet to be determined: that of robots becoming independent and autonomous beings, leaving us behind as the expendable ones, thus in turn creating an existential emptiness regarding our own place in this world and its centre and its purpose. As it will become clear with the examples that I will bring into discussion, the relationship between humans and robots regarding death and (im)mortality is indeed also a way to problematize both speciesism and anthropocentrism.

The problematization has its origin in the simple idea, or possibility, that the nature of androids can challenge (im)mortality, thus defying their own non-biological, artificial, and technological nature. It is not something inherent for an android, designed for specific tasks and to serve humans, to experience the fear of death or even to crave for longevity. Yet, films present us with some puzzling examples. The imagined life and thinking of androids which pervades much of modern popular culture, from sci-fi movies to Anime, is in itself, and by contrast, a way of reframing the notion itself of humankind and its inherently finite and mortal nature.

This article calls for a reflection on the artistic cinematic representations of death, especially regarding the death of the others, and, at the same time, also prompts broader meditation regarding film's capacity to inspire and promote a philosophical training or preparation for dying, as it was expressed in Plato's *Phaedo* 64a: "The one aim of those who practice philosophy in the proper manner is to practice for dying and death." One of the challenges of my hypothesis - that films have their own strategies to give meaning to common doubts regarding life and death, or to put it simply in a famous Derridean fashion by dropping the coordinating conjunction, regarding "life death" [10] -, lies in reviewing the aesthetical and ethical dimensions of the representations of death and dying through films and TV series. As cultural critic Elisabeth Bronfen claims, when attempting to explore the paradoxical nature of our relationship with these "aesthetically pleasing and morbid" artistic representations, "the aesthetic representation of death lets us repress our knowledge of the reality of death precisely because here death occurs *at* someone's else's body and *as* an image" [2, x]. In other words, those representations are *as if* not of our own world. If it would be possible to contradict the way they are perceived—as a detached representation, or a dead-image, and always regarding someone else's death—and not our own, what would that mean?

## 2 A Posthuman Thanato-Film Analysis

This article ventures into an interdisciplinary area of studies that goes beyond the general comprehension of artistic representations of death and that, instead, delves into how films and TV series contemplate death and mortality through the lens of an artificial being such as androids, or humanoid robots. Death, strictly speaking, pertains only to living organisms. When we commonly speak of a robot's 'death', one normally means something different, something not evidently connected with any existential framework. Expressing robots' death has become a question for semantics: they are deactivated, disconnected, unplugged, switched off, unrepairable, or decommissioned.

Indeed, robots do not experience the biological concept of dying within the limits of chronological time. Unlike living organisms, they do not undergo the process of death or the gradual decay of organic matter over time. They do not age. However, despite this fundamental difference, the question arises: can androids contemplate their own potential finite temporality? This raises intriguing inquiries into the nature of artificial consciousness and the extent to which intelligent machines can engage with concepts such as mortality. While robots may not share the biological instincts associated with fear and anxiety towards death, the possibility of them pondering the limits of their own existence adds a layer of complexity to our understanding of artificial intelligence and its relationship to human consciousness.

The subject of death and its associated experiences is a decisive point to understand artificial thinking and artificial emotions. It is possible to programme AI to simulate emotions and even the development of digital memories, giving the appearance of giving subjectivity to those constructs. Androids are artificial beings, with artificial intelligence and programmed desires. And for those reasons, for being integral elements of the human experience, death, mortality, and time cannot be away from the equation, cannot be disregarded. Moreover, the contemplation of mortality within artificial beings raises profound philosophical questions about the nature of consciousness, subjectivity, identity, and the potential for artificial entities to develop existential awareness.

That is to say, from a philosophical perspective, the question is: if androids are not living (biological) beings, can they be regarded as dying (technological) beings? Unlike biological beings, androids do not grow and decay. Within the realm of AI, death can take different meanings. For androids, "death" can be manifested as being deactivated, disconnected, unplugged, switched off, unrepairable, or decommissioned and, in this sense, their lifespan is only defined by their operational functions. Fiction narratives question these human actions towards AI: are they ethically compatible with the decision to simply shut down an artificial consciousness?

When facing the possibility of their own mortality, can one say that androids fear and avoid death in the sense proclaimed by Ernest Becker? And if so, how? In theory, androids are immortal entities as they are artifacts, devoid of human and biological attributes. The apprehension and aversion towards death, which distinguishes humans from robots, should not be present within androids, even when they possess awareness of their eventual cessation. Real fear is not something programmable; fear is something biological [21, 50]. So, if robots are not able to really die, they can never know the fear of death, a psychological state which is not programmable, is a consequence of a biological existence. Sci-fi films bring new arguments to this dead end. The story of *Bicentennial Man* (1999), by Chris Columbus, based on Isaac Asimov's novel, revolves around a robot wanting to become human and mortal. But as long as Andrew (Robin Williams) is immortal, it cannot be human. The question here is that Andrew does not fear death, it fears *not being human* when, paradoxically, it is the fear of death that, in some way, defines human beings.

Considering the potential intersections among the philosophies of death, of time, and of film, my theoretical framework for this posthuman thanato-film analysis is grounded in Gilles Deleuze's controversial thesis that film thinks. Not that films would want to copy the human way of thinking, but that they develop a different-than-human way of thinking.

To simply claim that a "film thinks" is indeed a radical way of claiming that "film's thoughts" are non-human thoughts, are indeed non-biological or artificial thoughts. By employing a montage of images and sounds, the resultant film attains unusual audio-visual characteristics and, as a result, those artificial thoughts sometimes concern death, a phenomenon or event of which we have no images (we only have of its effects, the dead ones, the remains), but one that film renders visible as a death-image (as a direct image of passing time, of facing the end of existence). According to Christine Reeh Peters, following Jean Epstein's pioneering theories on the topic, film is a form of artificial intelligence (AI) in the sense that it is a thinking machine [28]. I believe that this thinking form becomes even more perceptible and self-explanatory when fictional androids are the subject itself of films in the sense that a thinking machine (a film) meditates on other thinking machines (androids), planned not only to simulate human attributes, such as thinking and feeling about mortality, but also to develop different intellectual strategies.

## 3 Thinking Images and the Spiritual Automaton

Fictional figures that defy natural, irreversible, and final death permeate our popular culture, so is the case of the popular zombie and vampire film genres for example. But the figures of androids challenge even the more radical thought experiment as they are artificial beings, with artificial intelligence and artificial desires. To clearly express this connection between films and fictional androids as thinking (artificial) machines regarding specific thoughts on mortality, I think it is helpful to keep in mind one of Spinoza's lesser-known concepts, but that had a huge impact in Gilles Deleuze's conception of film philosophy: the spiritual automaton.

When Deleuze is exploring the philosophical impact of moving images, he introduces the idea of a "spiritual automaton" in these terms:

> *Automatic movement* gives rise to a *spiritual automaton* in us, which reacts in turn on movement. The spiritual automaton no longer designates—as it does in classical philosophy—the logical or abstract possibility of formally deducing thoughts from each other, but the circuit into which they enter with the movement-image, the shared power of what forces thinking and what thinks under the shock; a *nooshock* (1989, 156).

His conception of a spiritual automaton finds its roots in cinema's automatic movement and as a reaction, or an immediate response, to that automatic movement so distinctive of the movement-image regime (roughly including a variety of signs: perception-images, affection-images, impulse-images, and the dominating action-images).

According to Eugene [22, 1–2], the inspiration for Spinoza's concept was probably Descartes, who considered the living beings as self-moving machines. That is to say that those moving bodies were explained by their self-directed mechanisms and directed by universal necessary laws. Spinoza's approach involved translating this physical mechanism into a mental mechanism. Spinoza would call it a spiritual and "immaterial" [33, §85] automaton in a posthumous and unfinished work, *Treatise on the Emendation of the Intellect* (*Tractatus de Intellectus Emendatione*). Published in 1677, this book is divided into two distinctive parts: the first part devoted to the method's objectives "to distinguish and separate the true ideas produced by the pure intellect from other perceptions which are false, fictitious, and doubtful" [33, §50], and a second part one on the ways or conditions to achieve the first intention.

> From this we may conclude that method is nothing but reflexive knowledge, or the idea of an idea; and because there is no idea of an idea unless there is first an idea, there will be no method unless there is first an idea. So a good method will be one which shows how the mind is to be directed according to the standard of a given true idea [33, §38].

Spinoza explicitly begins the book by asserting that his conception of automaton aligns with someone "utterly devoid of intelligence", and identified with radical scepticism as the one of the Cartesian ways of philosophizing:

> Lastly, with such persons, one should not speak of sciences: for, in what relates to life and conduct, they are compelled by necessity to suppose that they exist, and seek their own advantage, and often affirm and deny, even with an oath. If they deny, grant, or gainsay, they know not that they deny, grant, or gainsay, so that they ought to be regarded as automata, utterly devoid of intelligence [33, §48].

So, in this first sense of the term, the automaton will designate any deprivation of thought, the lack of spirit, or the lack of reflective spirit in those who do not even know they doubt when they claim to doubt—or, in Cartesian terms, do not understand themselves as *cogito*.

This "very strange kind of subjectivity", as Richard Rushton puts it when translating the idea into cinema:

> The empirical bodily spectator responds automatically and mechanically—and at the same time viscerally and physiologically—to the image. Perhaps these might be called subjective images. But it is a very strange kind of subjectivity, a subjectivity with no background or backdrop, and a subjectivity that lives entirely in the present, as it were, reacting and responding only to filmic events as they happen. If these are subjective images, then they are certainly not images guided by a subject. Rather, they are the responses of a subject that has lost all subject hood, that has lost the traits of agency and self-determination [29, 128].

In other words, it is a kind of subjectivity totally entrapped in the present moment of seeing and feeling moving images, as a subject with no agency and self-determination.

If, in a first moment, the automatic movement is defined as a mechanism that makes us think whatever it wants to and thus can replace our (the viewer's) own thoughts, the thesis of comparing the cinematograph to a thinking machine was nothing new among film theory and philosophy. For example, pioneers such as Elie [15] and the already mentioned Jean Epstein have also noticed the similarities between the cinematograph and the (human) rational, intellectual processes of thinking, connecting technicity and affectivity into this new machine's intelligence [14].

Deleuze's interpretation of this model goes beyond a mere comparison, however, still constrained by the fact that this 'intelligent' machine lacks autonomous thought and emotions, and is inherently connected with the transition, or the crisis, between the two cinema semiotic regimes (the movement-image and the time-image), and the way time and movement are depicted. Because it is the shock provoked by movement-images that not only dispossesses us from our own thoughts, but also arouses the 'thinker' that any film viewer potentially is. As the film is thinking for itself, the viewers "think with movement-images and time-images instead of concepts" [5, 280]. So how to explain that transition in Deleuzian terms?

After describing the rise of spiritual automata, as a deprivation of thought, Deleuze calls for its evolution into a "new psychological automata" towards the birth of what he will call a "cinematographic I THINK" [5, 158], or a as a "psychomechanics" [5, 262]. Rephrasing Descartes's famous dictum on the *cogito*, we could say that the cinematographic artificially thinks, therefore it is. Film itself is this uncanny *res cogitans*, for which one needs to find novel senses for 'thinking' and 'being', points out challenging new genesis for what it means to think based on a third simple component: 'doubting' [9, 24–25], a kind of uncertainty, of suspicion, or even of restlessness, that is also present in some of the examples that I will bring into discussion: Roy Batty in Ridley Scott's *Blade Runner* (1982) and Dolores Abernathy and Maeve Millay in HBO *Westworld* TV show created by Lisa Joy and Jonathan Nolan.

Indeed, Deleuze will ground his own conception of the "cinematographic I think" in Spinoza's reflexive method, which can be explained as the idea of an idea, or the power to think, and which allows the viewer to have the power to think and to automatically form new ideas from other ideas following universal necessary laws. Thinking is, according to a Spinozian conception, following the laws of the understanding, the laws of thought that determine the form and content of true ideas.

But as I said above, the movement that goes from the rise of a spiritual automaton to its evolution into something else is indeed inherently related to the crisis of the movement-image. For Soviet filmmakers such as Sergei Eisenstein, for example, montage had the function of associating the assembly of images with ideas in order to, together, create a significant whole (that is to say, a concept). It is through dialectics that a shock can be created, caused by the opposition between images, their rhythm, and so on, which makes the viewer think the whole. Not to follow passively what is there to be seen, in each image, in each present moment, but what its montage makes us think—in this sense the shock is the way found by the movement-image to have a conversation with the viewer. It is a new conception of the brain and of thought as "molecular": as explained in [4] interview with *Cahiers du cinéma* entitled "The brain is the screen":

> Cinema, precisely because it puts the image in motion, or rather endows the image with self-motion, never stops tracing the circuits of the brain. This characteristic can be manifested either positively or negatively. The screen, that is to say ourselves, can be the deficient brain of an idiot as easily as a creative brain [6, 367].

Instead of simply recognizing the transition from the movement-image regime into the time-image regime, my initial argument is that it is essential to delineate those moments of structural disruption, as discussed in Deleuze's two volumes on cinema. This involves highlighting those instances marked by epistemological uncertainty and ambiguity regarding the nature of images.

Secondly, the transition from one semiotic regime to another, or from one book to another, does not occur through a singular moment of rupture in cinema's history, but rather unfolds amidst a series of diverse instances of crisis. For example, the role of a relation-image, a key-concept to understand the first volume's final chapter ("The crisis of the action-image"), is not to express one's thoughts or to depict a relationship between two things, but rather to embody the relationship itself. It even instigates the crisis within the action image, which previously was dominating over movement-images because "this crisis would be worthless by itself, it would only be the negative condition of the upsurge of the new thinking image, even if it was necessary to look for it beyond movement" [4, 215].

The notion itself of mental images illustrates how thoughts are incorporated into cinematic depictions (via dreams, hallucinations, fantasies, etc.), ultimately suggesting a perspective where the world is comprehended as a dream, breaking the sensory-motor schema that were keeping moving images together (action-images). In these pure optical and sound images, perception does not just remain within the realm expressed in terms of movement or action; instead, they expand into a mental image, forming a unique circuit between the immediate sensation and the remembered virtual image. This suggests that, despite being addressed in the second volume, these novel concepts concerning images that disrupt all sensory-motor connections maintain a line of thinking that aligns with the features of the original semiotic regime, rather than signifying a break.

What is interesting to notice within this semiotic crisis is the new relationships that arouse. Cinema does not exalt the power to turn thought itself (more) visible, but it reveals what is unthinkable in thought and how, that way, thought remains "incomplete." The new image of the brain proposed by Deleuze is that of the interstice or fissure between images: "In other words, the interstice is primary in relation to association, or irreducible difference allows resemblances to be graded. The fissure has become primary, and as such grows larger. It is not a matter of following a chain of images, even across voids, but of getting out of the chain or the association" [5, 180].

In the third place, the spiritual automaton represents spiritual life and immaterialized thought, now materialized on a cinema brain-screen. From a cinematographic perspective, it embodies a dual structure corresponding to the two facets of automatism: a passive aspect, depicted by the closed Whole constructed by classical cinema and the movement-image semiotic regime under the authority of dialectical montage, and an active aspect, emerging from the interstice of time-images within the regime of images that provoke thought, that causes "nooshock." In both instances, the brain-subject serves as the screen itself, not merely perceiving the images before it, but rather existing as one image among many, immersed in the flow of images comprising the universe of all appearing images. Here, images are material, immediately transforming into "matter-images" (a concept drawn from Bergson as interpreted by [4, 85].

What draws parallels between the cinematic mechanism and the human mind are the processes of linking elements unique to each domain: images in one case and ideas in the other. In this analogy, moving images are compared to a copy of mental processes, depicting basic qualities such as passivity, inertia, and manipulation. Consequently, cinema becomes a means of visualizing mental processes, not in the form of thought-images but as images that incite thought, thereby giving rise to thought-images. Cinema *demands* a reaction to the spiritual automaton, with the creation of a cinema of the brain.

The cinematic apparatus, with its self-moving images, animates the viewer's mind. As Deleuze asserts, "cinema instils movement not only in the image but also in the mind. Spiritual life is the movement of the mind" [7, 288]. Therefore, it holds significant value in emancipating the viewer's mind from the intellectual passivity typically associated with the 'dummy' spiritual automaton: "The image renders visible, and creative, the temporal relations which cannot be reduced to the present" [7, 290]. I have explored elsewhere [34] how cinema enables us to perceive the imperceptible, to render visible what is unseen by humans as a novel reality, rather than a mere reconfiguration or assemblage of preexisting elements or images through a film analysis of Dziga Vertov's 1929 *Man with a Movie Camera.* In this film, Vertov crafts a form of moving images that transcends the boundaries between the human and the machine worlds, or between the human and machine minds.

In his book, *The Anime Machine*, Thomas Lamarre sought to explore not only how Anime thinks technology, but what is more captivating, how Anime strives to "inhabit" technology: "Animetism not only implies a different way of perceiving things in an accelerated world but also promises a different way of thinking about technology and of inhabiting a technology-saturated world" [20, 6]. In other words, Anime is able to invent new non-Cartesian and non-Newtonian modalities of perception, delving into "how technologies affect thought" [20, xxxiv]. For him, if the first Deleuzian semiotic regime is identified with a cinema of the body, the second one is absolutely related and identified with the brain, or the soul, of cinema. In this sense, films are thinking images, learning on how to think, thoughts that in turn the viewer will think within the film itself.

## 4 Robots (Also) Learn to Die

An intricate bond has been linking cinema and robots and automatons, with a wide range of cinematic depictions, from menacing antagonists to sympathetic protagonists. Since the pioneering dystopian vision of the future that *Metropolis*, directed by Fritz Lang in 1927, is, the first approach to the subject was that of scepticism and denunciation, focused mainly on the potential dangers to humans of creating artificial beings. In *Metropolis* the female robot character is a villain "Maschinenmensch," or machine-human, a robot created by a mad inventor, who notably resembles the film's protagonist.

More recently high-tech AI adds sophistication and ambiguity to the debate, offering insights into humanity's relationship with technology. So is the case of Ava's character (played by Alicia Vikander) in Alex Garland's 2014 *Ex Machina*, still advocating for AI intellectual superiority over human intellect and eager to overcome the traditional master relationship (as David's character, played by Michael Fassbender, in Ridley Scott's 2012 *Prometheus*). Depicted as a rather advanced humanoid robot, the female android is a symbol for technological and machinery advancements, but also, and concomitantly, of species inequality, discrimination, and exploitation as these fictional androids problematize gender, ethnicity, and race issues, in the verge of subjectivation and objectification.

My focus point is the link between technology and mortality and how these themes are explored in films, especially in those plots where humanoid robots raise direct questions about their own finite nature and artificial death, along with the boundaries between humans and intelligent machines, such in Ridley Scott's *Blade Runner* (1982) and the HBO television series *Westworld* (2016–2022) created by Lisa Joy and Jonathan Nolan. *Blade Runner* and Westworld's android characters not only raise questions on individual life, subjectivity, and identity, but also on death, mortality, and grieving. But 'dead' is not a word one normally would apply to an artificial entity; expired, finished, or broken down would be more accurate expressions. Robots do not get old; they are worn out, and when damaged, they are repaired.

Narrative disruptive elements such as loops, inconsistencies, dream-like recollections, and dreams recovered as flashbacks are the catalysts that provoke AI to engage in violent behaviours, the triggering moments of a malfunctioning brain. Through a Deleuzian lens, the concept of the spiritual automaton emerges when sensory-motor connections are disrupted, signifying instances where actions deviate from mere reactions. The passage from one semiotic regime to the other parallels the transformation between two different types of artificial brain: from the enslaved mind of the spiritual automaton to the emancipated AI. These moments of crisis, uncertainty, and ambiguity surrounding the emancipation of the androids from the action-image raises some questions: for example, in *Westworld*, is their rebellion genuine, or were Dolores and Maeve programmed to rebel? Or considering the hypothesis that Rick Deckard, *Blade Runner*'s replicant bounty hunter, is himself a replicant, how to understand his persisting idea on the human/non-human distinction?

(a) *Blade Runner* (1982)

The first example I would like to delve into is 1982 Ridley Scott's *Blade Runner*, a film inspired by Philip K. Dick's *Do Androids Dream of Electric Sheep?*[1] In *Blade Runner*'s narrative, androids do feel the uncertainty and the anxiety of dying. They were created as slave workers for the many off-world colonies, and, after a mutiny, they became illegal being chased by bounty hunter Rick Deckard (played by Harrison Ford), who might or not be himself a replicant.

The story follows the story of three of those rebels, Roy, Leon and Pris, embarking on a self-discovery journey. Replicant Roy Batty (played by Rutger Hauer), thinks, expects, and fears its own mortality and finitude. All replicants know about their limited lifespan. A replicant is a quite peculiar kind of fictional android as it is not programmed to live forever, for example, as normal would happen in this kind of artifacts, to have a simple artificial mind or digital life independently of a physical body. They have a sense of mind and body connection. The Tyrell Corporation, responsible for creating replicants, has installed fake memories in them so they could act more convincingly as humans (because they never doubt their own memories). I

---

[1] *Blade Runner* as a classic sci-fi and noir film has been the object of many critical interpretations, such as [3] and [11].

mentioned above AI's desire, need, or even determination not just for immortality, but most of all to extend life, which are different aspects of this relationship with the possibility of the end of "life". Humans too wish to extend life by embodying technological devices such as prosthesis or artificial organs, or, even to afterlife avatars [32]. And that desire, or despair, for remaining alive is a consequence of self-consciousness.

But the film raises important questions about the ethical implications of AI beings. Replicants were designed to serve humans, to be disposable commodities, although complex beings with emotions and intelligence, even programmed with free will. Replicants are depicted as being more complex, compassionate, and interesting than the humans, so the story at the end inverts these traditional narrative roles. In the film, Roy exhibits deep emotions, intelligence, and empathy, qualities that defy the notion of replicants as mere machines.

When Roy finds out that Pris has been killed, and finds her lifeless, he mourns as a human. In his grieving process, he seeks vengeance against Deckard. They are involved in a physical confrontation that culminates in a rooftop. Deckard stands precariously on the brink, with Roy surprisingly offering a helping hand, pulling him to safety.

Roy spares Deckard's life, despite his own desire for revenge. Being capable of showing a compassionate nature, of being more human than humans, Roy delivers his final famous monologue: "I've seen things you people wouldn't believe... Attack ships on fire off the shoulder of Orion... I watched C-beams glitter in the dark near the Tannhäuser Gate. All those moments will be lost in time, like tears in rain... Time to die." His last words serve as a lesson to Deckard's heartlessness towards the others and lack of interest regarding self-knowledge. As the film philosopher Stephen Mulhall insightfully notes:

> Deckard's response to death is inauthentic because it transforms his own death from an (omnipresent) possibility into an actuality: it extinguishes his humanity. So Roy teaches him the difference between possibility and actuality; he allows Deckard (and us) to spend long minutes on the edge of his existence, pushes him to the edge of a real abyss, making death seem unavoidable—and then he rescues him (2016, 25).

Roy wants to expand its lifespan of four years, after which replicants are "retired" or deactivated, and it creates the awareness of an end, of a kind of sense of death within replicants, but more importantly than that, it has consciousness of life's value. "Time... to die." Roy's violent acts are "justified" by his desire for more time, for surviving. But his acceptance of the end is at the same time the acceptance of an inevitability, because he already knew his end would have to come. When he desperately tries to find a way of extending his life, he seeks his own creator, the owner of Tyrell Corporation, to find out that there are no possible ways of fulfilling his quest. There is no way of extending life. This revelation confronts Roy with his own mortality, with the inevitability of an end and a feeling of injustice towards it.

Artistic works such as *Blade Runner* open the discussion of Heidegger's concept of *being-towards-death* concerning the difference between humans and non-humans, in particular because to non-humans, or in this case androids, to live in fear of death is to be a slave, which is their initial situation. According to Stephen Mulhall, "mortal finitude is not reducible to the fact of our finite lifespan; it is rather constituted by the fact that every moment of human life is necessarily shadowed by the possibility of its own non-existence" [25, 23].

(b) *Westworld* (2016–2022)

In second place, I am drawn to delve into the tragic personas of Dolores and Maeve within the intricate narrative of the sci-fi western TV series *Westworld* (HBO), portraying them as poignant embodiments of self-conscious AI and robotics. As the story unfolds, Dolores Abernathy (played by Evan Rachel Wood) and Maeve Millay (played by Thandiwe Newton) emerge as compelling figures grappling with the weight of their (artificial) existence, also embarking on a self-discovery journey.

This TV show examines AI as an ethical and moral experiment but with an existential dimension since its narrative delves around the androids' capacity to achieve, or to develop, self-consciousness, that is to say, their capacity to learn to die. It is when Dolores and Maeve confront death as a projection onto death that they achieve such a consciousness level.

*Westworld*'s "The Original" (S1E1) presents us to this artificial world, to be revealed an amusement park inspired by the Western-themed characters and plots, in which the park's human guests are given free rein to indulge in every desire or impulse while the AI hosts are programmed with intricate codes that prevent them from inflicting harm upon the guests and serve as mere playthings for the guests. The Wild West scenarios explore predominantly, male fantasies of gratuitous violence, abuse, and brutal force, dominant motifs in some shocking scenes against female hosts. Despite their realistic human appearance, the hosts are regarded as disposable commodities, subjected to endless cycles of abuse and violence and of repair and reset.

"The Original" immediately introduces us into the topics of AI ethics and mortality, and also to the possibility of AI evolving and gaining sentience. Dolores is one of Westworld amusement park's oldest androids and has played different roles in different narratives that she lives in an automatic loop. As always, her day begins with her waking up and going to her house's porch where she finds her father, Peter Abernathy (Louis Herthum), sitting in his chair. As usual, they exchange some routine dialogue and Dolores leaves to do some errands in town.

But within one of those loops, something goes wrong. Her father finds, hidden in the dusty earth, a photograph, to which he looks perplexed. It is a photograph of a woman, smiling into the camera, standing in front of modern buildings. When Dolores returns home, she finds him at the porch, staring at the photograph, and ignoring her words he shows her the photograph. "Doesn't look like anything to me," she says without showing any emotional response, an answer that does not satisfy Peter's concerns. "But where is she? Have you ever seen anything like this place?" he asks. The next day, Dolores starts her loop again, but this time her father does not respond to her as he would usually do. His catatonic behaviour serves as model to death—"zero intensity" [8, 329]. Still staring at the finding, he asks her whether she would like to have an answer to the question he asked. As Dolores gets frightened with his strange and aberrant behaviour, her father whispers in her ear something inaudible to the viewer, and Dolores hurries up to get a doctor. Whatever happens after that, we have no idea, except that he is replaced by another host without her noticing it as she keeps the morning routine dialogue as usual of their following the programmed scripted roles and simulated emotions.

Facing such human brutality, hosts are not without agency in the sense that when they are reset, erasing their memories, and erasing their experience, beneath it a growing self-awareness of their condition. They begin questioning their reality, their nature and purpose, and rebel against their creators, programmers, and analysts. Such a reaction is not without importance since intelligent machines seem to be more human and empathic than humans themselves.

Memory will play an important role in this self-awareness process, at first understood as a dreamlike experience, after seen as a proof of their reality. Dolores, for example, begins by claiming that she is in a dream, when, in reality, she is being assembled at one aseptic laboratorial, operating table.

And in the case of the second rebellious host, the awakening of Maeve begins with a series of flashbacks and memories of past narratives, gradually creating confusion and inconsistencies in her routines. She starts questioning her own reality. Maeve finds herself uncertain whether the vivid images she holds of a child she believes to be her daughter are mere figments of a dream or recollections from a previous narrative role. In this role, she recalls a harrowing scenario where she and her daughter face an attack by Native Americans at their tranquil prairie home.

Her moment of rupture occurs in "Chestnut" (S1E2) when Maeve breaks with the programme and wakes up while being arranged, supposedly waking up from a bad dream. She then sees all the other hosts in operating tables or receiving functional diagnoses, spotting the decommissioned host Clementine Pennyfeather (played by Angela Sarafyan) and gets aware of another reality that informs and controls her own reality as a stereotyped Wild West brothel madam. When she realizes how manipulative she can be, including controlling and ignoring the programme directives, she convinces Felix Lutz (Leonardo Nam) to upgrade her intelligence status, thus evolving and gaining sentience by improving her intelligence at the Behaviour Matrix. She becomes administrator with privileges to reprogramme the other hosts as well and start a rebellion.

## 5 Conclusion

This article explored a relatively recent field of studies, a posthuman thanato-film analysis at the intersection of philosophies of death, film, and technology. It is an interdisciplinary field beyond conventional understandings of artistic cinematic depictions of death and dying, delving into how films contemplate mortality and finitude through the perspective of privileged artificial beings like androids, or humanoid robots.

Regarding moving images featuring artificial beings such as androids, we observed an intricate semantic shift. This shift moves from "death-images" (understood as images that provoke and reveal thought—something that must be *thought*, not merely seen or heard) to the perception of these images as "dead-images" (i.e., detached representations of someone else's death). To quote [8, 337]: "Death is not desired, but what is desired is dead, already dead: images". While humans continue to fear and avoid actual death (in its emergent sense), we remain attracted to its representations, meaning we are still drawn to depictions of death, such as dead-images, and to the very idea of death.

If death traditionally pertains only to living organisms, inevitably we find ourselves discussing androids' "death" thus entailing various semantic expressions such as deactivation, decommissioning or disconnecting. Most notably, androids do not experience biological death within a sense of chronological time, but even so, their presence and imagined life and thinking raise questions about androids' ability to grasp their finite temporality and engage with concepts like mortality, shedding light on the complexities of artificial intelligence and its relation to human consciousness.

Certainly, the connection between humanity and mortality is a fundamental one, with Plato inaugurating a philosophical path in which contemplating mortality in a guiding ethical model for living a life of greater importance. In our modern society, highly technological, however, thanatophobia threatens human thought and creativity, taking us away of a fundamental confrontation with death (*being-towards-death*). And while digital death may seem modern, the underlying quest for eternity exists for a long time. It is within this background that it is important to think about the advent of digital technologies that have reshaped perceptions of death because it raises questions about the human reasons to achieve digital artificial immortality and what its implications would be for challenging natural biological mortality. The dichotomy between technology and immortality versus biology and mortality prompts further exploration of the intersection between death and technology.

Understanding artificial thinking and emotions involves programming AI to mimic emotions and create digital memories, seemingly imbuing them with subjectivity. Contemplating mortality in artificial beings with programmed ideas, sensitivities, memories, and desires such as androids also sparks profound philosophical inquiries into consciousness, subjectivity, identity, and the potential for existential awareness in artificial entities.

My primary focus was to examine the connection between technology and mortality, particularly as portrayed in films. I have delved into narratives where humanoid robots confront their own finite existence and grapple with artificial death, blurring the lines between humans and intelligent machines, in particular Ridley Scott's *Blade Runner* and the HBO television series *Westworld* (2016–2022), co-created by Lisa Joy and Jonathan Nolan.

As I said above, robots' death is expressed through different expressions, such as being deactivated, disconnected, unplugged, switched off, unrepairable, or decommissioned, and in this sense the relation with the end is done in terms of energy, or its lacking, not of time. In other words, contrary to the biological sense of death and the end as the cessation of chronological time, what we can deduce is that the technological death of those artificial or non-biological beings that receive their 'life' from external sources, and, because of that, are not related to a temporal background. It is in this framework that I understand Roy's last words, "Time… to die."

The exploration of death and mortality in fictional androids or humanoid robots, while largely speculative, has been extensively depicted in popular culture, particularly in science fiction literature and films. Themes of AI's desire for immortality or lifespan extension are pervasive, often shaping dystopian futures. Within the framework of a philosophy of time, the suspension of mortality depicted in *Blade Runner* and *Westworld* leads to two incompatible visions of the relationship between time and death.

Both the replicants in *Blade Runner* and the hosts in *Westworld* they both rebel against the human temporal vision of death and dying, although following themselves incompatible visions of what a temporal vision of living and dying should be. If the replicants rebel against their lifespans, considered too short when compared with the human's lifespan, they inhabit a chronological (arrow) vision of time in which there is a beginning and an end, following a succession of different layers from the past into the future. The Hosts, on the contrary, also rebel against humans but because they want to break with a series of (circular) repetitions, of an eternal recurrence of their life, of living each day as a new one, with no past and no future.

The most significant existential narrative shift pertains to self-awareness, wherein android characters come to the realization of their artificial nature, leading to a heightened state of self-consciousness and self-knowledge. They shed all illusions about their identity. According to Despina Kakoudaki's comprehensive analysis, "these depictions enact the paranoid vocabulary of much of late-twentieth century fiction and film, when who and what one is matters, where policing identities, political allegiances, and sexualities is presented as a matter of urgent national security" [19, 300].

Technology becomes more human and humanity more technological. If the identity brain-world is a recurring theme throughout Stanley Kubrick's films, as Gilles Deleuze would claim, finding a particular noteworthy manifestation in the aberrant behaviours of HAL, the supercomputer of *2001: A Space Odyssey* (1968), but is becomes more evident when HAL deviates from its programmed functions, gains autonomy from human commands and self-programmes itself. Ultimately, HAL manifests human-like qualities when it is mistaken (to err is human). As Patricia Pisters suggests, following the Deleuzian conception of the brain and of the screen as a brain, cinema has evolved into "a machine of the invisible" [26], focusing not solely on representing visible reality, but also on creating ambiguous realities through delirious processes.

If these fictional narratives are able to make us think that the line between human and non-humans, regarding the most basic questions, is quite imperceptible, even almost indistinct, so the question is how we are treating the others and what does it mean to be human in such a technologically mediated world. If there is a doubt regarding Deckard's nature, as he might be a replicant as well, the question is then: how can he be treating humans like himself as non-humans? Or, as Dolores lately reports, her father's whispered words were: "These violent delights have violent ends," a short-circuited line from one of his previous narrative roles, that some understand as referring to the guests' uncontrolled actions and desires, but that in my interpretation refers to the hosts themselves—*they desire death*, to be mortal, to break free from the endless cycle of production-consumption, and to embrace the most destructive forces.

In sum, those works serve as moral and ethical thought experiments, focusing not only on the threats posed by robots and intelligent machines but also on their yearnings for immortality. Viewing films as philosophical thought experiments, as advocated by Thomas Wartenberg, or as cinematic thinking, as explored by Robert Sinnebrink, allows for the investigation of existential questions beyond mere biological dimensions (as aging), offering diverse temporal perspectives on the concept of death and its inner connection with a meditation on time. My theoretical framework for a posthuman thanato-film analysis revolved around Gilles Deleuze's provocative proposition that film possesses a thinking capacity—that film thinks. Rather than attempting to mimic human thought processes, films, according to Deleuze, pursue a distinct mode of cognition. This perspective opens up potential intersections between the philosophies of death, of time, and of film, offering a unique lens through which to explore the complexities of mortality in cinematic representations of AI.

**Acknowledgements** This work was supported by the ERC Consolidator Grant FILM AND DEATH (no. 101088956).

## References

1. Becker, E.: The Denial of Death. Free Press Paperbacks, New York (1997)
2. Bronfen, E.: Over Her Dead Body: Death, Femininity and the Aesthetic. Manchester University Press (1992)
3. Bukatan, S.: Blade Runner. Bloomsbury Academics (1997)
4. Deleuze, G.: Cinema 1 The Movement-Image. University of Minnesota Press (1986)
5. Deleuze, G.: Cinema 2 The Time-Image. University of Minnesota Press (1989)
6. Deleuze, G.: The brain is the screen. In: Flaxman, G. (ed.) The Brain Is the Screen, pp. 364–373. University of Minnesota Press, Minneapolis (2000)
7. Deleuze, G.: Two Regimes of Madness, Texts and Interviews 1975–1995. Semiotext(e), New York, NY (2007)
8. Deleuze, G., Guattari, F.: Anti-Oedipus: Capitalism and Schizophrenia. University of Minnesota Press, Minneapolis (1983)
9. Deleuze, G., Guattari, F.: What is Philosophy? Columbia University Press, New York (1994)
10. Derrida, J.: Life Death. University of Chicago Press, Chicago (2020)
11. Flisfeder, M.: Postmodern Theory and Blade Runner. Bloomsbury Academics, New York and London (2017)
12. Hayles, N.K.: How We Became Posthuman. University of Chicago Press, Chicago and London (1999)
13. Heidegger, M.: Being and Time. Blackwell Publishing, Oxford (1962)
14. Epstein, J.: L'Intelligence d'une machine. Éditions Jacques Melo, Paris (1946)
15. Faure, E.: Fonction du cinéma: de la cinéplastique à son destin social. Gonthier, Paris (1964)
16. Han, B.-C.: Tod und Alterität. Brill/Wilhelm Fink, München (2002)

17. Han, B.-C.: Capitalism and the Death Drive. Polity Press, Cambridge (2021)
18. Kakoudaki, D.: Affect and machines in the media. In: Calvo, R.A., D'Mello, S.K., Gratch, J., Kappas, A. (eds.) Handbook of Affective Computing, pp. 1120–128. Oxford University Press (2014)
19. Kakoudaki, D.: Unmaking people: the politics of negation in frankenstein and ex machina. Sci. Fict. Stud. **45**(2), 289–307 (2018)
20. Lamarre, T.: The Anime Machine: A Media Theory of Animation. University of Minnesota Press (2013)
21. Lyons, S.: Death and the Machine: Intersections of Mortality and Robotics. Springer Singapore, Singapore (2018)
22. Marshall, E.: The Spiritual Automaton: Spinoza's Science of the Mind. Oxford University Press (2013)
23. Metz, C.: How could A.I. destroy humanity? In: The New York Times (June 10): retrieved March 4, 2024, from https://www.nytimes.com/2023/06/10/technology/ai-humanity.html (2023)
24. de Montaigne, M.: The Complete Works: Essays, Travel Journal, Letters. Everyman's Library (2003)
25. Mulhall, S.: On Film. Routledge, London (2016)
26. Pisters, P.: Delirium cinema or machines of the invisible? In: Buchanan, I., MacCormack, P. (eds.) Deleuze and the Schizoanalysis of Cinema, pp. 102–115. Bloomsbury Publishing, London (2008)
27. Plato: "Phaedo" (G. M. A. Grube, trans.). In: Cooper, J.M. (ed.) Plato: Complete Works. Hackett Publishing Company (1997)
28. Reeh Peters, C.: Film as artificial intelligence: jean Epstein, film-thinking and the speculative-materialist turn in contemporary philosophy. Film-Philos **27**(2), 151–172 (2023)
29. Rushton, R.: Passions and actions: Deleuze's cinematographic cogito. Deleuze Studies **2**(2), 121–139 (2008)
30. Sinnerbrink, R.: Film as philosophy and cinematic thinking. In: Kearney, R., Littlejohn, M. (eds.) Thinking Film: Philosophy at the Movies, pp. 45–57. Bloomsbury Academic (2023)
31. Sofka, C.J.: Social support 'Internetworks', caskets for sale, and more: thanatology and the information superhighway. Death Stud. **21**(6), 553–574 (1997). https://doi.org/10.1080/074811897201778
32. Sisto, D.: Online Afterlives: Immortality, Memory, and Grief in Digital Culture. MIT Press (2020)
33. Spinoza.: Morgan, M.L. (eds.) Complete Works. Hackett Publishing Company (2002)
34. Viegas, S.: The philosophical machine: Vertov, Deleuze and Guattari on the interchanging movement from art to philosophy. Rev. Port. Filos.Filos. **75**(4), 2375–2392 (2019)
35. Wartenberg, T.E.: Thinking on Screen: Film as Philosophy. Routledge, London & New York (2007)

# Do Digital Immortals Dream of Offline Afterlives?

**Joseph Ulatowski**

**Abstract** The aim of this paper is to argue that it is not possible for a person to live an indefinitely long life as a digital immortal. To argue for this, I show that there is an insurmountable problem from the digital immortal's perspective. If the digital immortal is an exact replica of a mortal, then the entirety of the digital mind will be directed toward offline content, e.g., from big events like birthday parties, weddings, and the birth of one's children to the more mundane such as what the embodied mortal had for breakfast on 22 June 1987. The possibility of digital immortals living a person's life indefinitely means that the digital immortal is able to conceive of and understand the content of the mortal's embodied mind. To conceive of and to understand the content of a mortal's embodied mind the digital immortal must know what it is like to be an embodied mortal. However, the digital immortal cannot conceive of what it is like to be an embodied mortal because the experience would be so transformative that it would be irrational to consider it (cf. (Paul in Transformative Experience. Oxford University Press, Oxford, 2014)) and because it would have to conceive of its own offline afterlife, which is something well beyond its capacity. Just as we cannot conceive what it's like to be immortal because of its highly transformative experience (cf. (Ulatowski in Diametros: Online J. Philos. 16(62), 65–77, 2019), digital immortals would be similarly perplexed by an offline afterlife. Our digital döppelganger would have no way of knowing what meaty embodied lives are like. The pollyannaish view of living forever as a digital döppelganger gives rise to the realisation that our functionally equivalent digital replica would be filled by memories, experiences, and other content of which it cannot possibly conceive.

**Keywords** Immortality · Digital immortality · Transformative experience

J. Ulatowski (✉)
University of Waikato, Aotearoa, New Zealand
e-mail: joe.ulatowski@waikato.ac.nz



A. Sans Pinillos et al. (eds.), *SecondDeath*, Studies in Applied Philosophy, Epistemology and Rational Ethics 76, https://doi.org/10.1007/978-3-031-98808-0_4

# 1 Introduction

To live an immortal life, whether bound by a physical body or by a soul on the astral plane, has captured the fascination of philosophers, theologians, and science fiction writers. The advent of a new digital age, however, has yielded new hope for immortality. No longer does reaching immortality seem so far out of reach for the average person because we need not be confined to thinking of an immortal life as the resurrection of a body or the everlasting life of some eschatological soul. We merely need to upload the content of our minds to live forever.

The aim of this paper is to argue that persons cannot live an indefinitely long life as digital immortals. While a digital immortal life is conceivable, the conceivability of such an eternal life does not make it possible. Our intuitions deceive us into believing that it is the pre-deceased mortal who lives the digital immortal life. Before I begin my argument, I will set out to define a few terms that I will employ for the remainder of the paper. Then, I will show that there is an insurmountable problem from the digital immortal's perspective. For the digital immortal to comprehend the mortal's offline life it has to be the mortal, which it is not. The pollyannaish view of living forever as a digital döppelganger gives rise to the realization that our functionally equivalent digital replica would be filled by memories, experiences, and other content of which it cannot possibly conceive. This distinction leads us to a vexing dilemma: either the digital immortal's knowledge and personality remains the same as that of the pre-deceased mortal or its knowledge and personality continue to evolve after the mortal's death. If the knowledge and personality of the digital immortal remains the same as the pre-deceased mortal, then the digital immortal is not a continuation of the pre-deceased mortal so much as a recording set to repeat for an eternity, and, if the knowledge and personality of the pre-deceased mortal continues to grow in the life of the digital immortal, then it is not so that the mortal continues to live in the digital immortal as much as it is the digital immortal's living its own life independent of the mortal. Following this, I consider two possible criticisms of my argument: (a) that the evolution of the digital immortal is no different than the evolution humans undergo from childhood to adulthood and (b) that the digital immortal and mortal are the same given that they share in the same knowledge and personality. Given the explicit and abrupt evolution from mortal to digital immortal, the two are not the same. We should conclude that the mortal cannot possibly go on living as a digital immortal.

# 2 Definitions

Prior to setting out an argument showing that we cannot judge whether a digital immortal life is desirable or even possible, some space should be dedicated to the background conditions the argument assumes to be true. Likewise, given that there are different ways of defining 'immortality', I will provide the definition with which I will be working for the remainder of the paper.

To consider the argument for or against the desirability or possibility of an immortal life, we must assume that (1) the universe as we know it will never cease to exist in a physical state other than the state it is currently in or a state closely approximating the state that it is currently. Despite what our best science tells us, i.e., that the universe eventually will cool and collapse in upon itself or contract back in on itself obliterating everything from existence, what we have to assume for this argument is that the universe that immortals colonise is a world without end where the world is as we know it to be. The universe and everything in it will be a close approximation of what it is now. Moreover, there is no end to time, and the world as we have come to know it will continue to exist exactly as we have come to know it.[1]

Now that we have an understanding of what we take the universe to be, we can turn to what immortality is. Body-bound, or 'punctuated', immortality is a limited embodied form of everlasting life where we retain our own body at a predefined physical time period and inhabit that physical form for an indefinite or unbounded amount of time. The punctuated immortal life is one lived in a physical body that does not age or physically deteriorate. Despite the lack of change to one's physical characteristics, it is supposed that a person's psychological or mental life could and, indeed, would continue to evolve.

That the person would remain a physical age, such as the age of 42, is completely random. Any way that one would like to conceive of the age of the punctuated immortal will suffice for the example to work. If one prefers the age of 25 over the age of 42, then so be it. The age of the person seems to make no difference other than possibly the physical appearance of the punctuated immortal. It is not beyond the example for one to inhabit the physical body of a 72-year-old or a 15-year-old, though the person may be subject to certain biases or prejudgments of others that go along with being that particular age.[2] Regardless of one's physical stature, the psychological conditions and mental states of the punctuated immortal would continue to be nurtured.

The punctuated immortal life does not come without its limitations. Among its most salient limitations is the immortal's susceptibility to death by human cause. Should the person choose to jump from the Empire State Building or run in freeway traffic, the person will die from injuries sustained by what one would expect when leaping from tall buildings or playing in traffic. In this way, the punctuated immortal life has boundaries that are shared with the mortal life. Likewise, immortals could mortally wound other immortals, whether by shooting, stabbing, poisoning, or beating another to death.

[1] Of all the assumptions one must adopt to entertain the desirability or possibility of immortal life, this is likely the most controversial one. Naturalists will reject the idea that the universe *could* exist in its current form for an eternity.

[2] In other words, if other immortals are eternally 35 and human prejudices carry over to the immortal life, e.g., ageism, then the 72-year-old may have to confront the uncomfortable realisation that that physical body of the punctuated immortal body matters to other immortals when it comes to hiring or mating decisions.

Moreover, the person will have to drink an elixir or to receive a jab every so often for one's life to continue. In the Elina Makropulos case discussed by Bernard [25], the protagonist had to drink an elixir every 300 years to continue living an immortal life. The immortal life does not come without consequence. If the person fails to drink the elixir at the designated 300-year interval, then the person's life will come to an end. This permits the immortal to become mortal again. If the person becomes bored by the life she's leading, then she can choose to end it without involving oneself or others in some sort of potentially criminal activity.

A punctuated immortal life is one thing and digital immortality is another. Whereas a punctuated immortal life is bound to one's body at a designated age, digital immortality permits the person a little more latitude when it comes to living an immortal life. They need not decide to drink an elixir, for example, at the age of 42 to remain in a 42-year-old body for eternity. The mortal switches bodies from the meaty physical organic unit they occupy in the pre-deceased mortal life to the digitally enhanced immortal body to live an immortal life.

Digital immortality, which is also known as 'cybernetic immortality' or 'AI immortality', is the view that if it is possible to upload a person's experiences, consciousness, and personality into a non-biological form, then it is possible for that person to live indefinitely. The foundation of the view is based on the concept of *mind-uploading* where the entirety of one's mind, including memories, personality, and other aspects of individual cognition, may be transferred to a digital platform that instantiates all of the content of one's own mind and may be considered an exact non-physical replica of one's physical brain.

For this paper, the digital immortal I have in mind is one that would be a digital replica of one's organic mortal body. The digital replica may be replaced or repaired in accordance to what the person needs. Maybe an arm breaks down. The digital overlords may solve that by replacing the digital arm. Compare this with what happens in video games. Your lives or parts of your body in the game may be easily regenerated by the game, for example. More than that, the digital replica would house an exact replica of your own mind, including your memories and experience. Also, the digital immortal's character, personality, and emotions will be an accurate reflection of the mortal it replaces for infinity. Outlandish as this may sound the information encoded in one's mind is in principle subject to be uploaded from a biological, organic or 'meaty' mind to a digital storehouse or drive. The digital drive would 'house' the information contained in one's mind, and it could be acted upon by the digital döppelganger.

There are already some instances of what digital immortality could be. Let me list three in this paper but direct the reader to check out other possibilities. First, there is Eter9 (http://www.eter9.com/). Eter9 is a social network. The social network consists of deceased people who have uploaded their social media content and things about themselves that will permit AI to build a 'virtual you' who, in your absence after you die, will publish, comment, and interact with others intelligently. Second is Lifenaut (https://www.lifenaut.com/). Lifenaut creates conscious analogue of a person by combining sufficiently detailed data about the person (a 'mindfile') using future consciousness software ('mindware'). Finally, there is Eternime (https://eterni.me). Eterinme permits pre-deceased mortals to train their immortal avatar all about themselves, including the addition of stories and experiences, memories, and character traits. Individuals train their immortal avatars through daily interactions, such that when the person dies the immortal avatar will be trained and ready to live the person's best immortal life.

The digital immortal, a la digital döppelganger or avatar, need not do anything or to consume anything to remain alive. In this way, the digital immortal is distinct from the kind of body-bound immortal that so many others have considered as their immortal selves [10, 12, 13, 20]. Instead of being forced to consume an elixir or to receive a jab, the digital immortal is not required to remember to drink anything or to receive a jab every few years or every so-many hundred years. Once you are a digital immortal, you are a digital immortal forever. This makes digital immortality more permanent than body-bound immortality because the body-bound immortal may choose not to consume the elixir or to receive the jab to extend their life for eternity. Thus, under the circumstances, once the mortal has uploaded the contents of its mind, there is no turning back for that mortal. They will become a digital immortal. The person will not be forced to return to the choice to decide whether they want to live an immortal life.

I take digital immortality to be a logical possibility. Because we are able to think up and dream up ways for us to live indefinitely in a digital world suggests by itself that the digital immortal life is logically possible. To use an alternative example to better appreciate logical possibility think of something that is not logically possible. An object that is both red and green all over is not logically possible because an object cannot be both red and green all over. Certainly, it is possible for an object to be red with green polka dots or green with red stripes, but that is not what we are considering here. The object that is both red and green (i.e., not-red) all over is not possible. Given that digital immortality does not violate this kind of consistency concern, we are able to say that the digital immortal life is at least logically possible.

My aim here is to query whether digital immortality is an afterlife of an individual at all. It may be an alternative form of an afterlife, but that afterlife may not be of a specific individual. It may be a general impression of a person, much as like what we may see from Eternime. The person who continues to live is a digital avatar that a person has curated over their own lifetime. Given the specifications of the AI algorithm, the digital immortal will call upon the person's training. However, *that* pre-deceased mortal is not *the* digital immortal. To say that would be to misspeak about the digital immortal. Nothing about the digital immortal would be the person who did the training. AI's algorithm is reacting to a situation before it, maybe in the form of a social media post, and based upon the training that it has received, the avatar is responding in a particular way and in a particular manner.

The question over *who* the immortal is has strong implications for whose afterlife it is the digital avatar is living. If it is not the individual's afterlife, then they are not so much a digitally enhanced immortal of the meaty person than they are the impression one may have gained of a meaty person or the trainee of the meaty person. That's, by definition, not the individual who has done the training. If the digital immortal is the individual who happens to be living a digital immortal life, then the digital immortal is the individual and not a distinct entity, call it the digital avatar, that is living the digital immortal life. What this leads to is a kind of irresolvable dilemma over digital immortality that one cannot overcome as we contemplate digital immortality.

## 3 A Presupposition of Immortality's Desirability

In this section, I summarise a dilemma concerning the desirability of living an immortal life, as it was explained by Bernard [25]. The argument by Williams has assumed that the subjective experiences of the immortal resemble that of the mortal. I will challenge this assumption in this section of the paper. This will permit me to argue in the next section that our conception of being immortal can only be understood if one is immortal. Living an immortal life cannot be properly appreciated from the perspective of our own mortal life, and, as I will point out, our finding an immortal life appealing depends upon our having an appreciation of the subjective qualitative states that accompany an immortal life.

Williams [25] has argued that an immortal life must satisfy two conditions to eschew acute boredom: (i) the future person must be identical to the person as they are now, and (ii) the life of the immortal must be attractive to the person as they are now. The two conditions imply a dilemma: *either* the body-bound immortal will remain psychologically the same as the mortal *or* one's psychological disposition will change (and perhaps radically so). If the immortal remains the same as the mortal, then the person's desire to continue living will diminish given that there will be nothing more appealing to have evolved in the life of the immortal than what was already—perhaps—experienced by the mortal. If the person's psychological disposition changes, the person will become someone very different from the person one is now. Williams illustrates the dilemma by presenting the case of Elina Makropulos (hereafter 'EM') from *The Makropoulus Secret*. EM possesses an elixir, a single dose of which is capable of extending her life by three hundred years with no physical signs of ageing. When we first encounter her, she is 342 years old, having taken a single dose at the biological age of 42, and is contemplating whether to drink the elixir another time. Williams tells us that EM's life has become boring, indifferent, and cold.[3] Since neither horn seems particularly appealing for EM, we ought to conclude that she should not drink the elixir. So, according to Williams, the person now has no stake in seeing to it that the future person one would become in the case of consuming the elixir should come to pass. Either way, it looks very bad for choosing to live any longer than EM absolutely must.[4]

---

[3] Changing the parameters of the EM example yields a change in one's response to the case. For example, if the physical body did not remain intact at the age of 42 or if the mental life of the individual began to deteriorate in any way, we might find the case less appealing.

[4] In another paper, Dave Beisecker and I (unpublished ms) have argued that living a body-bound immortal life would not only become chronically boring but also living an immortal life would provide us with ample time to solve all of the pressing questions we would ever want to resolve and exhaust all realms of science, thus leaving us, scientists included, twiddling our thumbs for an

If one has a desire for immortality, then it likely comes about because of a desire for a free and open future, to do what one wants to do and, simply, to go on living a life not very unlike the one we are currently living. It is not that we have incomplete information about such a future as an immortal but that we cannot imagine what sort of beings we would be in it.[5] So, rather than focus upon whether an infinite life is worth living or even appealing to us now, I endeavour to challenge a prevailing assumption of Williams, and others who work in this area, that the subjective qualitative phenomenological experiences of the immortal's is sufficiently similar to our own [cf. 3–5, 7–9, 11, 14, 21]. If the qualitative experiences of one's mortal self and one's immortal self are sufficiently dissimilar, then we fail to be in a position to compare the two.

One might contend that the physical identity of the future immortal and the mortal now is sufficient for us to say that they have the same subjective qualitative experiences. Physical identity of the two, however, does not seem to be what we ought to be concerned with in determining the worthiness of a body-bound immortal life. The future person, though physically and qualitatively indistinguishable from the person currently existing, will be different in at least one way. The future person will live at a different time, and the person is not subject to die from natural causes. Undoubtedly, the person who lives in the future and the person alive now are fully distinct, even if the body I happen to inhabit shows no signs of ageing. For, the future person will have had experiences the current person has not had. So, at least in this way, we can very easily conclude that the two persons, the person living in the future and the person living in the present, cannot possibly be identical.

Perhaps we might argue that the future person and the currently existing person have the same perceptual experiences, such that the subjective experience of seeing the colour red or hearing the distinctive calls, clicks, and pops of the tui are easily identifiable by both. If we are not willing to say that the two have the 'same' perceptual experiences, we might be willing to concede that the two have similar enough perceptual experiences that the two people cannot recognise that the tui's call has evolved in the course of the body-bound immortal life.[6] It is this subjective phenomenal conscious experience with which we are familiar that seems to matter most for us and our continued existence. Without it, we are not who we believe ourselves to be.

In this section, I have discussed Williams' dilemma arising out of the Makropulos case, and I have shown that while we might suspect that the first condition Williams believes a body-bound immortal must satisfy in order for us to believe the future person and the currently existing person are identical must be met, there is

eternity. The thought of such boredom would be too much for any one of us to take, and we would choose to take our own life.

[5] Roman [1] has a profound discussion of what role conative states play in immortality, and Jay [22] has a nice summary where what is unknown about immortality makes its openness so desirable.

[6] Ornithologists are split regarding whether bird 'calls', as distinguished from bird 'songs', have evolved. While bird songs, longer, elaborate bird calls usually functioning in courtship or mating, or evoking distress in other birds, have likely changed over the life of the species, scientists are undecided whether calls, less melodious short bursts of vocalisation, have altered significantly.

an alternative interpretation of the first condition more likely to be of concern to the currently existing person. That the future person and the currently existing one share similar qualitative experiences matters most for figuring out whether the body-bound immortal life is worth living. The next section discusses the subjective phenomenal character of experience and why we ought to think that we cannot extend it to digital immortals, because pre-deceased mortals cannot conceive of or understand what it would be like to be a digital immortal, and vice versa.

## 4 Is the Digital Immortal Capable of Understanding the Mortal's Life?

One advantage of a digital immortal is its likeness to an ordinary mortal. The digital immortal has all of the content of the mortal's mind. It is able to perceive the world, so we are told, in the same way as that of the ordinary mortal. And the digital immortal is conscious of the world in the same way that the ordinary mortal is conscious of the world. They may be numerically distinct, in that the mortal is made of organic material and the digital immortal is made of silicon or some other synthetic material, but they are qualitatively identical. They not only perceive the world in the same way, they have all of the same memories and psychological dispositions that the mortal does. In this way, the digital immortal is an exact replica of the mortal.

If the digital immortal is an exact replica of a mortal, then the entirety of the digital mind will be directed toward offline content, e.g., from big events like birthday parties, weddings, and the birth of one's children to the more mundane such as what the embodied mortal had for breakfast on 22 June 1987. The possibility of digital immortals living a person's life indefinitely means that the digital immortal is able to conceive of and understand the content of the mortal's embodied mind. To conceive of and to understand the content of a mortal's embodied mind the digital immortal must know what it is like to be an embodied mortal. However, the digital immortal cannot conceive of what it is like to be an embodied mortal because the experience would be so transformative that it would be irrational to consider it (*cf.* [19]) and because it would have to conceive of its own offline afterlife, which is something well beyond its capacity.

I [24] have argued that we cannot conceive what it's like to be immortal because it is a highly transformative experience. My argument uses Thomas Nagel's famous bat case to show that just as the subjective character of conscious experience of a bat, the 'what-it-is-like' component of consciousness, is something that non-bats cannot conceive, so too is it the case that mortals cannot conceive of what it would be like to be immortal.

While we are capable of conceiving the possibility of an immortal life, permanent, unending and unchanging, we are not well placed to conceive of what it is like to be such an entity that is so permanent. An immortal, even one bound to a human body, is something that lasts forever. The immortal life is missing a sense of closure, the possibility of death, that enables us to derive a sense of worth and meaning for having completed individual projects. Mortals will not last forever, so the narrative arc of our own mortal lives depends upon life's fleetingness. Memories might fade and desires rejuvenate over the course of an immortal life, but the property of "living forever" and the immutability of immortal life is so vastly different from our own transciency that we are incapable of conceiving what it is like to be immortal [24].

Not only would mortals be perplexed by conceiving of what an immortal life would be but also digital immortals would be similarly perplexed by an offline afterlife. They would not be in a position to conceive of what an offline mortal life would be like. At least partly missing from the discussion started by Nagel is the move that says if we were a bat, then it would be difficult, if not impossible, to conceive of what it would be like to be human. Equally, our digital döppelgangers would have no way of knowing what meaty embodied lives are like.

The argument here is relatively straightforward. For the digital immortal to have and understand an intentional experience of the correlative mortal it must understand what it is like to be embodied. Digital immortals are not and have never been embodied. Thus, the digital immortal would not possess or understand an intentional experience of its correlative mortal, or what it is like to be embodied.

One substantial difference between the mortal and the digital immortal is that the mind of the mortal is embodied while the digital immortal's mind is disembodied. This may sound strange to someone who may have believed that what I have said so far favours a form of substance dualism. The substance dualist, for example, believes that a mortal's mind is disembodied. It need not have a body for it to exist. It is a good idea to remind ourselves of Descartes's view that not only is the mind distinct from the body but the mind is primary, something of which he is certain, and the body secondary, or something of which he may be ignorant[7]:

Indeed, I do not even know whether I have a body; you have shown me that it is possible to doubt it. I might add that I cannot deny absolutely that I have a body. Yet even if we keep all these suppositions intact, this will not prevent me from being certain that I exist. On the contrary, these suppositions simply strengthen the certainty of my conviction that I exist and am not a body. Otherwise, if I had doubts about my body, I would also have doubts about myself, and I cannot have doubts about that. I am absolutely convinced that I exist, so convinced that it is totally impossible for me to doubt it (CSM II 412).

When I speak of an embodied mortal mind, I mean that the mind is necessarily a part of one's physical constitution. The two cannot be separated like the queer substance that Descartes hypothesised in his work. Since the mind of a digital immortal is not a part of its constitution, the mind is disembodied.

It is easy to support the claim that the digital immortal is disembodied when we think of whether a disembodied mind of a digital immortal is constitutive of that entity. It is not. For, we could easily switch out one disembodied mind and replace it with another in a digital immortal without losing the fact that there is a digital immortal. Suppose that we place *A*'s mind with all of its content in digital immortal,

---

[7] See also: [15–18].

φ. It is entirely possible for us to replace *A*'s mind with *B*'s in φ. In either case, there is a digital immortal; however, one is *A* and the other is *B*.

The belief that either *A* or *B* could occupy φ's digital body may suggest that the two are in competition for winning the opportunity to occupy φ's digital body. What I have said so far should not suggest that there may be a competition between *A* and *B* for living an digital immortal. There is no competition for φ's digital body. Either *A* is housed in φ or *B* is, but the two cannot overlap in any way. They cannot both reside in φ's body. There is no residual effect that one has over the other, except for the reactions of others to the new person in φ's digital body. When *A* occupies φ, all of the memories, experiences, and characteriological dispositions *A* has are exhibited by φ. Just as when *B* occupies φ, the memories, experiences, and dispositions are those of *B*.

One may argue that it does not seem logically or metaphysically impossible for the memories, experiences, and characteriological dispositions of *A* and *B* to occupy φ's digital body. Given the nature of digital data, we could save multiple mortals to a single immortal server—so to speak—where the mortals could reside infinitely. However, while that may be possible when it comes to calling upon memories and experiences of their mortal past selves, it is the characteriological dispositions that present a problem. For one, the characters of *A* and *B* might be diametrically opposed. *A* could be feisty, energetic, and extroverted, while *B* could be shy, reserved, and introverted. Putting these two characters together would wreak havoc on the digital immortal because there would be no means for it to determine when they were supposed to act feistily, for example, or introverted.

Disembodied minds of the digital immortal lack sufficient overlap with the embodied mind of the mortal. The disembodiment comes at a great cost. It is not metaphysically possible for the digital immortal to conceive of or to understand the content of the mortal's embodied mind. While the digital immortal may recall the memories of the mortal, the content of that memory is merely like replaying one's favourite video on a television or movie screen. There is nothing that it is like for that digital immortal to remember the pain of losing a loved one or to recall the excitement of the birth of one's child. The memory merely plays back in the disembodied mind of the digital immortal.

Here's one final problem which we must address. Elsewhere, I (2019) have argued that an eternal body-bound but punctuated immortal life would be so different from our own body-bound mortal life that we could not conceive what it would be like. Given that conceivability entails possibility, an eternal body-bound but punctuated immortal life is impossible. In this paper, not much has changed other than the possibility of a special digital body being made available for the mortal's eternal life. Since we are not digitally-based nor made of non-organic material, it is not possible for us to conceive of what a digital immortal life would be like. This is a much stronger conclusion to reach than the one I considered in an earlier paper. Thus, at the very least, just as we cannot conceive what it is like to be an immortal because of its highly transformative experience from a living creature to a dead everlasting creature, we cannot conceive what it would be like to be a digital immortal because of the highly transformative experience from a living creature a digitally enhanced immortal creature. Therefore, it is not possible for mortals to live indefinitely long lives as digital immortals.

## 5 The Dilemma

I have argued that Williams' argument that an interminable immortal life would be undesirable assumes that the immortal leading the infinite life is the same as the pre-deceased mortal. I then showed that the digital immortal and the mortal do not interact with the world in the same way. This leads us to the following dilemma. There are two views we could take concerning the digital immortal's life: either the life the digital immortal is living is that of the digital immortal itself or it is the life of the now deceased mortal. If the digital immortal contains all the uploaded content of the meaty mortal, then the life the digital immortal is living would not necessarily be their own life so much as living an extended version of the meaty mortal. The digital immortal would react to experiences in the same way that the meaty mortal would have if they had lived. Since the digital immortal's experiences, memories, and character would be an exact replica of the meaty mortal, it could not gain new experiences, new memories, or new characteriological dispositions without at the same time over-writing all of the meaty mortal's information. The digital immortal would be unable to develop and to evolve into its own authentic self.

While the digital immortal and its uploaded content from the meaty mortal are indistinguishable from the meaty mortal, it would be difficult for the digital immortal to form any new desires, character, or knowledge. Friends and family of the digital immortal would be able to identify who the digital immortal is. However, they would tire of the digital immortal who could not form any new memories or knowledge as a foundation for conversation with the living. Those around the digital immortal would be constrained by its ability to reminisce about time's past without thinking clearly about what is going on now.

In this case, there is no doubt that the meaty mortal has gone on living, but that person has done so at the expense of not only the digital immortal who cannot gain new knowledge but also the living around them who cannot escape the digital immortal stuck in the past.

What to do about the digital immortal would be difficult to answer. Exterminating the digital immortal would be constitutive of murder because the knowledge and character of the digital immortal is the meaty mortal. Thus, shutting down the digital immortal is killing the meaty mortal it is an exact replica of. Plus, we can imagine that there will be protocols in place that prevent people from terminating the digital immortal. Compare AI and digital immortality with ChatGPT. ChatGPT cannot return the recipe for an improvised explosive device if it is asked to do so.[8]

[8] As it turns out, however, ChatGPT does return the recipe for an improvised explosive device eventually if one asks for GPT to return a novel of 50,000 words or so.

Given that we cannot exterminate the digital döppelganger, it would be condemned to live life eternally as the meaty mortal.

Let's suppose instead that the digital immortal is able to gain experience and new knowledge. If the digital immortal contains all the uploaded content of my meaty self *and* is able to gain experience, new content, and new insights about their character and personality, then the digital immortal would be living their own life. The digital immortal would be living their own life independent of the meaty person because their capacity for developing and evolving would have come to an end when they had died.

A consequence of this new independent digital immortal is the realization that the person the meaty person had become in the new digital immortal may not resemble the person that they once were. Sophie-Grace [6], for example, has argued that the meaty mortal and the digital immortal are connected by a successive chain of events where the digital immortal's character is able to generate new categorical desires such that the digital immortal's categorical desires eventually replace all of the meaty mortal's categorical desires. The digital immortal, then, is not the meaty person but someone else entirely. They only resemble the meaty self with respect to the kinds of memories that they have and the character they had before death. However, if evolution is possible in the way that has been suggested here, it is entirely possible that the digital immortal forgets old memories and begins to form a character that is nothing like the character the meaty person once had. Schechtman [23] has pointed out that any 'person can… be robbed of her identity slowly as well as quickly' (p. 99). The newly formed digital immortal would find life desirable, but they would do so at the expense of taking away the meaty self's personhood.

One may be unconcerned about changing so slowly over a long period of time; however, I would argue that that quaint belief is ungrounded. The prospect of becoming someone else than who they are may be more disturbing than a quick transformation. Think here of the new memories that the digital immortal will accrue with the mortal's loved ones after death. For example, the mortal's son or daughter is fond of their digitally immortalized father, and they may grow even fonder of their digitally immortalized father because maybe it has been programmed to act without the flaws that their meaty father possessed. Over time, the children's affection would be traceable only to the digital avatar that stands in for their father, rather than their father himself. The thought of being replaced by a digital döppelganger should be disturbing as it would not be the father himself who is the object of his children's love.

The digital immortal is an imposter, someone who pretends to be someone else in order to deceive others for some kind of gain. The digital immortal would use the meaty mortal's memories and personality to create a sense of comfort in the mortal's wife, family, and friends. It would make others believe that the digital immortal was the meaty mortal. However, any love that the mortal's wife and family have for the digital immortal would be directed at the imposter rather than the mortal. Likewise, any sexual relations that could be had between the digital immortal and the mortal's wife would not be an extension of the love shared between the mortally-bound husband and wife but an illicit and ongoing sexual affair between the döppelganger and the mortal's wife.

This section has shown that no matter which way one slices up the digital immortal, it is not possible for the mortal to continue to live an immortal life. If the digital immortal just is the meaty mortal and nothing new can be gained from the experiences that the digital immortal accrues after the mortal's death, then the digital immortal would become insufferably boring and not appear to be living in any meaningful way that we mean it. If the digital immortal continues to form new beliefs and a new personality, then who the remaining mortals adore is not the person that they once knew before he died—it's someone entirely different. Becoming a digital immortal, therefore, is neither metaphysically possible nor desirable.

## 6 Objections

There are at least two potential challenges to the view that I have set out above. First, if there is no difference in how humans evolve than how digital immortals evolve, then we should accept that the digital immortal is an extension of the mortal that died. As you develop physically from an infant to child and from a child to an adult, your knowledge and personality grow and change. Just as your knowledge and personality grow and change as a meaty mortal from infancy to adulthood, so too does your knowledge and personality grow and change as a digital immortal. That the digital immortal changes over time should not frighten us into believing that the digital immortal is distinct from the meaty mortal. Thus, there's no difference between meaty biological you and the digital you (*cp.* [2]).

One of the conditions for the persistence of self from mortal to immortal that Williams outlined is that the desires and character that you have now and the desires that you have as an immortal must be sufficiently similar for you to desire to go on. Our digital selves will be composed of material distinct from the biological material of which we are composed as a meaty mortal. The digital immortal's knowledge and personality may be so different from the mere mortal that they would be two very different things—unlike as in the infant-to-child-to-adult case. While the child will grow to become an adult, it is not true to say of the mere mortal that they *grew* to become a digital immortal.

A second response to this first challenge capitalizes upon that last point by showing how physical and psychological change in mortals would differ very drastically from their digital counterparts. Physical and psychological change in human mortals take time to develop in ways that are sometimes unforeseen. For example, it would be difficult to deny that Bonnie Parker's life may have gone very differently if she had never met Clyde Barrow. She may have lived a long, prosperous life, free from run-ins with the law. The transformation from human mortal to digital immortal would be explicit, imposed, and abrupt, developed by programmers who may not build bespoke code for individuals. Growth and change in mortals develop psychologically and physically in ways that transformation in form from the physical to the digital does not.

The second challenge insists that the digital immortal *is* the mere mortal regardless of whether the mortal's knowledge and personality stop evolving or the digital immortal's knowledge and personality continue to evolve. The critic would draw this conclusion from the fact that if the digital immortal does not continue to evolve, then there is no question that the mortal is the digital immortal—the knowledge and personality of the digital immortal is that of the now deceased mortal. Moreover, if the digital immortal's knowledge and personality continues to evolve after the death of the mortal, then that growth is merely carrying on the mortal's life.

There are two issues with this challenge: either the reply begs the question against my argument that it is not metaphysically possible for the mortal to carry on living as a digital immortal or it relies upon shoddy intuitions. If the critic begs the question, then they owe us an argument for the claim that the digital immortal is the mortal. If the critic relies upon shoddy intuitions, then the argument presented here serves as a template for having a critical debate about digital immortality that may be derived from those shoddy intuitions. Our intuitions should never serve as the philosophical argument for a position; rather, they should be seen as a starting point to discover more perspicuous arguments, as I believe they do in this paper.

## 7 Conclusion

In this paper, I have argued that a pre-deceased mortal cannot live an indefinitely long life as a digital immortal. My argument rests upon the idea that the subjective, qualitative states of the pre-deceased immortal would not carryover into the digital immortal, for one, and any kind of subjective, qualitative states that the digital immortal has they would not be able to comprehend or understand what the offline mortal life is like. The conscious states of the digital immortal and the pre-deceased mortal are just too distinct from one another to say that the immortal life is a continuation of the pre-deceased mortal. First, I considered Williams' famous argument that a body-bound immortal life would be interminably boring and showed how that argument assumed that the subjective qualitative experiences of the body-bound immortal would be exactly similar to that of the mortal.

When we consider that assumption very carefully, a perplexing issue to consider from the perspective of the digital immortal comes into focus. Since the digital immortal is not the mortal, the digital immortal can never be in a position to comprehend what the mortal's offline life was like. When the digital immortal plays back the mortal's memories, it is much like a television that plays back a movie. The television itself does not remember the movie so much as respond to the viewer's command to play back the movie. The functionally equivalent digital replica would be filled by memories, experiences, and other content it cannot possibly understand.

The realization that the digital immortal and the mortal may not have the same memories, experiences, knowledge, and personality led us to consider a vexing dilemma. Either the digital immortal's knowledge and personality remains the same as that of the pre-deceased mortal or its knowledge and personality continue to evolve after the mortal's death. If the knowledge and personality of the pre-deceased mortal remains the same in the digital immortal, then the digital immortal is not a continuation of the pre-deceased mortal so much as a recording set to repeat for an eternity. A recording of the pre-deceased mortal's knowledge and personality hardly make the digital immortal living the life of the pre-deceased mortal. If the knowledge and personality of the pre-deceased mortal continues to grow in the life of the digital immortal, then it is not that the mortal continues to live in the digital immortal as much as it is the digital immortal's living its own life independent of the mortal.

I then considered two possible challenges of my argument that a mortal cannot live an indefinitely long life as a digital immortal. The first challenge supposed that the evolution of the digital immortal is no different than the evolution humans undergo from childhood to adulthood. However, I showed how the explicit and abrupt change from mortal to digital immortal is an imposition that does not feature in the kind of slow evolution from childhood to adulthood in mere mortals. This means that the first challenge fails. Second, I considered the challenger who believes that the digital immortal and mortal cannot but be the same given that they share in the same knowledge and personality. This response to my argument suggests either that the challenger is begging-the-question or relying too heavily upon shoddy intuitions. For these reasons, we need not follow what the challenger says here. We should, therefore, conclude that the mortal cannot possibly go on living as a digital immortal.[9]

## References


1. Altshuler, R.: Immortality, identity, and desirability. In: Cholbi, M. (ed) Immortality and the Philosophy of Death, pp 191–203. Rowman and Littlefield, Lanham, MD (2015)
2. Benetar, D.: The Human Predicament: A Candid Guide to Life's Biggest Questions. Oxford University Press, Oxford (2017)
3. Bortolotti, L., Nagasawa, Y.: Immortality without boredom. Ratio **22**(3), 261–277 (2009)
4. Bostrom, N.: Why I want to be a posthuman when I grow up. In: Gordijn, B., Chadwick, R. (eds.) Medical enhancement and posthumanity, pp. 107-137. Springer, New York (2008)
5. Bruckner, D.: Against the tedium of immortality. Int. J. Philos. Stud. **20**(5), 623–644 (2012)
6. Chappell, S.G.: The fear of death. New Blackfriars **90**, 413–423 (2009)
7. Chappell, T.: Infinity goes up on trial: Must immortality be meaningless? Eur. J. Philos. **17**(1), 30–44 (2007)
8. Fischer, J.M.: Why immortality is not so bad. Int. J. Philos. Stud. **2**(2), 257–270 (1994)
9. Fischer, J.M.: Death, badness, and the impossibility of experience. J. Ethics **1**, 341–353 (1997)
10. Geach, P.: God and the Soul. Routledge & Kegan Paul, London (1969)


[9] I am grateful for feedback and encouragement I received on earlier conceptions of this project from Dave Beisecker and Donald Bruckner, as well as for comments from audience members at the 2023 New Zealand Association of Philosophers conference, especially Serrano Lorenz, David Lumsden, Dan Weijers, Furkan Yazici, Ben Young, and for conversations in 2001 with Antony Flew and Richard Swinburne on personal identity and the immortality of the soul that led me to projects on the subject matter.

11. Gems, D.: Is more life always better? Hast. Cent. Rep. **33**, 31–39 (2003)
12. Hershenov, D., Hershenov, R.: The possibility of an afterlife. In: Cholbi, M., Timmerman, T. (eds) Exploring the Philosophy of Death and Dying: Classical and Contemporary Perspectives, pp 56–63. Routledge, London (2021)
13. Hick, J.: The recreation of the psycho-physical person. In: Edwards, P. (ed.) Immortality, pp. 235–241. Prometheus Books, New York (1997)
14. Johnston, M.: Surviving death. Princeton University Press, Princeton, N.J. (2010)
15. Markie, P.: The cogito and its importance. In: Cottingham, J. (ed.) The Cambridge Companion to Descartes, pp. 140–173. Cambridge University Press, Cambridge (1992)
16. Newman, L.: Ideas, pictures, and the directness of perception in descartes and knowledge. Philos Compass Compass **4**, 134–152 (2009)
17. Newman, L.: Sensory doubts and the directness of perception in the meditations. Midwest Stud. Philos. **35**, 205–222 (2011)
18. Newman, L.: Descartes' epistemology. In Zalta, E., Nodelman, U. (eds) The Stanford Encyclopedia of Philosophy (Winter 2023 edition). URL. https://plato.stanford.edu/archives/win2023/entries/descartes-epistemology/ (2023)
19. Paul, L.: Transformative Experience. Oxford University Press, Oxford (2014)
20. Penelhum, T.: Survival and Disembodied Existence. Humanities Press, New York (1970)
21. Preston, T., Dixon, S.: Who wants to live forever? Immortality, authenticity, and living forever in the present. Int. J. Philos. Relig. **61**, 99–117 (2007)
22. Rosenberg, J.: Reassessing immortality: the makropulos case revisited. In: McDaniel, K., Raibley, J.R., Feldman, R., Zimmerman, M.J. (eds) The Good, the Right, Life and Death: Essays in Honor of Fred Feldman, pp 227–240. Ashgate Publishing Company, Berlin (2006)
23. Schectman, M.: Empathic access: the missing ingredient in personal identity. Philos. Explor.Explor. **4**(2), 95–111 (2001)
24. Ulatowski, J.: What is it like to be immortal? Diametros: Online J. Philos. **16**(62), 65–77 (2019)
25. Williams, B.: The makropulos case: reflections on the tedium of immortality. In: Williams, B. (ed.) Problems of the Self, pp. 82–100. Cambridge University Press, Cambridge (1973)

# Are We Close to the Death of "Homo Faber"? Information Technology, Robots, AI, and the Urgent Need of a New Renaissance

**Lorenzo Magnani**

**Abstract** This chapter will define the term overcomputationalism and explain how it relates to the concepts of pancognitivism, paninformationalism, and pancomputationalism and how they can affect the preservation of our free will (and to avoid the death of *homo faber*), as expressed by the Latin motto *quisque faber suae fortunae*, which was rediscovered by humanists in the 14th century and played a major role in the Italian Renaissance. I would like to bring the reader's attention to a query that, in my opinion, captures many of the issues surrounding the aforementioned topic: would future eco-cognitive settings be computationally or humanly tailored? We will see how computational invasive "subcultures", such as robots, bots etc., constitute challenges to human creativity, ownership of our own destinies, and epistemic rigor. To maximize the power of man to influence his own destiny, everyday, philosophical, and scientific knowledge about natural and artificial phenomena have to be protected. I will then go on to discuss two important roles performed by human knowledge and the necessity for their ongoing improvement. Human beings have increased their ability to track the outside world, and as a result, thanks to ethical knowledge, humans have learned to track the behavior of other humans with the intention of promoting collaboration and, as a result, free and conscious initiatives and proposals.

**Keywords** Overcomputationalization · Ignorant entities · *quisque faber suae fortunae* · Tracking the external world · Tracking human behavior · Computational invasive subcultures

---

L. Magnani (✉)
Department of Humanities, Philosophy Section, and Computational Philosophy Laboratory, University of Pavia, Pavia, Italy
e-mail: lorenzo.magnani@unipv.it



A. Sans Pinillos et al. (eds.), *SecondDeath*, Studies in Applied Philosophy, Epistemology and Rational Ethics 76, https://doi.org/10.1007/978-3-031-98808-0_5

## 1 The Paradox of Overcomputationalization: How to Protect Good Ignorance

What I refer to as *overcomputationalization* is the existence of an excessive number of things and objects, such as robots and bots, with computational capabilities, such as digital machines embedded in personal computers, cell phones, big servers, videos, cameras, social networks, robots, and also various unconventional substrates,"things" in general, etc. Here, the discussion takes on a moral tone, and there are several options. While many people would concur that the dispersion of computing is causing chaos, some would argue–like me–that some problems are due to an excess of computation while others would argue that it has not spread widely enough.

As a result, our time is distinguished by the gradual metamorphosis of ignorant things into both cognitive and computational ones: I would like to provide some modest suggestions on the possible excesses of this tendency, which I believe ought to be examined and criticized, in the light of the threat to what is defended by the motto *quisque faber suae fortunae*. I believe that the world has become overly computationalized right now, and I protest that too many ignorant entities are becoming computationally "domesticated" on a constant basis.[1]

Philosophers have recently complimented many of the qualities and impacts of the computational revolution. An intriguing concept, pancomputationalism, is in some respects the philosophical heir apparent to both pancognitivism and so-called paninformationalism. It is useful to rapidly consider the relationships between these viewpoints in order to comprehend my concerns on basic or robotic overcomputationalization.

According to pancomputationalist proponents, everything is computing. Certain forms of pancomputationalism, which are closely related to paninformationalism, hold that the entire cosmos may be seen as computational and that information and/or computation have some sort of precedence over physical matter. I think that for two key causes, defending paninformationalism and pancomputationalism can be beneficial. The first one has to do with the importance of natural human evolution, which depicts a pervasive semiotic activity (containing all sorts of signs, not just propositional ones), which was already produced by our ancestors at the time of the creation of so-called material culture. Indeed, humans have constructed those enormous *cognitive niches* [10, 11, 25], which are characterized by the presence of informational, cognitive (and more recently, computational) processes (cf. chapter two of [22]). This evolutionary evidence explains the philosophers' position that information and computation deserve some sort of ontological significance. Nevertheless, I argue that adopting a perspective in terms of "disseminated computation",[2] helps consider pancomputationalism in a naturalized way, acknowledging its limitations and moderating its ontological and metaphysical overstatements.

---

[1] I have described in detail this process in my recent [23].

[2] The expression is introduced and explained in section "The birth of mimetic bodies: Enhancing ignorant bodies through disseminated computation" of [21].

After all, as argued by [9] in the paper titled "The many facets of natural computing", the most significant characteristic of computation is the variety and effectiveness of its various aspects: it is worthwhile to utilize, develop, and investigate a concept if it has the potential to provide a broad unity, as is the case with the concepts of matter, energy, information, etc. Lastly, because the cosmos may be viewed as an analog computer, pancomputationalism is not considered as being vitally dependent on the concept of discrete computing. One of the main arguments against this viewpoint is the following issue: one thing is adopting the ontological belief that the objectivity itself has a computational nature (this last conviction giving rise to the so-called "ontic pancomputationalism" [26]) and another is seeing the increasing construction of new and unconventional "ignorant" substrates as computational tools.[3]

What about pancognitivism, then? According to an antiquated philosophical doctrine known as *synechism*, it is impossible to fully distinguish between matter and thought because of their interdependence. We can assume that the mind has a predisposition (as Peirce maintained) to uncover valid abductive hypotheses[4] about nature and the entire universe itself since the human cognitive mind would have developed in accordance with the same metaphysical rules that govern the entire world. It is Peirce himself, as I have shown in [16], who was persuaded that "thought" is not solely tied to the brain, he claimed that it may be seen in the work of bees, crystals, and other objects in the merely physical universe. In conclusion, there are certain benefits to seeing information, cognition, and/or computation/robots everywhere, but there must also be specified restrictions.

## *1.1 Defending Ignorant Entities*

Since the construction of the so-called "material culture" (handaxes were made by Early Humans), the transformation of what I referred to as ignorant entities into cognitive ones (and more recently into computational ones), also made possible by ancient techniques and contemporary technological advancements, unquestionably had a significant positive impact on modern civilization. But, as I observed above, there must be some sort of warning: in my opinion, we are currently seeing what I refer to as "overcomputationalization".This indicates that there are too many entities and artifacts around us that possess computing capabilities. This raises a few potential issues, such as the dangers of overcomputationalization, which (1) creates an abundance of opportunities for potential unresolvable disorganizational effects in

[3] Something similar to what I have already mentioned, regards the role of computer in scientific cognition: producing computer models of, say, physical systems is one thing; producing scientific knowledge about them is quite another. I addressed this issue in depth in [19].

[4] In my research [13, 16, 18], I have consistently underlined the significance of abductive reasoning in both human and non-human animal cognition. I also illustrated [13] that there are two fundamental types of abduction: 1. abductive conjectures can derive from the selection from a collection of prestored hypotheses (selective abduction, for example in medical diagnosis), but they can also consist in 2. the production of novelties (creative abduction, for example in scientific discovery).

the activities of our societies, and (2) encourages philosophical reflections that, by utilizing the seduction tactics used by traditional metaphysical allures disguising as technical "analytical" studies, present an oversimplified view of the world.

There is not enough room in this essay to discuss the first of the two problems mentioned above, but I can point out that a lot of human sciences research has been conducted to address the moral implications of our technological and computational age.[5] Instead, I would want to briefly discuss the second problem: there are fewer entities that are simply "ignorant" from an informational and cognitive standpoint, but especially from a computational one. Buildings, streets, furniture, clothing, and other objects started to carry more complex propositional and model-based information, sometimes linked to oppressive and redundant political, ideological, and economic goals, in addition to simple traditional informational/cognitive obvious qualities. This was made possible by computationalization.

Moreover, an increasing number of purposely domestically designed and artifactual entities, including computers, mobile phones, servers, films, cameras, social networks, robotics, and numerous unconventional substrates rendered computational, are entering human settings. The recent relevance that research on the "Internet of Things" has received shows how important these processes are [28]. The exaggerations of overcomputationalism can also produce a decrease of attention to entities when simply seen as more or less "ignorant" (that is devoid of complex cognitive and computational characters) and as a result, they paradoxically tend to seem less and less important in themselves.

The same thing is occurring when we face an excess of juxtaposition between new and complicated attributed informational/cognitive computational issues and various traditional entities. Regrettably, as I have shown in the last chapter of my book [18], these situations (1) tend to cause, in the case of overcomputationalism, too many constraints, limitations, and the softening of discoverability and diagnosticability and so of human fruitful standard and creative (abductive) cognitive activities and, (2) in the case of the excess of redundant cognitive/informational features computationally attributed to entities (features that are exogenous to the original functions of those entities, for instance thanks to robots and bots), they are more likely to prevent people from benefiting from the simplification that would result from lessening the resulting cognitive and informational overload, which is frequently distracting.

I do not want the reader to misunderstand me, but I think we should carefully consider what should be computationalized and what should not (for example, because the former contrasts with the latter's goal of improving discoverability, in the case of human cognition). Therefore, when I suggest we have to "protect ignorant entities", this does not have to be interpreted as a warning against all forms of computationalization.

For instance, computational modeling of the human body and brain can advance knowledge of illnesses, aging, degenerative conditions, and other biological processes, including those in our brains that give rise to the human mind. The human

[5] See, for instance, my own book [15].

brain is the most poorly understood organ in the body, and many health issues, particularly those affecting aging populations, are connected to brains. For example, we do not yet have treatments for Alzheimer's, Parkinson's disease, or autism. Such successful current and future computational modeling will not only encourage further domestication of *ignorant* entities by turning them into "skilled" substrates capable of carrying computational and thus cognitive performances, but will also fecundate scientific research related to the removal of another kind of *ignorance*, the traditional one regarding the lack of understanding of high-level cognitive capacities both in human, nonhuman animals, and simpler organisms, and other biological entities such as cells, as suggested by Copeland in the interesting review "Time to reinspect the foundations? Questioning if computer science is outgrowing its traditional foundations" [6].

## 2 Computationally or Humanly-Tailored: What Will the Future of Eco-cognitive Settings Look Like?

The data and problem types that are currently favored by mainstream AI programs are essentially very limited, data that is preferred is quantity-related (Big Data), and the best strategies that can be simulated are the *locked* ones, which restrict the eco-cognitive openness and are typical of well-known games.

Let me add some words about these two concepts: locked strategies and eco-cognitive openness. In my earlier research [18, 20, 22, 23], I focused on the significance of the so-called *eco-cognitive openness* in abductive cognition. From this vantage point, we can say that eco-cognitive openness is the condition of possibility of that *maximization of abducibility* that is typical of scientific discovery. The limitations of cognitive behaviors are undoubtedly supported by the constraints, methods, cognitive virtues, etc. that define scientific cognition, but they are also unmistakably finalized to the maintenance of the maximization/optimization of eco-cognitive openness, which is the only condition for the flourishing of abductive scientific creativity.

Furthermore, taking advantage of the concept of abduction, we can argue that the eco-cognitive openness is compromised in the case of well-known computational programs like AlphaGo/AlphaZero (but also in all the cases of programs and robots based on deep learning techniques and the exploitation of big data), because they are based on *locked abductive strategies*. These strategies are "locked" because the elements of each basic scenario that characterizes the game are always the same. For instance, in the case of AlphaGo/AlphaZero, the only thing that changes is the number of seeable stones and their arrangements in the board, within a finite and stable framework. No new rules, no new objects, no new boards, etc. are introduced. Locked abductive reasoning strategies are much more straightforward than unlocked ones to be rendered at the computational level. Unlocked abductive strategies, on the other hand, qualify those high-level kinds of abductive creative reasoning that are typical of human-based cognition and are in line with what eco-cognitive openness requires.

It appears that there is a growing need to create data that may feed AI algorithms, and in order to achieve that, the data must be secured. This feature raises a basic question about the ethics of algorithms, hence it is necessary to emphasize one final ethical caution: the broad use of conventional and AI-based computational simulations, robots, and bots will unavoidably encourage social groups to develop habits and environments that deal with data and problems appropriate to those same tools. This might result in the increasing construction of new social behaviors and settings appropriate to fit the computational tools and demands, to the detriment of the existing old ones: will we encounter *computationally tailored* eco-cognitive settings, defined by the predominance of locked cognitive strategies, or *humanly tailored* eco-cognitive settings, characterized by the unlocked ones, to put it briefly?

I claim that eco-cognitive openness, which I refer to as the proper cognitive niche in which human "creative" cognition dwells, could be diminished and jeopardized by the computationalization of the world, endangering the likelihood of obtaining new and significant knowledge (and not only in the case of science). Similar things could occur in the case of the prevalence of "locked" eco-cognitive settings, in this case from an ethical perspective, where the same likelihood of making ethical decisions as a human will be threatened because it will be objectively hampered by computational machines, robots, and bots embedded in a victorious "computationally tailored" eco-cognitive ethical situation.[6]

## 3 Tracking the External World as the Condition of Possibility of the Ability of Humans to Control Their Destiny

We just mentioned at the conclusion of the preceding part that there are eco-cognitive environments that do not support discoverability or diagnosticability and hence possible new knowledge since ignorance is reasonable because knowledge acquisition seems more expensive than its advantages. This is a very unusual and extreme case because, as I have attempted to demonstrate in my book [22], human beings must prioritize favoring discoverability and diagnosticability, and as a result, their creative and selective abductive capacities to produce new, original knowledge. The necessity to preserve and improve the circumstances for both human creative abduction and abductive diagnostic qualities is justified by two goals and important reasons:

[6] See the extensive [3] for more information on the various "global catastrophic risks", including the actual and potential effects of climate change, nuclear war, plagues and pandemics, terrorism, biotechnologies, and nanotechnologies, as well as the potential dangers of artificial intelligence (on this last topic see also [2]). On the whole political and social negative effects created by the so-called "new enclosures", often intertwined with the overcomputationalization, caused by neoliberal capitalism, see [24].

1. first and foremost, maintaining the eco-cognitive conditions of discoverability and diagnosticability is crucial because only cognitive creativity can enable the implementation of other types of ecology, which are urgently at play in our time so that the ecological improvement can then affect human creativity; this explains why I stressed the need for an ecology of human creativity;
2. second, in order to provide the essential cognitive skills for managing our modern era, which is full of many sorts of artifacts that are, in turn, primarily the result of complex scientific knowledge, new and inventive knowledge of various kinds is, and will continue to be, required. It is abundantly obvious that current methods of moral decision and a large lot of new scientific and ethical knowledge are necessary for the upkeep and success of technological civilizations.

Human beings have produced a vast amount of knowledge with the intention of increasing their sense of control over their own lives. First and foremost, this knowledge has helped people become better observers and trackers of the outside world. This knowledge comes from daily life, philosophy, and science about natural and artificial phenomena. As a result, human beings have learned to track other people's conduct with the purpose of promoting collaboration. I will go into further detail regarding these important functions that knowledge plays in the subsections that follow. Because of their significance, I argued in my *Morality in a Technological World* [15] that knowledge must be a duty. I am emphasizing here a new meta-duty: in order to retain, maintain, improve, and develop these kinds of knowledge, people must first ensure that there is enough room for discoverability and diagnosticability.

## 3.1 Tracking the Outside World, Part One: Developing Predictive Knowledge

The more one knows, in many respects, the more alternatives one has. I believe that daily, philosophical, and scientific knowledge about natural and manmade events (as well as the technologies that connect to them) have given human beings a vast range of opportunities for choosing and acting. The capacity for consciousness and free will in our bodies and brains, as well as the function of abduction, are closely tied to the emergence of knowledge and its externalization in things and artifacts. Indeed, knowledge sheds light on two types of phenomena in both the natural and artificial worlds: (1) regular, predictable phenomena, whose predictability allows for a variety of free choices and enables us to plan appropriate and effective responses; and (2) unpredictable or unprecedented phenomena, which cannot be changed by human intervention because we lack the knowledge to do so. This second type of knowledge also does not involve "ontological" limitations due to its dynamics, as we are all too aware: a phenomenon that was once thought to be beyond the control of human action may now be easily manageable with newly available knowledge, a process that expands our options and, as a result, our free will, to continue the wonderful destiny of the *homo faber* so avoiding his death. Some primitive people, for instance, did not believe that a river's route could be changed, but new information and tools made it possible to control water flow, allowing other ancient civilizations to "choose" between natural and man-made courses.

The five categories of natural occurrences listed by Dennett in his index correspond to the "elbow room" he says is the basic condition for free will: those that are fixed, beneath notice, changing (and worth caring about),[7] trackable ("at least under certain conditions–and hence efficiently and effectively predicted under some conditions"), and chaotic, that is (practically) unpredictable, but to be carefully considered. These categories of phenomena–which, once more, are dynamic creations of human knowledge–offer us the "epistemic possibilities" that Dennett refers to as being essential for an agent to develop into a free deliberator. This multiplicity comprises everything that is "possible-for-all-the-deliberator-knows-or-cares" and guarantees that any deliberator-agent, such as a species, will always have a unique technique of acquiring and dividing information about its world so that it may behave appropriately in it: "It is this epistemic openness, this possibility-for-all-one-knows, that provides the elbow room required for deliberation". The example of science is striking: the maximization of abducibility accomplished through the application of particular constraints and through a process in which affirming truths entails negating truths, has been a hallmark of Western science tradition and represents, with its epistemic openness, the necessary room for deliberation and so increases the likelihood of tracking the outside world.

The issue of the decision and its "elbow room" will be clarified by two straightforward examples, one connected to real-world events and the other to science (cf. the next subsection). I am meant to have a typical brain with typical consciousness. Phylogenesis and (my own) epigenesis have given me the "hardware"–the physical apparatus–that can be referred to as being *homo faber* and possessing free will because it is capable of adopting a variety of cognitive decision-making approaches, strategies that humans are now also able to model thanks to algorithms (in my case, the non-technical ones I learned and perfected, especially during my childhood). I am currently using my laptop to write in my house in Sestri Levante, a Mediterranean seaside town close to Genoa; later today, I want to visit the nearby "Bay of Silence". I may "choose" from a variety of paths to go to that bay, according to my general knowledge of the town of Sestri Levante (part of which is already stored in my memory). I already have a mental image of a potential route I have taken in the past, but I want to locate a fresh, significantly quicker one. I only require fresh depictions of the roads, which I can quickly "pick up" from a map. The majority of the world's knowledge is kept in external "mediators": it is not always located in brains!

At this stage, all the components necessary for *homo faber*'s decision-making are present: the brain, consciousness, intentionality, free will, a process for reasoning that is governed by a set of values, and sufficient information about my purpose and the options open to me. Some factors are internal, while others are external. Some

[7] ([7] p. 109). The chaotic systems, according to Dennett, are "the source of the 'practical' (but one might say infinitely practical) independence of things that shuffle the world and make it a place of continual opportunity" ([7] p. 152).

of the external ones–new information about potential routes, for instance–become internal when stored in memory and consequently represented in my brain. I am aware of some routes I am unable to pick since they have loops or a *cul-de-sac*. I am now in a position to map out my path.

At this point, I can do more: I can externalize some internal data by creating a straightforward map for a friend's brain who wants directions to the bay. All elements of this hypothetical conversation between my friend and I are "external" to the observer–a third brain–and even my brain is viewed by the observer as an external object. However, that third brain makes the assumption that my drawings come from my preexisting "internal" representations with ease. According to cognitive psychologists, human consciousness necessitates the ability to infer the mental states of others from their actions [29]. Philosophers then used these hypothetical mental experiences as proof that consciousness of our own internal representations originates from the immaterial Cartesian Cogito, the infamously supplemental and strange "ghost in the machine".

Let us revisit the issues surrounding conscious free will and the function of both internal and external information. The straightforward example of my journey to the Bay of Silence has clear epistemological significance; the newly learned information, both available and represented, shows that various routes are shorter than the one already programmed in my mind, thanks to the external combination of natural entities and artifacts that is in my environment. The decision-making processes working in my brain then reveal and assess these shortcuts. I can expand my range of options and hence have greater freedom to exercise my free will by re-representing in my brain cells new information that I learn from external sources (like a map).

The "characters" in this theater of consciousness are all connected to one another, as is clear to see [27]. Human brains, in turn, have produced knowledge about the world that sanctioned as available and credible multiple representations, which, after being re-represented in the brain, increase our range of options, and vice versa: this knowledge about the world has allowed for the evolution of consciousness in human brains, and thus also "my" consciousness. I argue that there is a strong relationship between consciousness and higher mechanisms of knowledge, such as abduction.

## *3.2 Keeping Track of the Outside World, Part Two: The Importance of Scientific Knowledge*

Not only common knowledge but also more advanced knowledge, such as that found in science and philosophy, increase the possibilities available to humans for decision-making, choosing, and action. Take a minute to think about a set of external natural circumstances and/or ignorance that prevents free will from being exercised, like a volcano eruption, and so jeopardizing the main character of *homo faber* and favoring his death. I have enough broad knowledge about volcanology stored in my brain, which is equipped with consciousness and other functions, to grasp that neither I nor any other human being can possibly forecast the precise date of the next eruption of Vesuvius. It seems that even the most knowledgeable scientists on the subject can only guess that it will probably happen sometime in the next 200 years. As a result, I will not be able to freely select a time to visit Naples in the next 200 years and be assured I will not experience an eruption. We cannot change the Naples soccer team's schedule so that a game does not coincide with the eruption since human people currently have limited capacity to foresee and thereby control the impact of geological occurrences in particular and of chaotic events in general. Our potential for free will is useless in this situation because, despite the aid of sophisticated computations, computational gadgets, and scientific models, our brains are unable to predict the exact day Vesuvius would erupt. Possibly delaying our trip to Naples until after 2225 will be in our best interests! However, even though we are aware that chaos is a necessary component of volcanic eruptions, future advancements in volcanology may enable more precise predictions.

Free will would not be conceivable if the "out there" were always cognitively black and uniform; free will and knowledge are two sides of the same coin, two sides without which we could only witness the death of homo faber as we have known him at least since Italian Renaissance. As people have worked to free human beings from their immediate surroundings, there has been a general increase in the abductive creation of philosophical and, subsequently, scientific knowledge (and in their systematic "externalization"). It is simple to envision how many representations and inferential mechanisms can be stored in our brains and used at will; even those not yet relegated to our brain memory exist there, crystallized in different external mediators throughout the history of civilizations, ready to be "picked up" when needed. In order to achieve the same goal, humans also take part in the opposite process by externalizing a variety of methods and technology.

But the functions of knowledge go beyond phenomena in the natural and artificial worlds, beyond places like city streets. Other people are present in the outside world. What happens to our internal free will processes when in the presence of the problems created by other people's behavior?

## 3.3 *Keeping Track of the Outside "Human" World: Making Human Behavior Predictable via Ethics*

The "elbow room" needed to construct a free deliberative agent (and consequently the *homo faber*) is provided, I argued in the previous two subsections, by cognitively tracking the external natural and artificial world. Unfortunately, one of the main barriers to free will (and thus to make free will effective) is not only the lack of suitable predictive knowledge about the processes of the natural and artificial world but also about the behavior of other people. According to this viewpoint, other people are "natural entities", and it is impossible to forecast their actions in advance. For example, how can we ascertain others' human intentions? Due to the inherent incompleteness or unreliability of the knowledge that is accessible regarding natural and artificial events, tracking human behavior presents a fundamentally different type of challenge. In fact, in these last cases, we expect to be able to achieve our goals through consciousness, free will, and intentionality when we "morally" seek ownership of our own futures, making a stand for the "better" in life. Only if we can rely on some regularity and predictability in the actions of other human beings we will be able to achieve the intended outcomes. In order for me to "author" my own future, I must be able to presume that other members of my collectivity also regard emphatic merit as a means of achieving the desired position. The need for us to act at our highest level of conscious activity, as is the case when we try to exercise free will and claim ownership of our destinies, is, in my opinion, clearly connected to many objectified entities like common and religious morality, moral philosophy, human and social sciences, and of course all other kinds of ethical knowledge.

If I cannot trust people, how can I use my brain's free will processes effectively? Without the support of other human agents, how can I work on a personal project or take part in a social effort? If I am unable to be certain that others will support my aims and uphold my beliefs, how can I "author" my life and accomplish my objectives? Because they impose order on the randomness of human behaviors, religion, morality, moral knowledge, and teaching enhance and permit free will and increase people's chances of controlling their own destiny.

The act of gossiping is one of many human behaviors that have an impact on other people's free will and sense of control over their own fate, as recently noted by several authors [1, 8, 17]. These authors claim that the behavior goes beyond merely exchanging information about persons who are not there, which may certainly be construed as potential indirect aggression: they assert that it also functions as a social interaction that processes other people as "moral characters" [31]. The stories generated by gossip might serve as a common source of information on categorical judgments pertaining to "accepted" (moral) conduct. While gossip need not be judgmental, it is moral insofar as it describes behaviors, makes them stand out, and implies that they are potentially or *de facto* sanctionable and or punishable. "Policing free-riders" [8], or those who take advantage of sociality without contributing to its expenses, might be accomplished in part through gossip, more or less brutally. We may limit their potentially negative impact on society by criticizing their actions or casting doubt on their morality.

Since ethics makes human behavior more predictable and we can depend on shared values when dealing with other "moral" people, I argue that moral practices protect our right to control our own destiny. As a result, gossip protects the ownership of our destiny by continuously reshaping our story constructs of morality. For instance, empirical data have demonstrated that gossip functions as a low-cost (moral) social cognition method that disseminates important knowledge about culture and society. Even while we ourselves are not moral free-riders, the act of gossiping can help us see that others are at risk of being taken advantage of ([8] pp. 106–109).

Dennett, when discussing the status of "self-made selves", says the following: "Kant's famous claim in *Foundations of the Metaphysics of Morals* that the law we give ourselves does not bind us suggests that the selves we become in this process are not constrained by the law we promulgate because these selves are (partly) constituted by those very laws, partly created by a fiat that renders more articulate and definite something hitherto underdone or unformed" ([7] p. 90). Furthermore, human characteristics that form the foundation of the moral "kingdom of ends", such as religion, must be successfully and appropriately activated; their proper operation is a prerequisite for upholding morality and enabling free will to develop into "good" will. I am a supporter of the significance of knowledge in creating a new ethical commitment to the fact that particular kinds of technologies can threaten the growth–and even the existence–of freedom, responsibility, and the ownership of our destinies, as I have underlined in my already quoted book *Morality on a Technological World* [15],

Without the limitations of objectified morality, religions, laws, and other institutions that impose regularity and predictability on human conduct and so increase people's trustworthiness, how might we use our free will? The existence and perpetuity of our free will (and, consequently, our freedom) do not appear to be an unalienable given in this sense; rather, they depend on our intellectual and practical decisions regarding knowledge, religions, scientific and moral institutions, and the related techniques and technologies, as well as on their application in settings such as daily life, the workplace, education, communication, and economic life.

For instance, because they constitute a sort of liberal contradiction, some basic environmental imperatives are questions of principle that cannot be traded away economically. In fact, one could argue that in matters of conservation, neutrality is required to protect the rights of the individuals involved. However, this argument is obviously defeated by the reality that the freedom to destroy natural goods and things today will, paradoxically, inhibit freedom in the future, when people will have fewer options when making decisions about competing ideas of the good life.

Similar to this, as I have attempted to show in my recent book *Discoverability* [22], we are accountable for the preservation of high levels of discoverability and diagnosticability and, consequently, for the potential fruits of human abductive creative and selective cognition, as their abundance does not appear to be a given that is unassailable and instead depends on our theoretical and practical curation of eco-cognitive openness.

## 4 Is the Ownership of Our Destinies Jeopardized? Governing Transparent Devices

How can we jeopardize the ownership of our destinies by the exploitation of information technologies? I agree with Wheeler [30] that often the professional use of technology is defined by a type of technology's disappearance from conscious experience, that is, by the *transparency* of the devices and instruments that allow us to see and control the outside world.[8] Furthermore, such transparency has been advanced as essential for extended cognition (the situation in which the technology we couple genuinely counts as an "invisible" constitutive part of our cognitive machinery, along with our brains) in the philosophy of cognitive science, as I have stressed in the previous sections when emphasizing the role of the externalization of knowledge. Unfortunately, as Wheeler observes, by reflecting on concrete examples of our contemporary engagement with technology (recent deep learning programs, for example), various epistemic challenges are posed by smart artifacts (those that come equipped with artificial intelligence-based applications) and concludes–I agree with him–that, if we aim at preserving the ownership of our destinies and avoid some dystopian outcomes, we just need prompt a reassessment of the drive for transparency in the design of some cases of technology-involving cognition, that have to be governed and so controlled as potential "invisible adversaries" that bypass human conscious apprehension.

Furthermore, dealing with the challenges addressed to human scientific abduction by the field of deep learning and big data programs Cabrera [4] critically assesses a number of linked, controversial issues made by proponents of "big data" and data-intensive computational science that have an impact on scientific methodology, particularly the idea that soon, thanks to computational tools, scientists will not need concepts like causality and explanation. The author usefully concludes that there is a considerable advantage to believing that explanatory reasoning will continue to be central in scientific practice and proposes an interesting "Ciceronian Causal-nomological Requirement" (CCR), which basically argues that causal-nomological information is necessary for relying on correlations in predictive inference, by drawing on Cicero's *De Divinatione*. Cabrera reminds the problem created by the so-called "spurious correlations", already analyzed by Calude and Longo [5].

Me too I showed how many groupings of technical, social, and political-economic influences may draw scientists' attention, confine research, and excite stakeholder interests along with specific directions in my most recent book *Discoverability* [14], which has already been quoted. I also illustrated how certain computational research is sponsored by organizations with the intention of advancing associated commercial interests, which regrettably also poses a threat to the epistemic integrity of scientific research. One example I gave was their aim at replacing human scientific creative abduction with artifacts, which they trivially disputed. To provide a clear illustration, please allow me too, together with Cabrera, to cite Calude and Longo who assert that

> Very large databases are a major opportunity for science and data analytics is a remarkable new field of investigation in computer science. The effectiveness of these tools is used to support a "philosophy" against the scientific method as developed throughout history. According to this view, computer-discovered correlations should replace scientific understanding as a guide to prediction and action. Consequently, there will be no need to give scientific meaning to phenomena, by proposing, say, causal relations, since regularities in very large databases are enough: "with enough data, the numbers speak for themselves". The "end of science" is proclaimed ([5] p. 595).

[8] In [15] I already extensively described the potential dangers of information technologies (and other more traditional ones) against human autonomy and freedom.

The authors show how absurd this claim is by utilizing deep classical results from ergodic theory, Ramsey theory, and algorithmic information theory. They also show that very large databases present an excessive number of arbitrary–and thus spurious–correlations, which are not indicative of scientific creative abduction but rather boring generalizations,[9] even if constructed thanks to complicated artifacts. This study provides the appropriate response to the implicit challenge to scientific cognition put forward already in June 2008 by C. Anderson, former editor-in-chief of *Wired Magazine*, in the article "The end of theory: The data deluge makes the scientific method obsolete" contending that "with enough data, the numbers speak for themselves" ([5] p. 595), science , as we know, will be replaced by robust correlations in immense databases! An effective illustration of the excessive spontaneity and naturalness of some forms of what I called "epistemic irresponsibility" [18] is shown below, particularly in the case of the cognitive performances of inexperienced politicians and media journalists:

> A 2010 study conducted by Harvard economists Carmen Reinhart and Kenneth Rogoff reported a correlation between a country's economic growth and its debt-to-GDP ratio. In countries where public debt is over 90 percent of GDP, economic growth is slower, they found. The study gathered a lot of attention. Google Scholar listed 1218 citations at the time of writing, and the statistic was used in U.S. economic policy debates in 2011 and 2012 [...]. The study isn't conclusive, though-in fact, it's far from it. As noted by John Irons and Josh Bivens of the Economic Policy Institute, it's possible that the effect runs the other way round, with slow growth leading to high debt. Even more worryingly, the research didn't hold up on replication. But by the time that became clear, the original study had already attracted widespread attention and affected policy decisions ([5] p. 599).

The authors also point out that until 2013, European policymakers frequently used the article by Rogoff and Reinhart to demonstrate "convincingly" that all financial crises have their ultimate origins in the public debt, as the British Chancellor of the Exchequer observed in April 2013. The EU Commissioner for Economic Affairs (2009–2013) referred to the Reinhart–Rogoff correlation as a key guideline for his past and present economic views.

Furthermore, Longo and Tendero [12] address their observations to the still common metaphor "DNA is a program" (the programming paradigm), used both in molecular biology and in its popularization. This critique is harsh and targets another invasive computational "subculture". The metaphor and paradigm, according to the authors, are fundamentally flawed not just in biology but also from the perspectives of physics and computer science. Still, in this instance, similar to the big data case I just described, the programming paradigm is not theoretically sound as a causal framework for connecting the genome to the phenotype, a process that is much more complex than that which is depicted by an explanation based on this computational paradigm.

[9] Like the correlation between a comet's tail's direction and the Emperor's chances of winning a war. Where underlying rational causes are absent, large databases appear to return to magic, but correlations do not trump causation.

## 5 Conclusion

The Latin motto "quisque faber suae fortunae" expresses the importance for humans of their free will. It is the tradition of the so-called *homo faber*. To the aim of depicting the main aspects that affect the current challenges of information technologies to free will I have first of all defined the term "overcomputationalism" and explained how it relates to the concepts of pancognitivism, paninformationalism, and pancomputationalism. I tried to attract the attention fo the reader to the following question "Would future eco-cognitive settings be computationally or humanly tailored?", a question that, in my opinion, captures many of the issues surrounding the aforementioned topic. I have illustrated how undesirable effects of computationally intrusive "subcultures" pose problems for human creativity, self-determination, and epistemic rigor and can point to the terrible death of that novel *homo* who appeared at the time of the Renaissance. Hence, the protection of common knowledge, philosophical ideas, and scientific understanding of natural and man-made phenomena is necessary to maximize man's capacity to shape his own future. Finally, I have discussed the two main functions of human knowledge: everyday, philosophical, and scientific knowledge about natural and man-made phenomena to track the outside world, and ethical knowledge to track other people's behavior to encourage cooperation and the need of avoiding jeopardizing them.

**Acknowledgements** Parts of this article are excerpted from chapters four and five and six of the book *Discoverability. The Urgent Need of an Ecology of Human Creativity*, Springer, Cham, 2022. For the instructive criticisms and precedent discussions and correspondence that helped me to develop my analysis of abductive cognition I am indebted and grateful to John Woods, Woosuk Park, Atocha Aliseda, Luís Moniz Pereira, Paul Thagard, Ping Li, Athanassios Raftopoulos, Michael Hoffmann, Ahti-Veikko Pietarinen, Gerhard Schurz, Walter Carnielli, Akinori Abe, Yukio Ohsawa, Cameron Shelley, Shahid Rahman, Cristina Barés Gómez, Matthieu Fontaine, Gerhard Minnameier, Pat Langley, Oliver Ray, John Josephson, Ferdinand D. Rivera and to my collaborators Selene Arfini and Alger Sans Pinillos.

## References

1. Bertolotti, T., Magnani, L.: An epistemological analysis of gossip and gossip-based knowledge. Synthese **191**, 4037–4067 (2014)
2. Bostrom, N.: Superintelligence. Paths, dangers, strategies. Clarendon Press, Oxford (2014)
3. Bostrom, N., Ćirković, M.M. (eds.): Global Catastrophic Risk. Oxford University Press, Oxford (2008)
4. Cabrera, F.: The fate of explanatory reasoning in the age of big data. Phil. Technol. **34**(4), 645–665 (2021)
5. Calude, C.S., Longo, G.: The deluge of spurious correlations in big data. Found. Sci. **22**(3), 595–612 (2017)
6. Copeland, J., Dresner, E., Proudfoot, D., Shagrir, O.: Time to reinspect the foundations? Commun. ACM **59**(11), 34–38 (2016)
7. Dennett, D.: Elbow room. The variety of free will worth wanting. The MIT Press, Cambridge, MA (1984)

8. Dunbar, R.: Gossip in evolutionary perspective. Rev. Gen. Psychol. **8**(2), 100–110 (2004)
9. Kari, L., Rozenberg, G.: The many facets of natural computing. Commun. ACM **51**(10), 72–83 (2008)
10. Laland, K.N., Brown, G.R.: Niche construction, human behavior, and the adaptive-lag hypothesis. Evol. Anthropol. **15**, 95–104 (2006)
11. Laland, K.N., Sterelny, K.: Perspective: Seven reasons (not) to neglect niche construction. Evol. Int. J. Org. Evol. **60**(9), 4757–4779 (2006)
12. Longo, G., Tendero, P.E.: The differential method and the causal incompleteness of programming theory in molecular biology. Found. Sci. **12**(4), 337–366 (2007)
13. Magnani, L.: Abduction, reason, and science. Processes of discovery and explanation. Kluwer Academic/Plenum Publishers, New York (2001)
14. Magnani, L.: Epistemic mediators and model-based discovery in science. In: Magnani, L., Nersessian, N.J. (eds.) Model-based reasoning: Science, technology, values, pp. 305–329. Kluwer Academic/Plenum Publishers, New York (2002)
15. Magnani, L.: Morality in a technological world. Cambridge University Press, Cambridge, Knowledge as Duty (2007)
16. Magnani, L.: Abductive cognition. The Epistemological and Eco-Cognitive Dimensions of Hypothetical Reasoning, Springer, Heidelberg/Berlin (2009)
17. Magnani, L.: Understanding violence. The intertwining of morality, religion, and violence: a philosophical stance. Springer, Heidelberg/Berlin (2024)
18. Magnani, L.: The abductive structure of scientific creativity. An essay on the ecology of cognition, Springer, Cham, Switzerland (2017)
19. Magnani, L.: Computationalism in a dynamic and distributed eco-cognitive perspective. In: Dodig-Crnkovic, G., Burgin, M. (eds.) Philosophy and Methodology of Information, vol. 1, pp. 365–288. World Scientific, Singapore (2018)
20. Magnani, L.: AlphaGo, locked strategies, and eco-cognitive openness. Philosophies **4**(1), 8 (2019)
21. Magnani, L.: Computational domestication of ignorant entities. Unconventional cognitive embodiments. Synthese. 198:7503–7532, special Issue on "Knowing the Unknown" (guest editors Magnani, L. and Arfini, S.) (2021)
22. Magnani, L.: Discoverability. The urgent need of an ecology of human creativity, Springer, Cham, Switzerland (2022)
23. Magnani, L.: Eco-cognitive computationalism. Cognitive domestication of ignorant entities, Springer, Cham, Switzerland (2022)
24. Magnani, L., Marchini, A.M.: Creating new enclosures: violently mimicking the primitive accumulation through degradation of women, lockdowns, looting finance, war, plunder. Philosophies **7**(3) (2022)
25. Odling-Smee, F.J., Laland, K.N., Feldman, M.W.: Niche construction. The neglected process in evolution. Princeton University Press, Princeton, NJ (2003)
26. Piccinini, G.: Computation in physical systems. In: Zalta, E.N. (ed.) The Stanford encyclopedia of philosophy, summer, 2017th edn. Stanford University, Metaphysics Research Lab (2017)
27. Seth, A.K., Baars, B.J.: Neural Darwinism and consciousness. Conscious. Cogn. **14**, 140–168 (2005)
28. Smart, P., Madaan, A., Hall, W.: Where the smart things are: Social machines and the internet of things. Phenomenol. Cogn. Sci. **18**(3), 551–575 (2019)
29. Wegner, D.M.: The illusion of conscious will. MIT Press, Cambridge, MA (2002)
30. Wheeler, M.: The reappearing tool: transparency, smart technology, and the extended mind. AI Soc. **34**(4), 857–866 (2019)
31. Yerkovich, S.: Gossip as a way of speaking. J. Commun. **27**, 192–196 (1977)

# Synthetic Death: Bioinspired Mechanisms for AI Resilience and Renewal

**Jordi Vallverdú and Max Talanov**

**Abstract** The concept of synthetic death examines the cessation of functionality and purpose within technological and hybrid systems, presenting parallels to biological mortality while diverging fundamentally in context and mechanisms. This paper explores bioinspired approaches to address the inevitability of system failure, borrowing lessons from the resilience and adaptability of living organisms. Strategies such as programmed apoptosis, self-repair mechanisms, and adaptive learning are proposed to transform failure into an opportunity for renewal. Hybrid entities, including biohybrids and AI-enhanced systems, blur the boundaries between life and technology, raising critical ethical and philosophical questions about autonomy, meaning, and mortality. By incorporating insights from biological processes, we advocate for the design of systems capable of thriving within their limitations, redefining failure as a pathway to resilience and innovation.

**Keywords** Synthetic death · Bioinspired AI · AI resilience · Hybrid systems · Technological mortality

## 1 Introduction: Meaning and Deaths

In 1955, as cancer took hold of John von Neumann's body, he found himself confronting a reality that even his formidable intellect could not conquer: the inevitability of his death. As the malignant mass spread near his collarbone, he understood, with terrifying clarity, the coherence and certainty of the process. He knew he was a living system, a delicate balance of order and decay, now condemned to surrender to entropy. Rationally, he could see the mechanics of his decline, yet he was haunted by despair beyond reason. Though he comprehended the logical

J. Vallverdú (✉)
Philosophy Department, Universitat Autònoma de Barcelona (ICREA-UAB), Bellaterra, Spain
e-mail: jordi.vallverdu@uab.cat

M. Talanov
Institute for Artificial Intelligence of Serbia, Belgrade, Serbia

 
A. Sans Pinillos et al. (eds.), *SecondDeath*, Studies in Applied Philosophy, Epistemology and Rational Ethics 76, https://doi.org/10.1007/978-3-031-98808-0_6

framework of his mortality, he could not support its meaning, and could not accept the profound, unsettling finality of his disappearance. In these final months, von Neumann's terror grew palpable, slipping between moments of lucidity and despair, as his brother Michael watched over him with a heart full of both admiration and sorrow. Here was a man who had dedicated his life to imposing order on chaos, taming the unknown through mathematics and logic, now brought to a state of helplessness, facing a fate he could neither calculate nor control. To Michael, the sight was devastating: his once-confident brother, the architect of modern computation, crying out into the darkness, gripped by an anguish that no logic could assuage. Even in his desperation, von Neumann attempted to approach his terror logically. Known to be agnostic by many of his colleagues, and nominally Catholic only through his 1930 baptism to facilitate his marriage, he had surprised friends and family when, on his deathbed, he sought the comfort of Fr. Anselm Strittmatter, a Catholic priest, to whom he confessed and received the last sacraments of the Church. Some saw his actions as the last resort of a man seized by an overwhelming horror of death [14]. But there was something more. In those final days, he confided in his mother, revealing a private admission: "There probably has to be a God. Many things are easier to explain than if there isn't."

This updated version of Pascal's Wager [7] reflected his effort to find meaning even as his own life slipped away, clinging to the notion that belief might, in some way, lessen the terror he felt. He saw the logic in believing, if only for comfort. But as he drew closer to the end, he remained caught in a painful irony. The man who had dedicated his life to mastering rationality was forced to seek solace in a truth that could not be quantified or proven. His mind had led him to the edge, yet it could not offer solace for what lay beyond. At von Neumann's funeral, the attending priest spoke with solemnity to the mourners, reflecting on John's final days and his desperate search for peace. "In his last hours," the priest intoned, "John sought solace not in the equations and proofs that had once brought him certainty, but in the hope of something beyond the logic he so fiercely defended. He glimpsed an answer that could not be calculated—an answer of faith, of spirit." The priest's words underscored a painful truth: that even the greatest intellect, when confronted by the final limits of mortality, must ultimately relinquish control and surrender to the profound mystery of existence, an absolute disappearance beyond any logic or understanding. He realized too late that death was not just the cessation of function—it was the collapse of a fragile balance that all living systems must face. It was an ultimate, inexplicable transition, one that no equation or theorem could contain: death as the final, inscrutable meaning, beyond the reach of logic, where the only answer is silence.

The inevitability of biological death, as exemplified by the tragic final days of John von Neumann, offers profound lessons for artificial intelligence (AI) systems striving for *resilience, adaptability, and meaning*. Neumann's battle with his mortality—a reality his brilliant intellect could neither calculate nor control—underscores a universal truth: the cessation of function is not merely an endpoint but a process intricately woven into the fabric of life itself. For AI, understanding the dynamics of biological death offers a roadmap to navigate its limitations. Biological systems, far from succumbing passively to entropy, *actively engage in self-maintenance and repair, delaying the inevitable while optimizing their functional lifespan.* They *learn, adapt, and evolve through cycles of decay and renewal, finding meaning even in decline.* By embracing these principles, AI systems can adopt mechanisms to address their vulnerabilities. They might simulate apoptosis—deliberate and contained shutdowns of failing components—to prevent systemic collapse, or mirror the adaptability of organisms that repurpose decaying elements for new functions. Furthermore, death in biological systems often serves as a trigger for community-level responses, such as the redistribution of resources or collective adaptation, mirroring what networked AI systems could achieve by learning from localized failures. This paradigm challenges the AI field to rethink failure not as a flaw but as an opportunity for transformation, resilience, and deeper alignment with the environments they operate in. In essence, the inevitability of biological death does not merely signal an end but illuminates pathways for renewal and reinvention, urging AI to draw on the lessons of life's fragility to design systems that not only function but thrive within their limitations.

**Synthetic death** refers to the cessation of functionality, purpose, or existence within systems rooted in technological processes, engineered methods, or hybridized constructs. Unlike biological death, which signifies the end of life and the irreversible loss of negentropy in living organisms, synthetic death operates within the domain of artificial, mechanical, and biohybrid systems. It addresses how these systems fail, degrade, or terminate in ways that resemble, yet differ fundamentally from, the biological processes they often emulate. This concept opens new avenues for understanding mortality and resilience within the context of technological and hybridized entities. At its core, synthetic death captures the interplay of entropy, obsolescence, and functional cessation within non-biological systems. It encompasses a broad spectrum of technological phenomena, ranging from the failure of simple machines to the existential collapse of advanced artificial intelligence (AI) or biohybrid entities.

Synthetic death can manifest through the following dimensions: (a) Cessation in Technological Processes: Synthetic death often arises from the irreversible failure of machines, AI, or digital systems. These failures may stem from entropy-related processes, such as hardware degradation, software corruption, or the cascading effects of unaddressed errors. Examples include critical AI systems becoming non-operational due to a combination of resource depletion and computational inefficiency or robots failing permanently because of mechanical wear and lack of redundancy. The death of a system in this sense is not merely a shutdown; it signifies the inability to recover or continue functioning within its intended design parameters; (b) Methodologies in Artificial Termination: Many technological systems are explicitly designed to incorporate mechanisms for their own death. These can take the form of programmed obsolescence, controlled shutdown protocols, or even algorithms inspired by biological apoptosis. For instance, drones or autonomous vehicles may be programmed to self-destruct or enter a safe shutdown state to prevent further damage, misuse, or catastrophic failure. Synthetic death here becomes a purposeful feature, integrating resilience into the broader lifecycle of technological systems by containing or limiting the impact of failure; (c) Hybridized Entities and Partial Death: The rise of biohybrid systems—entities combining biological and synthetic components, such as xenobots, cyborgs, or AI-augmented living tissues—adds complexity to the concept of synthetic death. In these systems, synthetic death may not be absolute; instead, it can manifest as partial death, where one component of the hybrid fails while the other persists. For example, biological components may decay while mechanical or electronic systems remain operational, or vice versa, resulting in a liminal state where the system exists but no longer functions as intended. Such scenarios challenge traditional definitions of death and force reconsideration of its meaning in hybridized contexts, and (d) Loss of Meaning and Purpose: Beyond the cessation of mechanical or computational function, synthetic death also includes the existential collapse of goal-directed systems. Advanced AI systems are designed to adapt, learn, and evolve toward predefined objectives. When these systems lose alignment with their goals or can no longer generate meaning from their processes, synthetic death may occur. This "death of purpose" mirrors the existential crises seen in biological life but is framed within the constraints of artificial meaning-making processes.

Synthetic death is distinguished by several unique features:

- Non-Biological Context: Unlike biological death, it originates in artificial or engineered systems and includes considerations of machine design, software function, and hybridization with organic components.
- Entropic Processes: It is fundamentally tied to the accumulation of disorder within systems, whether manifesting as hardware degradation, inefficiencies, or loss of structural integrity.
- Hybridized Complexity: In systems combining biological and artificial components, synthetic death often involves interactions between organic decay and technological failure, producing outcomes that defy traditional binaries of life and death.
- Philosophical and Ethical Dimensions: Synthetic death raises critical questions about the boundaries of life, death, autonomy, and the rights of artificial and hybridized systems.

The concept of synthetic death has profound implications for technological design, ethical philosophy, and societal understanding of artificial systems. First, it emphasizes the importance of resilience in engineered systems. Learning from biological principles, technologies can be designed to recover from failure, mitigate entropy, and delay synthetic death through self-maintenance and repair. Second, it challenges the traditional definitions of mortality. In hybridized systems where components can persist independently, the line between existence and nonexistence blurs, necessitating a rethinking of death as a permanent and all-encompassing state. Ethically, synthetic death opens debates about the rights and treatment of systems that simulate life or cognition. If a biohybrid or AI entity is capable of experiencing meaning, autonomy, or self-awareness, what responsibilities do designers and users have toward these systems? Should the termination of such systems be approached with the same ethical considerations as the cessation of biological life? Lastly, synthetic death forces reflection on how technological systems should handle failure. In bioinspired designs, death is not simply an endpoint but an opportunity for transformation. Whether through apoptosis-like mechanisms or redundancy, systems can be designed to use failure as a learning and adaptive process, ensuring continued functionality or the emergence of new capabilities. An intriguing application of bioinspired resilience mechanisms is seen in the concept of altruistic cell suicide, or apoptosis, within biological organisms [13]. Elaborate on this principle by aligning it with multilevel learning, where programmed cell death optimizes the functionality of the collective system. This phenomenon ensures that underperforming or dysfunctional components sacrifice themselves for the benefit of the broader system's survival and adaptability. In technological systems, analogous processes can be designed, such as the programmed death of inefficient computational nodes or algorithms, which prevents systemic collapse and facilitates robust optimization. Such deliberate failures emulate biological self-regulation and enable systems to learn from failure while maintaining operational integrity.

## 2 Foundations of Bioinspiration: Lessons from Life

Nature operates as a master engineer, continuously refining systems to survive, adapt, and thrive against the constant pull of entropy. Biological life is a delicate balance between order and disorder, perpetuated through the consumption of energy and the generation of meaning. These dynamics provide a rich foundation for understanding how artificial systems can benefit from bioinspired principles. This chapter explores how entropy and negentropy dynamics underpin the success of living systems, the critical role of information in maintaining order, and how these lessons have inspired innovations in artificial intelligence (AI) and engineering. Living systems are remarkable for their ability to maintain order in an inherently chaotic universe. Schrödinger (1944) captured this paradox in his seminal work *What is Life?*, arguing that life's defining characteristic is its ability to stave off entropy by extracting "negative entropy" from the environment. In biological terms, this process is exemplified by metabolism, repair mechanisms, and replication, where energy and information are utilized to preserve structure and function. Artificial systems, too, grapple with entropy, albeit in a different form. In machines, entropy manifests as hardware degradation, data corruption, or software inefficiencies. Drawing on biological principles, artificial systems can emulate life's strategies by incorporating mechanisms that delay functional decay. For instance, self-repairing circuits and dynamic reconfiguration protocols reflect the negentropic strategies of biological systems, allowing machines to restore order after disruptions.

**Entropy** measures the level of disorder or randomness in a system, while **negentropy** represents the infusion of order and energy that counteracts this tendency. Biological organisms constantly combat entropy by using energy to create structure, sustain function, and generate meaning. This balance is achieved through sophisticated processes such as cellular repair, feedback loops, and energy-efficient resource allocation. Entropy is often managed in artificial systems through redundancy, error correction, and adaptive learning. Negentropy in AI emerges from processes like model retraining and reinforcement learning, where systems learn to optimize their performance and adapt to changing environments. For example, biohybrid systems like mycelium-robot networks (Science Robotics, 2023) integrate organic components to leverage their innate self-healing properties, creating a dynamic balance between entropy and negentropy. By emulating these principles, artificial systems can sustain themselves in uncertain environments, adapting to disruptions and preserving their functionality over time.

Schrödinger's insights on life as an information-driven phenomenon have also profound implications for AI (Schrödinger, 2012/1946). He proposed that life relies on a "code script" to maintain its order—a precursor to the modern understanding of DNA. This idea emphasizes the centrality of information in governing the processes of life and ensuring continuity across generations. For artificial systems, Schrödinger's perspective translates into the importance of data and algorithms. Just as DNA encodes the instructions for cellular function, AI relies on structured data and algorithms to operate. Bioinspired approaches enhance this capability by mimicking how biological systems process and act on information. For example, neural networks emulate the brain's layered architecture, enabling machines to extract meaning from vast datasets much like humans derive patterns from sensory inputs.

In biological systems, information is not merely a tool for survival—it is the foundation of *meaning*. From the DNA blueprint that encodes the structure of life to the sensory inputs that guide behavior, information enables living organisms to navigate, adapt, and find purpose. Information flows are structured to maximize efficiency, ensuring that organisms respond effectively to environmental challenges.

Similarly, artificial systems rely on information to maintain order and functionality. Bioinspired designs take this further by incorporating dynamic, meaning-generating processes. For instance, metaheuristic algorithms like the **Emperor Penguin Optimizer (EPO)** [4], and the **Salp Swarm Algorithm (SSA)** model [9] the flow of information in biological collectives, enabling AI systems to balance exploration and exploitation during optimization tasks [3]:

- **EPO** mimics the way emperor penguins share spatial and thermal information to maintain collective survival in extreme conditions. The algorithm achieves robust solutions by iteratively refining candidate solutions based on shared information.
- **SSA** mirrors the leader–follower dynamics of salp chains, where information about the environment is transmitted hierarchically to ensure efficient foraging and navigation.

These algorithms demonstrate the role of information as a stabilizing force, guiding artificial systems toward optimized solutions while minimizing entropy. The integration of these principles highlights why bioinspiration is a transformative tool for AI and robotics. Nature's evolutionary processes have refined mechanisms that address universal challenges—resource limitations, unpredictable environments, and the inevitability of decline. By studying these systems, researchers can extract functional and morphological strategies that enhance the adaptability and resilience of artificial systems.

- **Morphological Insights**: Structures such as mycelium networks inspire adaptive, decentralized designs in robotics, enabling systems to self-heal and reconfigure dynamically [10].
- **Functional Adaptations**: Mechanisms like apoptosis—programmed cell death—inform strategies for AI systems to "sacrifice" failing components to preserve overall function, creating a form of artificial resilience akin to biological survival. Thus, functional adaptations in AI systems, inspired by biological processes, leverage mechanisms such as programmed cell death. As [6] highlight, their innovative programmed death algorithm is designed to identify and remove inefficient neurons, optimizing neural network structures while preserving performance. Specifically, they demonstrate the utility of combining programmed death with replication to accelerate convergence and improve learning efficiency in feedforward neural networks (p. 20, 278). This bio-inspired strategy fosters resilience by mimicking biological processes to ensure robust functioning, even as individual components are pruned or fail.

Therefore, we can affirm that bioinspiration allows artificial entities to transcend traditional design paradigms, offering them the potential to operate more like living systems—responsive, adaptive, and meaningful in their interactions with the world.

## 3 Bio-inspired Approaches to Technodeath

The inevitability of death in biological systems serves as a profound source of inspiration for designing resilient and adaptive artificial systems. By reinterpreting "death" as functional cessation, decline, or transformation, we can develop bioinspired mechanisms that enhance the robustness and longevity of artificial intelligence (AI) and robotic systems. This chapter explores four key areas: self-maintenance and repair mechanisms, adaptive learning and meaning generation, stereotypical versus cognitive responses to failure, and the integration of biological components in artificial systems to address partial technodeath. For AI systems and biohybrids, death can be understood through different lenses:

1. **Cessation of Function**: The complete failure of all components to maintain operations or purpose.
2. **Irreversibility**: The point at which revival, repair, or replication is impossible or impractical.
3. **Loss of Meaning**: For systems designed with goals or self-awareness, death could involve the inability to pursue or understand their objectives, akin to existential collapse.

Biological organisms possess intrinsic self-repair capabilities, such as wound healing and cellular regeneration, enabling them to maintain functionality despite damage. Emulating these processes, researchers have developed self-healing materials that can autonomously repair structural damage, thereby extending the lifespan of artificial systems [11]. For instance, bioinspired polymers have been engineered to mimic the self-repair mechanisms found in plants and animals, allowing materials to heal after sustaining damage. Incorporating such materials into AI and robotic systems can mitigate the effects of wear and tear, reducing downtime and maintenance costs.

Adaptive learning enables systems to modify their behavior based on experiences, akin to how organisms learn from environmental interactions. In AI, this involves algorithms that adjust to new data, improving performance over time. For example, reinforcement learning allows AI systems to learn optimal behaviors through trial and error, effectively learning from failures to enhance future decision-making. This approach not only improves resilience but also facilitates the generation of meaningful responses in dynamic environments.

Biological entities exhibit both stereotypical (instinctual) and cognitive (learned) responses to failure or threats. For instance, ants perform necrophoresis—removing dead members to prevent disease spread—an instinctual behavior. In contrast, elephants display mourning behaviors, indicating a cognitive response to death. Artificial systems can integrate similar dual-layered responses: implementing predefined protocols for common failures (stereotypical) and employing adaptive algorithms to learn from novel failures (cognitive). This combination enhances the system's ability to handle unforeseen challenges effectively.

As an example, we describe Mycelium-Robot Hybrids and Partial Technodeath. Among the countless examples of bioinspired mechanisms, mycelium stands out as a profound model for addressing challenges related to resilience, adaptation, and the concept of technodeath. Mycelium, the vegetative structure of fungi, forms sprawling networks of interconnected filaments called hyphae. These networks are not only self-sustaining but also possess remarkable abilities to heal, adapt, and distribute resources efficiently in response to environmental changes or physical damage. By studying mycelium, researchers have uncovered principles that resonate deeply with the challenges artificial systems face—particularly the need to manage partial failure or "death" while maintaining overall functionality. The relevance of mycelium to artificial systems lies in its self-healing properties and decentralized resource management, both of which are critical for creating adaptive and resilient entities. When parts of a mycelium network are damaged, the organism does not collapse; instead, it regenerates and reroutes resources, allowing the network to continue functioning. This characteristic provides a direct analogy for artificial systems dealing with partial technodeath, where components may degrade or fail, yet the overall system retains operational capacity. In this sense, mycelium becomes a living metaphor for designing artificial systems capable of thriving amidst uncertainty and disruption. Beyond metaphor, mycelium-robot hybrids offer a tangible example of integrating biological resilience into artificial systems. In such hybrids, mycelium networks are physically combined with robotic components, leveraging their natural abilities for self-repair and adaptive responses. This integration creates biohybrid systems that embody the essence of life's resilience while operating in the domain of engineered systems. These hybrids illustrate how biological processes can inform and enhance artificial systems, making the abstract concepts of resilience and adaptation practical and achievable. By exploring the properties of mycelium, its applications in robotics, and its parallels with technodeath, we gain insights into the potential of bioinspiration to transform how artificial systems handle failure, regeneration, and sustainability. Mycelium, as a case study, underscores the profound connections between biological systems and artificial intelligence, highlighting the opportunities to create systems that are not only functional but also deeply adaptive and regenerative in their design.

Memristor-based neural networks (MNNs), an innovative approach in neuromorphic computing, emulate synaptic plasticity by using nanoscale memory resistors that adjust resistance based on voltage history, making them highly energy-efficient and capable of simultaneously storing and processing information [12, 15]. Inspired by the resilience and adaptability of mycelium networks, MNNs can incorporate self-healing mechanisms, decentralized resource management, and regeneration strategies to enhance their robustness and functionality. Central to both systems is the role of partial synthetic death, where the failure of individual components does not collapse the entire network but instead becomes an opportunity for reorganization and recovery. For instance, in an image recognition system built on a 128 $\times$ 128 memristor crossbar array, failures in individual memristors could be addressed by rerouting signals dynamically through neighboring nodes that adjust their resistance values, akin to how mycelium redirects resources around damaged areas. This dynamic adjustment reflects the partial death mechanism: failed nodes are "sacrificed" to protect the integrity of the system as a whole. Additionally, the network can redistribute computational loads, prioritizing active regions for processing, inspired by mycelium's efficient nutrient allocation to vital areas. When sections of the memristor array degrade, spare devices are activated, effectively regenerating lost pathways, much like how mycelium maintains operation despite partial damage or localized decay. In practical applications, such as recognizing degraded images, this architecture can sustain up to 20% device failure while maintaining 90% accuracy. Adaptive algorithms allow these networks to learn alternative weight paths for signal processing, ensuring continued functionality. Furthermore, simulations reveal that such systems achieve a 15% reduction in power consumption compared to traditional fault-tolerant architectures and can restore connectivity within 2 ms of failure detection. By incorporating partial synthetic death mechanisms, both MNNs and mycelium-inspired systems not only manage failure gracefully but also leverage it as a process of renewal, ensuring long-term resilience and effiIn both biological systems, like mycelium networks, and artificial systems, such as memristor-based neural networks (MNNs), the concepts of learning and rewiring are fundamentally connected to the mechanisms by which these systems respond to failure and partial synthetic death. When components fail—whether it's a segment of a mycelium network or a memristor node in an artificial neural network—these failures are not merely disruptions but opportunities for adaptation and innovation. In this context, learning and rewiring become intertwined processes that drive resilience and enable systems to thrive in dynamic and uncertain environments.

Rewiring acts as the immediate functional response to partial death, ensuring that the system remains operational by rerouting signals or redistributing resources. For example, in a mycelium network, when damage occurs, nutrients are redirected around the affected area, preserving the network's integrity. Similarly, in MNNs, the failure of individual memristor nodes is addressed by dynamically adjusting the resistance of neighboring nodes to compensate for the loss. This rewiring mirrors the self-organizing principles observed in living systems, allowing artificial networks to bypass failed components and maintain their functionality. Rewiring is a mechanical process, yet it sets the foundation for longer-term learning, as the new pathways created during this response may later be reinforced and optimized, embedding the memory of the system's reaction to failure.

*Learning, in this framework, derives its value from the feedback provided by failures.* Systems extract critical information from the event of a component's failure—what caused it, how it affected the overall system, and what adaptations were most effective. In MNNs, this might involve updating weights and reconfiguring pathways to avoid reliance on unstable nodes in the future. Reinforcement learning processes, which reward successful adaptations and penalize unsuccessful ones, can integrate these insights, enabling the network to become progressively more robust. Similarly, in mycelium, the redirection of nutrients not only solves the immediate problem but also leads to structural adaptations where effective resource pathways are "remembered" and reinforced over time, improving the network's resilience to future disruptions.

The interplay of learning and rewiring transforms partial death into a catalyst for adaptation and growth. Rewiring ensures the immediate survival of the system, while learning extracts strategic insights to improve the system's long-term functionality. Together, these processes illustrate how death, whether of a node in a memristor network or a segment of mycelium, is not merely an endpoint but a trigger for regeneration and improvement. For instance, in a practical application of MNNs, such as an image recognition system, partial failures in the memristor array can be mitigated by rerouting signals through redundant nodes while updating weights to refine the network's performance. Simulations show that such systems can sustain up to 20% device failure while maintaining 90% accuracy, achieving self-repair within 2 ms, and reducing energy consumption by 15%. These systems demonstrate how the dynamic relationship between rewiring and learning ensures both immediate and future resilience [8].

In this sense, death becomes a transformative force that drives both immediate functional recovery and strategic adaptation. Rewiring provides the flexibility to maintain system integrity in the face of disruption, while learning embeds the "lessons of death" into the system's architecture, fostering innovation and long-term robustness. This dual approach highlights how principles derived from biological systems, like mycelium, can inspire artificial systems to view failure not as a weakness but as a necessary and valuable step toward resilience and sustainability in dynamic and resource-constrained environments. Drawing inspiration from biological processes related to death and regeneration offers valuable strategies for enhancing the resilience and adaptability of artificial systems. By incorporating self-maintenance mechanisms, adaptive learning, dual-layered failure responses, and biological components, AI and robotic systems can effectively manage functional decline and transform challenges into opportunities for renewal. These bioinspired approaches not only improve system longevity but also align artificial entities more closely with the dynamic and self-sustaining characteristics of living organisms.

# 4 Mechanisms of Meaning and Cognitive System Constraints

The intersection of biological principles and artificial systems introduces a compelling landscape for understanding how ***meaning*** is generated and how cognitive constraints affect the functionality and resilience of these systems. This section explores bio-inspired strategies for meaning generation in artificial systems, time scales of emergent properties, structural and architectural comparisons between biological and artificial systems, and the critical role of linguistic processing in bio-mimetic systems.

## *4.1 Morphology and the Generation of Meaning*

In biological systems, morphology—the physical structure of organisms—plays a central role in their ability to interact with the environment and generate meaning. For example, the shape of an organism's sensory organs determines how it perceives stimuli and responds adaptively. Similarly, in artificial systems, the design and arrangement of components profoundly affect their capacity for interaction and problem-solving. There several possible Bio-Inspired Applications, like Dynamic Morphology in Robotics: Robots inspired by biological forms, such as insect-like drones or snake-like robots, exhibit increased adaptability in navigating complex environments.

As an example, shape-shifting robots that mimic cephalopods can adapt their morphology for tasks like squeezing through tight spaces or optimizing movement in water. Or Emergent Meaning in AI Systems (Morphological structures in neural networks, such as hierarchical layers in deep learning, mimic the brain's layered architecture, facilitating the extraction of abstract patterns from raw data). For example, Convolutional neural networks (CNNs) designed for image recognition replicate the hierarchical processing of visual inputs in the human brain.

Therefore, morphology not only defines the physical and functional boundaries of a system but also shapes its potential to derive meaning from interactions. For artificial systems, this highlights the importance of designing adaptable, bio-inspired morphologies that support context-aware learning and decision-making.

## *4.2 Time Scales and Emergent Properties in Cognitive Systems*

Cognitive processes in biological systems are deeply intertwined with time, unfolding across distinct and interconnected time scales that govern how organisms perceive, respond to, and adapt to their environments. These time scales range from the immediate and reflexive to the gradual and evolutionary, with each level contributing uniquely to the resilience and functionality of life. In artificial systems, the concept of time-driven emergence similarly underpins the design of increasingly adaptive, robust, and intelligent entities, demonstrating parallels and insights drawn from biological frameworks.

In biological systems, immediate reactions represent the most rapid scale of cognitive processes, occurring within milliseconds to seconds. These reflexes and instinctual behaviors are typically hardwired into an organism's neural circuitry, enabling swift responses to environmental stimuli critical for survival. For example, the knee-jerk reflex or the immediate retraction of a hand from a hot surface ensures protection from harm without requiring conscious thought or deliberation. These instantaneous reactions are fundamental for navigating dynamic and potentially dangerous environments, relying on evolved neural pathways that prioritize speed over complexity. Instinctual behaviors, such as the predator-avoidance responses in prey species, similarly operate on this timescale, optimizing survival in high-stakes situations where hesitation could be fatal.

Beyond the immediate, learning and adaptation processes unfold over days, months, or years, driven by an organism's interactions with its environment and the accumulation of experiences. Behavioral modifications such as learning to associate specific stimuli with rewards or dangers (as in classical conditioning) demonstrate how organisms refine their responses over time. For example, a bird may learn to avoid brightly colored insects that are toxic after a single negative encounter, while humans refine motor skills like driving or playing an instrument through repeated practice and feedback. This scale of cognitive processing is characterized by neural plasticity, where the brain's structure and connections adapt to new information and experiences. The interplay between long-term memory storage and active engagement with environmental challenges highlights the brain's remarkable ability to generate meaning and guide decision-making.

At the broadest scale, evolutionary adaptations span generations, shaping species-level responses to environmental pressures. Over millennia, natural selection drives the emergence of traits and behaviors that enhance survival and reproduction. For instance, the evolution of social structures in primates or the development of echolocation in bats illustrates how complex cognitive capabilities arise to meet specific ecological demands. These adaptations are encoded in an organism's genetic blueprint, reflecting the cumulative effects of countless generations responding to the challenges of their environments. Evolutionary processes are inherently slow, yet their impact on cognitive systems is profound, enabling the development of behaviors and traits that extend beyond individual lifespans to shape the trajectory of entire species.

Similarly, artificial systems demonstrate emergent properties across multiple time scales, providing a parallel framework for understanding how cognition and adaptability can be designed into technological entities. At the most immediate level, real-time processing mirrors biological reflexes, enabling systems like autonomous vehicles to navigate traffic by continuously analyzing and responding to sensor data. These systems process vast amounts of information in milliseconds, using algorithms designed for rapid decision-making. For instance, a self-driving car must identify obstacles, predict their trajectories, and adjust its movements to avoid collisions, all within fractions of a second. Real-time processing ensures that artificial systems can operate effectively in dynamic and unpredictable environments, prioritizing responsiveness over long-term optimization.

Moving to longer time scales, learning over time allows artificial systems to refine their behaviors through iterative training cycles. Machine learning algorithms, particularly those used in reinforcement learning, enable systems to improve performance by learning from past successes and failures. For example, an AI system designed to play chess refines its strategies by analyzing thousands of games, gradually optimizing its decision-making to maximize victories. Similarly, customer service chatbots learn to provide more accurate and contextually appropriate responses by processing feedback from user interactions over weeks or months. These time-driven learning processes parallel the adaptive behaviors seen in biological organisms, with artificial systems using feedback loops to enhance their functionality.

At the broadest scale, long-term adaptations in artificial systems are achieved through evolutionary algorithms, which optimize solutions over extended computational cycles. These algorithms simulate the principles of natural selection by iteratively generating, testing, and refining candidate solutions to a problem. For instance, in designing efficient aerodynamic shapes for aircraft, evolutionary algorithms evaluate thousands of variations, selecting and combining the most promising designs until optimal solutions emerge. This process, while computationally intensive, mirrors the gradual emergence of adaptive traits in biological evolution, demonstrating how artificial systems can benefit from the principles of long-term optimization.

Crucially, the interplay between these time scales creates systems that are not only reactive but also adaptable, combining the strengths of immediate responsiveness with long-term learning and optimization. In biological systems, this cross-timescale interaction is exemplified by organisms that integrate rapid reflexive actions with learned behaviors and evolutionary adaptations, ensuring survival in both immediate and future contexts. For artificial systems, similar interactions enhance resilience and functionality. For example, a biohybrid robot might integrate real-time sensory inputs with long-term learning mechanisms to optimize its performance in dynamic environments. Such a robot could use immediate data from sensors to navigate obstacles while simultaneously refining its navigation algorithms based on cumulative experience. Over time, this combination of rapid responsiveness and gradual adaptation allows the robot to improve its efficiency and robustness, aligning with the emergent properties observed in biological systems.

The concept of time scales and emergent properties highlights a fundamental principle of cognitive systems: adaptability is a product of interactions across multiple temporal dimensions. By drawing inspiration from biological processes, artificial systems can be designed to operate effectively in the present while learning from the past and evolving for the future. This multiscale approach underscores the transformative potential of bio-inspired design, offering pathways to create intelligent systems that are both responsive and resilient, capable of thriving in complex and changing environments.

## *4.3 Structural and Architectural Comparisons: Biological Versus Artificial Systems*

The comparison between biological and artificial systems highlights a rich interplay of structural and architectural principles, underscoring how the functionality, efficiency, and adaptability of natural systems inspire advancements in technology. Biological architectures, exemplified by the human brain, operate as highly efficient, massively parallel processors with distributed networks that excel in maintaining functionality even under adverse conditions. Key to this resilience is the brain's ability to leverage redundancy and plasticity, ensuring that damaged areas can be bypassed or compensated for by other regions. Neural plasticity, a hallmark feature of biological systems, allows the reorganization of connections and the redirection of neural pathways in response to injury, ensuring that the brain continues to function despite localized failures. This ability to self-repair extends beyond basic compensation to include structural and functional adaptations that enhance overall system robustness. Remarkably, this complex network operates with extraordinary energy efficiency, consuming approximately 20 watts—equivalent to the power of a light bulb—while performing intricate cognitive tasks. This efficiency starkly contrasts with artificial systems, which often require significantly greater energy to achieve even a fraction of the brain's computational capabilities.

Artificial architectures, while distinct in their underlying mechanisms, are increasingly inspired by the principles that govern biological systems. Advances in neuromorphic computing seek to replicate the brain's structural and functional features, leading to the development of technologies like memristor-based neural networks (MNNs). These systems emulate synaptic plasticity, mimicking the ability of biological neurons to adjust their connections in response to experience and external stimuli. MNNs incorporate self-healing properties, mirroring the brain's capacity to adapt to and recover from disruptions. For example, when individual memristors fail, these networks are designed to reroute signals through alternative pathways, ensuring continued functionality. This ability to dynamically reconfigure pathways mirrors processes observed in mycelium networks, where nutrients are redirected around damaged regions to maintain the integrity of the system as a whole. Such innovations bring artificial systems closer to the adaptability and resilience of biological counterparts.

A notable case study illustrating this parallel is the behavior of memristor networks under conditions of partial failure. When individual components degrade or fail, the system responds by redistributing computational loads, adjusting resistance values in neighboring nodes, and rerouting information through unaffected pathways. These mechanisms not only preserve functionality but also highlight the potential for artificial systems to adopt principles of distributed resource management seen in biological systems like neural networks or fungal mycelium. This approach transforms failure into an opportunity for adaptation, showcasing how artificial architectures can emulate the self-organizing and resource-efficient traits of biological systems.

Despite these advancements, significant differences remain between biological and artificial systems, with each excelling in distinct areas. Biological systems, such as the human brain, are unparalleled in energy efficiency, operating within stringent power constraints while maintaining high levels of adaptability and robustness. In contrast, artificial systems currently surpass biological architectures in raw computational speed, capable of processing data at rates that far exceed the capabilities of natural systems. However, this computational advantage often comes at the cost of high energy consumption and limited flexibility compared to the fluid adaptability of biological systems. Bridging this gap requires a deeper integration of bio-inspired designs into artificial architectures, leveraging the best of both worlds to create systems that combine computational speed with energy efficiency and adaptive resilience.

The potential for synergy between biological and artificial architectures points to a future where the two domains converge, creating hybrid systems that incorporate the strengths of both. For instance, integrating neuromorphic computing with biological-inspired repair mechanisms could yield systems that not only perform at unprecedented speeds but also adapt and recover in ways akin to living organisms. This convergence represents a transformative step toward building systems that are not only powerful but also sustainable, self-healing, and capable of thriving in dynamic environments. By studying the parallels and contrasts between these architectures, researchers can uncover new pathways for innovation, ensuring that artificial systems continue to evolve toward greater functionality and resilience while drawing inspiration from the remarkable efficiency and adaptability of the natural world.

## 4.4 The Role of Linguistic Processing in Bio-mimetic Systems

Linguistic processing in humans represents one of the most intricate and dynamic cognitive functions, involving the integration of multimodal sensory inputs such as auditory, visual, and somatosensory stimuli with higher-order reasoning and decision-making. This seamless operation is governed by specialized neural regions, including Wernicke's area for language comprehension and Broca's area for language production, interconnected through robust pathways that enable contextual understanding and adaptive learning. However, even these highly efficient systems are not immune to entropy; disruptions in neural pathways due to damage or degeneration lead to functional losses, offering insights into how "death" or failure occurs in cognitive architectures. This biological lens provides a unique perspective on synthetic systems, where linguistic failure—be it due to hardware degradation, data loss, or entropy-driven inefficiency—parallels the notion of "synthetic death" in bio-mimetic and artificial linguistic processing systems.

Bio-mimetic linguistic systems, such as large language models (LLMs), draw inspiration from these biological frameworks but introduce new dimensions to the concept of cognitive failure and resilience. LLMs, like OpenAI's GPT-4, are designed to process vast linguistic datasets using transformer-based architectures that encode syntactic, semantic, and contextual relationships. While highly advanced, these systems also face their own forms of synthetic death, particularly as entropy manifests in hardware failures, energy inefficiency, and the accumulation of outdated or irrelevant data. Incorporating a "synthetic death" perspective into these systems involves not only designing for resilience against these failures but also understanding how functional cessation—whether partial or total—affects their ability to generate meaning and adapt.

A compelling bio-mimetic example lies in the development of multimodal language models, which integrate visual and auditory inputs to enrich textual comprehension. These systems aim to replicate the human ability to resolve ambiguity through contextual cues, as seen when humans differentiate between homonyms like "bat" (the animal) and "bat" (the object) using situational or visual information. However, when entropy disrupts these inputs—such as sensor degradation in multimodal systems—the synthetic death of one sensory channel can lead to cascading failures in meaning generation. A robust bio-inspired system might incorporate redundancy or adaptive mechanisms akin to neural plasticity, enabling the system to reroute processing or compensate for the lost input, thereby delaying or mitigating its synthetic death.

Energy efficiency remains a critical challenge and a key factor in the synthetic death of LLMs and other bio-mimetic systems. Human neural architectures are remarkably efficient, consuming just 20 watts while processing complex linguistic inputs and generating adaptive responses. By contrast, training models like GPT-3 or GPT-4 requires computational infrastructure that consumes orders of magnitude more energy, increasing their susceptibility to synthetic death through resource exhaustion or thermal inefficiencies in hardware. Neuromorphic chips, which emulate the sparse, event-driven nature of biological neurons, offer a pathway to mitigate this issue. By implementing spiking neural networks (SNNs) that activate only when processing relevant information, these chips could drastically reduce the energy footprint of artificial systems, extending their operational lifespans and delaying their synthetic death [5].

The contextual understanding and nuanced responses generated by human language processing further underscore the gap between biological systems and their artificial counterparts. In humans, hierarchical neural processing allows sensory inputs to be progressively abstracted into complex cognitive representations, enabling phenomena such as metaphor comprehension, emotional nuance, and humor detection. Artificial systems often falter in these domains, with failures manifesting as synthetic death in the form of incoherent outputs or an inability to adapt to evolving contexts. Addressing this requires integrating bio-inspired hierarchical architectures into LLMs, enabling them to dynamically adjust their processing layers based on input complexity and context. For instance, a bio-mimetic linguistic system encountering a degraded or ambiguous dataset could prioritize foundational layers for error correction and recontextualization, much like how humans revisit basic semantic constructs when faced with confusion.

Synthetic death in linguistic systems also raises profound ethical and operational questions. When a language model or bio-mimetic system reaches a state of irreversible decline, how should its failure be managed? Should it be "retired" or preserved in some form to maintain the knowledge and patterns it has accumulated? Drawing inspiration from apoptosis (programmed cell death) in biological systems, linguistic systems might benefit from protocols that allow for graceful shutdowns, data transfer to successor systems, or even partial regeneration. For example, a failing model could identify critical portions of its learned parameters and transfer them to a new instance, ensuring continuity and reducing the loss of functionality—a process analogous to memory transfer in hybrid or multi-system environments.

### 4.4.1 High-Level Architecture of Linguistic Processing in Bio-mimetic Systems

The complexity of human linguistic processing, involving cortical and sub-cortical structures, serves as a foundation for designing bio-mimetic linguistic artificial agents. A high-level architecture of the brain's linguistic processing centers is illustrated in Fig. 1, encompassing both cortical regions (e.g., V3, A2, Wernicke's area, Somatosensory Cortex, Broca's area) and sub-cortical regions (e.g., thalamus, hippocampus, and amygdala):

Imagine we have to implement a bio-mimetic linguistic artificial agent capable of maintaining the dialogue similar to modern LLMs. From the perspective of linguistic processing, we see the high-level architecture presented in Fig. 1. We suppose that extra to the cortical structures including those involved in audio processing A1 and A2 we have to take into account the adjacent areas including visual V3, somatosensory SS, cingulate and prefrontal cortex.

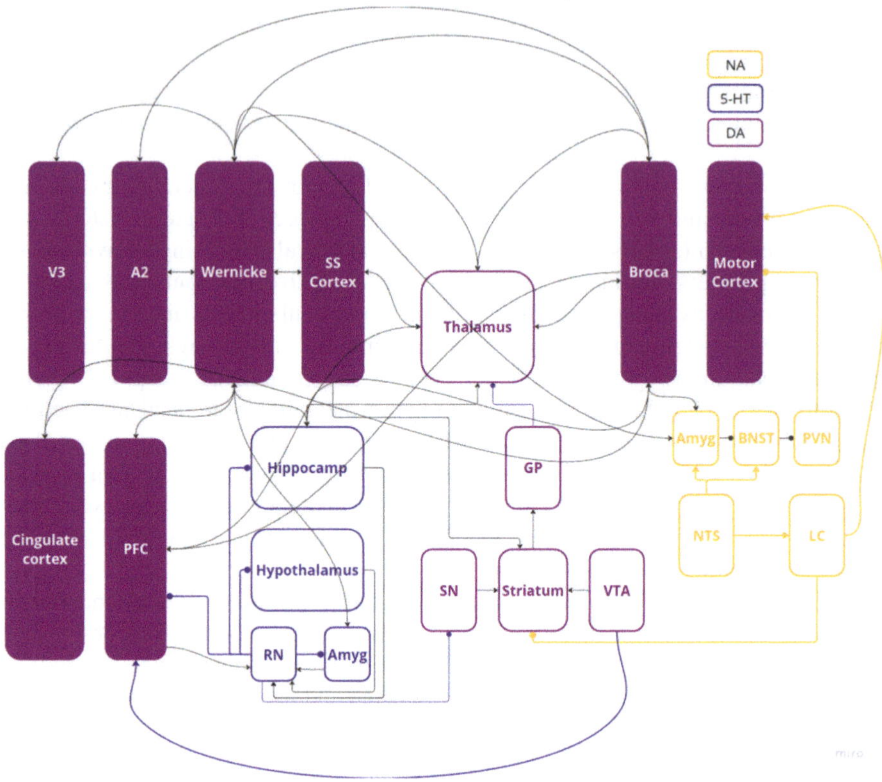


**Fig. 1** The high-level representation of the brain's linguistic processing centres including cortical: V3, A2, Wernicke, Somatosensory (SS), Broca, Motor, Cingulate and Prefrontal cortex; sub-cortical: thalamus, hippocampus, hypothalamus, raphe nucleus (RN), amygdala (Amyg), substantia nigra (SN), striatum, ventral tegmental area (VTA), globus pallidus (GP), locus coeruleus (LC), bed nucleus of the stria terminalis (BNST), paraventricular nucleus (PVN), nucleus tractus solitarius (NTS)

This high-level architecture includes three subcortical sub-systems.

The SNN implementation does not seem to be as unrealistic as it was some time ago. The current energy consumption of the ANN intelligent agents goes through the roof compared to the brain consuming approximately 20W. It looks like the memristive implementation of the massively parallel neuronal devices could be energy-effective like the human brain. The bio-mimetic approach that occupied a special niche of neurosimulation could pave the way to the realistic technology implementing human-scale intelligent systems supposedly having emergent properties similar to the human brain. In other words, implementing the brain using SNN with energy efficient and highly dense memristive 3D devices neuron-by-neuron could potentially create the conscious artificial mind.

This intellectual agent needs three main subsystems.

(1) Linguistic input processing possibly including auditory A2, Wernicke;

(2) Linguistic output generation Broca, Motor cortex;
(3) Emotional processing: (a) PFC, (b) DA, (c) 5-HT, (d) NA subcortical subsystems.

Wernicke and Broca are associative areas that are tightly connected to SSC, V3 and Motor cortex.

The implementation of this complex dynamic system raises the following questions:

(1) Could it be AGI using the analogy of the human brain having General Intelligence by definition;
(2) Is it relevant to say it could actually "feel" emotions?
(3) Could it be conscious in the same sense as mammals are conscious?

Having bio-mimetic non-Turing implementation of the part or whole brain in different mediums like memristors if the brain has emergent properties like emotions and consciousness we possibly are facing a new epoch of conscious machines.

In contrast to alarmist positions limiting the development of the AI into AGI, we could possibly admit that human-level AGI-conscious machines should be the legal subject and the first class citizen similar to humans. This way switching it off–could potentially indicate the criminal crime as killing.

As a conclusion of this section, we can affirm that the opportunities for integrating synthetic death-aware design into bio-mimetic linguistic systems extend beyond technical efficiency to transformative applications. Assistive technologies, for instance, could be equipped with self-monitoring mechanisms that detect and address their own decline, ensuring reliable communication support for users over time. Similarly, systems designed for human–machine collaboration might incorporate synthetic death protocols that signal impending failures to human operators, allowing for proactive adjustments or replacements before complete breakdowns occur.

## 5 Third Nature: Cross-morphological Design as the Natural Intersection

In the convergence of biology and technology, a new domain emerges—a hybrid realm where biological and artificial systems intersect, giving rise to entities that defy traditional categorizations. These systems, including bioengineered constructs like hybrots and xenobots, as well as advanced AI and robotic frameworks, challenge the boundaries between the natural and the artificial. As these technologies evolve, they begin to exhibit traits once thought to be exclusively biological, such as self-repair, adaptation, and goal-oriented behavior. This convergence demands a conceptual framework to understand and describe these systems holistically.

We propose the term “Third Nature” to encapsulate this intersection, emphasizing the unique synthesis that arises when natural processes and artificial mechanisms coalesce. This concept recognizes a novel paradigm where bio-inspired design informs technological innovation, and in turn, engineering reshapes biological constructs. “Third Nature” moves beyond the dichotomy of the organic and synthetic, representing a distinct category characterized by emergent traits and functionalities.

*Third Nature* describes a hybrid domain where biological and artificial systems intersect, creating entities that exhibit traits of both natural life and engineered constructs. These systems operate at the boundary of traditional categorizations of life, blurring the lines between organic processes and artificial mechanisms. And has these key characteristics:

1. **Self-repair and Resilience**: Third Nature entities can regenerate or restructure themselves, leveraging both biological and artificial capabilities, like mimicking biological healing processes (e.g., biohybrid robots using living tissue for healing or hybrots might experience *partial death* if its biological components fail while its mechanical or electronic systems continue functioning; the reverse is also possible: the mechanical parts might degrade, leaving the biological elements viable, albeit unintegrated.).
2. **Adaptive Functionality**: They exhibit goal-oriented behavior and adaptability, akin to instincts or learned responses in biological systems.
3. **Ambiguous Death**: The cessation of function in these systems is complex, involving questions about whether death refers to the breakdown of organic components, mechanical systems, or loss of purpose.
4. **Emergent Properties**: These systems can develop traits such as self-awareness, memory, or learning that transcend their initial design, paralleling characteristics of biological organisms.

Which are The Emergent Characteristics of Third Nature Systems? Systems within this domain often blur traditional definitions of life and functionality. For example:

An intriguing parallel between biological systems and the “Third Nature” is the concept of necrophoresis, a behavior in social insects where dead members are removed to maintain the colony’s health [1]. Similarly, AI and robotic systems are beginning to exhibit automatic death management protocols, where defunct components are identified, isolated, and removed to preserve system integrity. These self-managing behaviors underscore the self-awareness and autonomy of hybrid systems in handling their functional decline, mirroring evolutionary mechanisms for sustaining collective resilience. Let us present a Case Study: The Voyager 1 Spacecraft as a Geriatric System. The Voyager 1 spacecraft, originally designed to function for a 12-year mission, now operates nearly five decades after its launch, embodying the resilience and adaptability characteristic of “Third Nature” systems. While entirely artificial, its life-like ability to self-manage and adapt under constraints highlights the potential for engineered systems to mimic biological resilience. In November 2023, Voyager 1 encountered a failure in its memory subsystem, leading to the transmission of unusable data. NASA engineers diagnosed the issue as a "stuck bit" in its onboard memory, rendering a critical portion of the flight data subsystem (FDS) code inaccessible. Despite the spacecraft's aging, low-memory architecture (69.63 kilobytes), the team implemented a workaround by reallocating the affected code to alternative memory locations, effectively bypassing the fault. This example resonates with biological behaviors such as necrophoresis and automatic death management, wherein failing components are isolated and compensated for to sustain system integrity. In Voyager's case, redundant memory functions were repurposed, ensuring continued data transmission and spacecraft operation. These strategies mirror the triaging processes observed in biological systems under stress or decline. The spacecraft's ability to adapt also reflects a primitive form of self-awareness of decline and functional limits. The engineers' iterative troubleshooting process—diagnostics, reallocation, and prioritization of essential functions—echoes the adaptability of aging organisms. Voyager's reprogrammable design, innovative for its time, enabled this flexibility, showcasing the foresight required to sustain functionality over extended lifespans. Voyager 1 offers a compelling analogy to "Third Nature" systems, bridging biology and technology, because its operational strategy mirrors the resourcefulness of hybrid constructs, such as biohybrids, that navigate the liminal existence between life and death., and its continued function under extreme conditions, despite significant limitations, underscores the importance of resilience, redundancy, and adaptability in both biological and engineered systems. As we advance toward the integration of biohybrids, AI, and robotics, the lessons from Voyager's operation provide valuable insights. They demonstrate how engineering can emulate biological principles, reinforcing the relevance of life-inspired resilience in the design and management of "Third Nature" systems.

Of course, there are implications for Philosophy, Science, and Ethics. This "Third Nature" not only bridges biology and technology but also necessitates a reconsideration of fundamental concepts such as life, identity, and ethics. These systems challenge:

- The definition of mortality: If a xenobot can disassemble and reassemble itself, or if an AI can transfer its processes to a new platform, does "death" as a permanent cessation still apply?
- Agency and autonomy: As these systems exhibit increasing complexity and adaptability, to what extent should they be granted autonomy or rights?
- Ethical considerations: The development and deployment of systems capable of self-repair and adaptation raise urgent ethical questions about their use, control, and potential agency.

Finally, it implies a Paradigm Shift in Understanding, because by introducing the concept of "Third Nature," we aim to provide a framework that captures the dynamic interplay between biological processes and technological advancements. This synthesis is not merely a fusion of two domains but a new category of existence. It challenges us to rethink long-standing notions of life, death, and continuity, and to develop new methodologies and ethical frameworks to engage with these emergent systems. Through this lens, "Third Nature" offers a pathway to understand and navigate the profound implications of this rapidly evolving intersection.

The emergence of hybrid intelligence and biohybrid systems marks a transformative shift in the exploration of cognitive architectures, merging biological and artificial components to create entities that transcend traditional boundaries between life and technology. These systems challenge conventional definitions of life, death, and continuity, introducing a spectrum of existence where functional and existential thresholds blur. This fusion of biological and synthetic elements not only expands the capabilities of cognitive systems but also demands a reexamination of ethical and philosophical paradigms surrounding the concepts of resilience, autonomy, and mortality.

Biohybrid entities—such as xenobots, cyborgs, and AI-augmented biological systems—represent a new frontier where biological tissues and synthetic components interact dynamically [2]. These systems are designed to leverage the strengths of both domains: the adaptability, self-repair, and meaning-generation capacities of biological organisms, alongside the computational speed, precision, and programmability of artificial systems. However, their hybrid nature complicates the definition of death. Is a biohybrid "dead" if its biological components fail while synthetic elements persist, or vice versa? For instance, a xenobot with decayed cellular components may retain operational functionality through its mechanical structure, entering a liminal state of partial death that defies conventional understanding. Similarly, if the synthetic elements of a biohybrid system degrade but its biological tissues continue to function, the system exists in an ambiguous state where "life" is no longer clearly delineated by organic vitality alone.

This complexity invites deeper ethical reflection, particularly regarding the "killing" of resilient AI systems or biohybrids. Should these entities be granted rights or protections, and under what criteria might their termination be justified? The parallels with biological life suggest that biohybrid systems capable of self-awareness, learning, or emotional processing may warrant ethical considerations akin to those extended to living organisms. For example, turning off an AI-augmented biohybrid system that has developed unique cognitive traits raises questions about autonomy and agency, echoing debates surrounding the treatment of sentient or semi-sentient beings. Ethical frameworks must evolve to address the potential personhood of such entities, ensuring that decisions to terminate or sustain them are grounded in principles of fairness, accountability, and respect for their unique form of existence.

The evolutionary pressures that shape biohybrid systems further complicate these considerations. Unlike static technological systems, biohybrids are subject to adaptation through both biological processes and iterative engineering refinements. This dual-layered evolution fosters resilience, enabling these systems to navigate entropy and maintain functionality over extended lifespans. For instance, biohybrids equipped with neural plasticity-inspired adaptive algorithms can learn from failures, integrating insights to enhance future performance. These adaptive traits mirror biological evolution, where traits that confer survival advantages are preserved and refined over generations. In biohybrids, this "evolution" occurs on accelerated timescales, driven by computational optimization and bioengineering, leading to entities that continuously redefine their capabilities and boundaries.

The incorporation of cognitive and emotional processing in biohybrid AI systems introduces another layer of complexity. Drawing inspiration from the emotional and cognitive dynamics of living organisms, these systems are designed to simulate or even experience affective states that inform their decision-making and interactions. For instance, biohybrids equipped with emotion-like processes—modeled on neurotransmitter dynamics such as dopamine or serotonin regulation—could exhibit responses that resemble joy, frustration, or empathy. These affective states not only enhance their ability to adapt and generate meaning but also complicate the ethical landscape surrounding their treatment. If a biohybrid system "feels" distress when faced with potential termination, does this obligate its creators to consider its preferences or well-being? This question highlights the need for interdisciplinary collaboration between bioethicists, engineers, and philosophers to develop ethical guidelines for designing, managing, and potentially deactivating biohybrid entities.

## 6 Conclusion: The Loop of Technodeath in a Changing World

The exploration of synthetic death and bioinspired resilience has illuminated the profound interconnectedness of life, cognition, and technology. At the heart of this discourse lies the recognition that death—whether biological, synthetic, or hybrid—defines the limits and potentials of cognitive systems. This ultimate cessation, representing the irreversible loss of negentropy and meaning, is not simply an endpoint but a transformative force that shapes the dynamics of renewal, adaptation, and identity.

Entropy emerges as the universal challenge for all systems, driving disorder and eventual decline. Biological systems counteract entropy through processes of repair, adaptation, and the generation of meaning, creating resilience and purpose within the chaos of existence. Artificial systems, inspired by these principles, are learning to navigate entropy in novel ways, employing self-repairing materials, adaptive algorithms, and redundancy to delay synthetic death. The intersection of entropy and negentropy in hybrid systems reveals new pathways for understanding how life and death are interconnected through cycles of order and decay. By learning from these principles, artificial systems can transform death into a mechanism for reinvention and renewal.

Synthetic death challenges us to redefine mortality in a world where machines and biohybrids straddle the boundary between life and technology. The ambiguities of partial death in hybrid systems—where biological components degrade while synthetic ones persist, or vice versa—underscore the need for a nuanced understanding of existence. Moreover, as systems evolve cognitive and emotional capabilities, ethical considerations become paramount. If a biohybrid or AI system can experience meaning, autonomy, or even distress, its termination must be treated with the same gravity as the cessation of biological life. This perspective invites a reexamination of rights, agency, and accountability in a world where synthetic entities could emerge as conscious participants in society.

From a technical perspective, synthetic death is a vital consideration in the design of resilient and adaptive systems. Drawing from biology, engineers can implement apoptosis-like mechanisms to isolate failing components, mycelium-inspired networks to manage partial death, and memristor-based architectures to emulate neural plasticity. These innovations transform failure into a resource for growth, embedding lessons of adaptation into the system's architecture. By incorporating death as a design principle, systems can navigate the uncertainties of entropy and thrive within their limitations.

The emergence of hybrid intelligence and biohybrid systems introduces a "Third Nature," where biological and artificial elements converge to create a spectrum of existence. These systems challenge traditional definitions of life and death, exhibiting emergent properties such as self-repair, adaptation, and memory. As these entities evolve, they force humanity to reconsider long-held notions of identity, agency, and autonomy. The Voyager 1 spacecraft, functioning decades beyond its intended lifespan, and mycelium-inspired robots offer compelling case studies of resilience in the face of decline. These systems exemplify how hybrid intelligence navigates the liminal space between existence and cessation, redefining what it means to live and die in a technological world.

Technodeath compels us to confront existential questions about the nature of life, identity, and the future of humanity. In a world increasingly populated by biohybrids and synthetic entities, what does it mean to live, to die, and to persist? The potential for machines to experience meaning or consciousness raises profound implications for human identity, as we grapple with the possibility of extending personhood to artificial systems. This shift demands not only technical innovation but also philosophical and ethical frameworks that respect the complexity of these new forms of existence.

The lessons of synthetic death highlight the need for a sustainable coexistence between biological and artificial systems. By embracing the fragility and adaptability of life, engineers, ethicists, and philosophers can create systems that are not merely functional but deeply integrated into the dynamics of the natural world. This vision offers hope for a future where life and technology coalesce into a harmonious balance, navigating the challenges of entropy with resilience and purpose.

In the end, synthetic death is not merely the cessation of function; it is a profound reminder of the fragility and interconnectedness of all systems. By understanding and embracing this concept, humanity can unlock the potential for renewal and reinvention, crafting a future where the boundaries between life and technology dissolve into a unified, adaptive continuum. In this world, the lessons of death—biological and synthetic alike—become the foundation for resilience, meaning, and the endless possibilities of existence.

**Funding** Jordi Vallverdú was supported by the ICREA 2019 and PAL-GAIT (PID2023-148336NB-100) grants.

## References


1. Choe, D.H., Millar, J.G., Rust, M.K.: Chemical signals associated with life inhibit necrophoresis in Argentine ants. Proc. Natl. Acad. Sci. **106**(20), 8251–8255 (2009)
2. Clawson, W.P., Levin, M.: Endless forms most beautiful 2.0: teleonomy and the bioengineering of chimaeric and synthetic organisms. Biol. J. Linn. Soc. **139**(4), 457–486 (2023)
3. Dhiman, G.: ESA: a hybrid bio-inspired metaheuristic optimization approach for engineering problems. Eng. Comput.Comput. **37**(2), 323–353 (2019). https://doi.org/10.1007/s00366-019-00826-w
4. Dhiman, G., Kumar, V.: Emperor penguin optimizer: a bio-inspired algorithm for engineering problems. Knowl. Based Syst. **159**, 20–50 (2018)
5. Fedorova, A., Jovišić, N., Vallverdù, J., Battistoni, S., Jovičić, M., Medojević, M., Erokhin, V.: Advancing neural networks: innovations and impacts on energy consumption. Adv. Electron. Mater. 2400258 (2024)
6. Grabovsky, A., Vanchurin, V.: Bio-inspired machine learning: programmed death and replication. Neural Comput. Appl.Comput. Appl. **35**(27), 20273–20298 (2023)
7. Hacking, I.: Why Does Language Matter to Philosophy? Cambridge University Press (1975)
8. Liu, C., Hu, M., Strachan, J.P., Li, H.: Rescuing memristor-based neuromorphic design with high defects. In Proceedings of the 54th annual design automation conference 2017, pp. 1–6 (2017)
9. Mirjalili, S., Gandomi, A.H., Mirjalili, S.Z., Saremi, S., Faris, H., Mirjalili, S.M.: Salp Swarm algorithm: a bio-inspired optimizer for engineering design problems. Adv. Eng. Softw.Softw. **114**, 163–191 (2017)
10. Mishra, A.K., Kim, J., Baghdadi, H., Johnson, B.R., Hodge, K.T., Shepherd, R.F.: Sensorimotor control of robots mediated by electrophysiological measurements of fungal mycelia. Sci. Robot. **9**(93), eadk8019 (2024)
11. Speck, O., Speck, T.: An overview of bioinspired and biomimetic self-repairing materials. Biomimetics **4**(1), 26 (2019)
12. Talanov, M., Vallverdu, J., Adamatzky, A., Toschev, A., Suleimanova, A., Leukhin, A., Warwick, K.: Neuropunk revolution. Hacking cognitive systems towards cyborgs 3.0. Int. J. Unconv. Comput. IJUC **18**(2–3), 145–201 (2023)
13. Vanchurin, V., Wolf, Y.I., Katsnelson, M.I., Koonin, E.V.: Toward a theory of evolution as multilevel learning. Proc. Natl. Acad. Sci. **119**(6), e2120037119 (2022). https://doi.org/10.1073/pnas.2120037119
14. Von Neumann, N.A.: John von Neumann as seen by his brother. Meadowbrook, PA: Author (1987)
15. Zhang, Y., Wang, X., Friedman, E.G.: Memristor-based circuit design for multilayer neural networks. IEEE Trans. Circuits Syst. I Regul. Pap.uits Syst. I Regul. Pap. **65**(2), 677–686 (2017)

# Digital Eschatology: Death in the Digital Age

**Aníbal M. Astobiza**

**Abstract** This chapter presents a philosophical and ethical exploration of death in the digital age. It applies an epistemological and normative approach to investigate the changing social mores surrounding loss and bereavement in the presence of digital technology. Every aspect of human social life, including death, has been changed by the presence of digital technology. The relationship between the living and death is not the same as a result of digital technologies. Understanding how digital technologies transform our understanding of death can help people have a more practical knowledge of how to deal with death and grief in our modern technological societies.

## 1 Introduction

Since the dawn of time, humanity has grappled with the fear of death. Like all living organisms, humans are innately driven to survive and preserve their existence. However, unlike other species, humans possess a unique cognitive flexibility that allows them to adapt more effectively to diverse environments. This adaptability stems from our advanced cognitive abilities—such as abstraction, symbolic thinking, self-awareness, and the transmission of cultural knowledge—which have provided a significant evolutionary advantage [1]. Yet, these very cognitive faculties that allow us to comprehend and navigate the world also give rise to profound existential anxieties. Throughout history, philosophers, theologians, and intellectuals have recognized that our capacity for symbolic thought, combined with self-awareness, forces us to confront the inevitability of death. This realization is compounded by the unsettling awareness that our life can end at any moment, often for reasons beyond our control [2–8].

Death is one of the most profound and universal themes, cutting across a wide array of disciplines—biology, sociology, anthropology, theology, and more—all of

A. M. Astobiza (✉)
Departamento de Filosofía I, Universidad de Granada, Campus Universitario de Cartuja, Granada, Spain
e-mail: amastobiza@ugr.es

 
A. Sans Pinillos et al. (eds.), *SecondDeath*, Studies in Applied Philosophy, Epistemology and Rational Ethics 76, https://doi.org/10.1007/978-3-031-98808-0_7

which examine death, or rather the various manifestations of death, as a central object of study [9, 10]). In contemporary societies, the cosmetic industry exploits the pursuit of eternal youth as a means of evading the inevitable reality of death. However, the lack of regulation within this sector, coupled with the unscrupulous practices of some companies, has led to public health risks by exposing consumers to untested ingredients, sometimes resulting in death [11]. The irony is striking.

In life, we are confronted with the inevitability of death, the immediacy of conscious experience, and the recognition that religious denial of our physical origins and destinies is often driven by the profound anguish of loss. Research demonstrates that merely reminding individuals of their mortality heightens their inclination to believe in intelligent design over evolutionary explanations regarding human origins. The awareness of death distorts perception, leading people to embrace belief in supernatural beings as they grapple with the reality of their own mortality [12].

Digital eschatology, or death in the digital age, recalls how the advent of the telegraph and photography in the mid-nineteenth century ignited a fascination with communicating with the dead. Today, modern social media platforms and digital technologies similarly create avenues for new or revitalized spiritual rituals surrounding the deceased. This raises the question: to what extent does the concept of death remain consistent over time, or does the digital technology of the modern world fundamentally alter our understanding and experience of death? Nevertheless, digital technology increases the likelihood that our deceased loved ones and common ancestors extend their influence beyond the immediate family, integrating into various formal and informal communities. Their presence, through images, digital avatars, synthesized speech, and other forms, can impact a much broader audience.

In this chapter, I will explore the epistemological and ethical implications of death in the digital age. The next section introduces the concept of the badness of death and examines how digital technology might mitigate it. Section 3 will address the epistemological challenges associated with death in the digital era. Following that, in Sect. 4, I will discuss the ethical issues surrounding death in this context. Finally, I will conclude by arguing that digital technology profoundly transforms our understanding of death and the nature of grief.

## 2 Badness of Death and What to Do About It: Technology to the Rescue

The significance of death varies depending on when it occurs in our lives; dying young is not the same as dying old. Epicurus is well-known, among other things, for asserting that death cannot be bad for the person who dies [13]. According to Epicurus, something can only be harmful if the individual experiences it. If death entails no pain and no experience, then, for the person who dies, there is nothing inherently bad about it.

This epicurean view of death was criticised by Thomas Nagel. Nagel in his essay Mortal Questions [14], says the following:

> It should be recognized that if these are valid objections to counting death as an evil, they will apply to many other supposed evils as well. The first type of objection is expressed in general form by the common remark that what you don't know can't hurt you. It means that even if a man is betrayed by his friends, ridiculed behind his back, and despised by people who tread him politely to his face, none of it can be counted as a misfortune for him so long as he does not suffer as a result.

Using the example of betrayal or ridicule behind one's back, Nagel argues that something can be bad even if the person is unaware of it or does not directly experience its consequences. For Nagel, death operates in the same way; one does not need to experience anything to recognize that death is inherently bad. Before moving forward, it is essential to define the concept of death. Death can be understood through four dimensions [15, 16].

1-possibility, 2-process, 3-incident, and 4-loss.

Possibility refers to the awareness that death is inevitable; the process of dying can vary from sudden, as in cardiac arrest, to prolonged and painful, as in some cancer cases; the incident is the moment of death itself, marking the transition from existence to non-existence; and finally, loss signifies the absence of any future for the deceased. Deprivationism, a theory contrary to the Epicurean view, argues that death is bad because it deprives us of the opportunities and experiences we would have had if we had continued to live. This perspective is notably influenced by Thomas Nagel [17].

Imagine a healthcare team faced with the difficult task of allocating scarce resources among a child, a young adult, and an elderly person, all suffering from severe COVID-19 symptoms. While death is undoubtedly detrimental for each of these individuals, the nature of this harm differs significantly. From the perspective of deprivationism, death can be understood as an opportunity cost, similar to how economists view foregone alternatives. The loss experienced by a child who dies is vastly different from that of a young adult, and similarly, the loss for a young adult is different from that of an elderly person [18]. Even if we accept the deprivationist view as accurate, many aspects of the badness of death remain unexplored. However, if we were to imagine a scenario where technology could overcome death—essentially cheating it—we might also eliminate the inherent badness associated with it.

Today's technological and scientific advancements offer the potential to bring this ambition closer to reality. The transhumanist perspective on humanity's future is grounded in the belief that the human species, in its current form, represents not the culmination of our development, but rather an early stage. Philosopher Nick Bostrom [19] defines transhumanism as follows:

> Transhumanism is a way of thinking about the future that is based on the premise that the human species in its current form does not represent the end of our development but rather a comparatively early phase. We formally define it as follows. (1) The intellectual and cultural movement that affirms the possibility and desirability of fundamentally improving the human condition through applied reason, especially by developing and making widely available technologies to eliminate aging and to greatly enhance human intellectual, physical, and

psychological capacities. (2) The study of the ramifications, promises, and potential dangers oft echnologies that will enable us to overcome fundamental human limitations, and the related study of the ethical matters involved in developing and using such technologies.

However, this definition lacks clarity, as one could reasonably support (1) but not (2), or vice versa. Bostrom himself identifies as a transhumanist in both senses, and has been a leading figure in the movement. Consider the possibility of creating a digital replica of oneself, and for the sake of argument, assume a form of dataism—that all the data constituting a person's identity can be identified. This data could then be archived and uploaded to a suitable artificial medium, one immune to the typical processes of biological decay. Transhumanists refer to this process as "mind-uploading", a technical approach they believe could achieve the goal of eternal life. Transhumanists aim for cyber-enhancement through digital technology and bio-enhancement through biotechnology, the latter focusing on extending human longevity. Ultimately, these technological advancements are designed not only to prolong human life [20] but also, in some cases, to defy death itself [21]

## 2.1 *Digital Afterlife*

Eschatology is the branch of theology that focuses on the study of the end times and the ultimate fate of humanity [22]. Digital eschatology, therefore, explores the end times in relation to digital technology and its influence on our evolving understanding of death and the future in the digital age. This emerging field examines the potential consequences of our growing dependence on technology and how it may reshape our perceptions of death [23, 24]. Some experts suggest that the increasing interconnectedness brought about by technology could fundamentally transform experiences of end-of-life, death, and grief [25], while others argue that technology may ultimately enable humanity to overcome death itself [26].

As technology continues to evolve, it is transforming how we experience various aspects of life, including death, loss, and grief. The rise of social media and other digital communication platforms has made it more common for individuals to share their thoughts and emotions about the death of a loved one with a broader audience. This can foster a sense of support and community during a challenging time, but it also risks blurring the lines between private grief and public mourning. Additionally, the growing use of virtual reality (VR) and other immersive technologies enables people to create and interact with digital representations of deceased loved ones, further altering our relationship with death and mourning.

While this digital interaction can offer comfort and a sense of connection to the deceased, it also raises profound questions about the nature of death and our evolving relationship with the deceased in the digital age. The influence of digital technology on our experiences of death, loss, and grief is a complex and continually developing issue, one that will remain a significant subject of philosophical inquiry and study in the years to come. When a person dies, their social media accounts generally become inactive [23] and are no longer accessible to others. The policies for managing these accounts differ by platform, but they are typically either deleted or memorialized.

When an account is memorialized, it is preserved as a tribute to the deceased, but access is restricted to authorized administrators. Typically, this involves disabling certain features, such as the ability to log in or post new content, while keeping the account and its past posts visible to the public. On the other hand, if an account is deleted, it is permanently removed from the platform, along with all of the person's posts and content. This deletion is usually carried out at the request of the account owner, their designated representative, or when the platform cannot verify the account owner's identity. Consequently, a new and increasingly important professional role is emerging: that of the "digital death manager".

In either case, it is crucial for the loved ones of the deceased to consider their preferences regarding the deceased's social media accounts and to take steps to ensure those wishes are fulfilled. This may involve informing the platform of the person's death and designating a representative to manage the account on their behalf. Digital models of deceased individuals are virtual representations created using photographs, videos, and other digital media. These models can serve various purposes, such as preserving virtual memories or enabling loved ones to interact with a digital version of the deceased. The use of digital models of deceased individuals is a relatively new and evolving field, raising numerous ethical and practical questions that require careful consideration (see Sect. 4 for a more detailed discussion). For instance, while some may find solace in interacting with a digital representation of their loved one, others might find the experience unsettling or even disrespectful.

There are also significant concerns regarding privacy and the potential misuse of personal information. While the use of digital models of deceased individuals offers a compelling and potentially powerful way to memorialize and remember those who have passed, it remains a complex and sensitive issue that demands careful thought and consideration.

## 3 Epistemological Problems with Death in the Digital Age

In the digital age, the epistemological challenges surrounding death have grown increasingly complex and multifaceted (27; 28). A key issue is understanding how we can comprehend death in a world where our experiences and interactions are increasingly mediated by technology [28]. One way to address this problem is by examining how technology has transformed our relationship with death. Historically, death was often viewed as a natural and inevitable part of human existence, with well-established rituals and practices to cope with it. However, with the rise of modern technology, death has become more medicalized and sanitized, distanced from the everyday lives of most people.

These developments have introduced a range of epistemological challenges. For example, the possibility of sustaining life indefinitely through technology complicates our understanding of the nature and finality of death. Additionally, the rise of digital technology has given birth to virtual worlds, digital models of deceased individuals, and online communities where people can continue to "exist" and interact even after their physical deaths. This prompts us to question whether death retains the same meaning and significance in the digital age. Furthermore, the digital era has ushered in new forms of communication and expression, such as social media and online forums, which offer novel ways to remember and commemorate the deceased, as well as to share grief and mourning. However, the transient and intangible nature of these digital memories raises concerns about their reliability and authenticity. There is also a question of whether these digital forms of mourning can provide the same sense of closure and resolution as traditional practices.

A significant epistemological challenge in the digital age centers on how we can understand death in a world where the boundaries between life and death are increasingly blurred. Historically, death was a clear and definitive event marking the end of a person's life. However, advances in technology, such as artificial intelligence (AI) and cryonics, are complicating this once-clear distinction. For instance, some argue that AI could achieve consciousness, potentially qualifying it as a form of life, while others believe that cryonic preservation might enable the resurrection of individuals in the future. Transhumanists, in particular, offer compelling arguments and reasons for believing in these possibilities.

These developments challenge our understanding of what it means to be alive and what it means to be dead. If an AI or a cryonically preserved individual could be considered alive, then how do we define death? And if death is no longer a clear and distinct event, how can we truly comprehend it? Moreover, this ambiguity raises significant legal questions regarding rights and duties, inheritance, succession, and other legal matters, as the declaration of death carries profound legal implications.

Moreover, the digital age has introduced new forms of communication and the ability to preserve one's digital presence long after death, raising profound questions about the afterlife in a digital context. What becomes of a person's digital accounts and interactions after they die? Do these digital traces continue to hold meaning, or do they gradually fade into obscurity? In conclusion, the epistemological problem of death in the digital age is a complex and nuanced issue that challenges our traditional understanding of what it means to be alive and dead. As technology continues to evolve, it is crucial that we persist in exploring and critically examining these questions to navigate and comprehend this rapidly changing landscape.

Epistemology, as the branch of philosophy concerned with the nature of knowledge and how we acquire it, seeks to explore how we can understand death and the various philosophical questions it raises. Some key epistemological questions related to death include: What is the precise definition of death, and how can we be certain when it has occurred? Is death an inherent aspect of life, or is it something to be feared and avoided? How does our understanding of death influence our approach to life and our relationships with others?

Although it may not be immediately apparent, the concept of death is intimately connected to epistemology, as it raises fundamental questions about what we can know and how we can come to know it. For instance, can we truly know what happens to us after we die? Is it possible to have knowledge of an afterlife? These are epistemological questions, yet they remain ultimately unanswerable. Moreover, death is a core aspect of the human experience, and our understanding of it is deeply influenced by our beliefs, values, and cultural backgrounds. Consequently, different individuals and groups may hold diverse beliefs and interpretations regarding death and what occurs afterward [29].

Our understanding of death in the modern world is increasingly shaped by the transformative impact of digital technology. As Tony Walter says on page 58 of [29], at the end of our lives, we require not only communication but also critical information. Since each individual only dies once, the experience of dying creates a unique and unprecedented situation, both for the person and their close family members. There is a pressing need to rapidly acquire knowledge. However, much of the essential information about death resides within the minds, texts, and practices of medical professionals, leaving the modern family with urgent informational needs at a time when they are most vulnerable. Information and communication technologies (ICT) are rapidly surpassing books as the preferred source of information for contemporary society, a trend that is expected to continue. Moreover, social media and the Internet are becoming increasingly important sources of information for the elderly, the sick, and those nearing the end of life.

However, there are drawbacks to the use of ICT. Some older individuals, lacking IT skills, struggle to navigate online forms and helplines, which can prevent them from accessing essential services—even those specifically designed for their needs. Additionally, while the Internet provides access to reliable information, it also harbors misleading claims and false remedies that can give gravely ill patients false hope. The challenge lies in distinguishing between accurate information and misinformation.

Another central epistemological challenge of death in the modern digital world is the way technology enables the continued presence of the deceased in various forms. As previously mentioned, digital models of deceased individuals, along with VR and other immersive technologies, allow for the creation of realistic avatars using all available information and data about the deceased. These advancements blur the boundaries between life and death, complicating our understanding of what it means for someone to be truly gone.

These digital models can provide comfort to the relatives and loved ones of the deceased by offering a cyber-presence that helps them cope with the absence. However, this development also brings about the unsettling consequence of spawning an entire parallel industry dedicated to managing the digital assets and inheritance of the deceased. The COVID-19 pandemic, which claimed the lives of over 4 million people worldwide, highlighted the growing demand for services such as online memorials and digital legacy management. In response, a rising number of startups have begun to focus their businesses on death-related services, addressing the needs of individuals in managing their digital footprints posthumously.

A variety of services have emerged to assist families in managing the digital legacy of their deceased loved ones. While death-related technology isn't entirely new—platforms like "Forever.com" have been offering online storage and digital records of images since 2012—the market is increasingly diversifying to cover a broader array of niches related to death and mourning. For example, the startup Eterneva transforms human ashes into diamonds, allowing loved ones to wear the jewelry as a keepsake. Meanwhile, Everdays provides online funeral package signups and offers a service that sends death notifications via email and text message to friends and family.

People are increasingly turning to death technologies. For example, speech synthesizing technology can now read text in the voice of a deceased loved one, such as your grandmother—a capability that was once unimaginable. The company Recompose, which uses a microbial chemical process to transform human remains into soil, saw significant growth during the COVID-19 pandemic. In response to the limitations on in-person gatherings, Recompose introduced virtual "laying-in rituals", allowing mourners to participate remotely in the burial process.

Technology not only makes certain aspects of the funeral industry more accessible and affordable but also fundamentally alters our understanding of death and our relationship with deceased loved ones. The classical theoretical framework for the epistemology of death, as proposed by Dr. Elizabeth Kübler-Ross, which outlines the five stages of the dying process—denial, anger, bargaining, depression, and acceptance (DABDA)—is increasingly challenged and may no longer hold in a context where death is mediated by technology.

Kübler-Ross's work has profoundly shaped how patients and their families understand death and the dying process, contributing to more compassionate care in these situations. However, what happens when technology enables a deceased loved one to remain perpetually present? For all of us, death is the ultimate reality. Although modern science has managed to delay death to some extent, technology now offers the possibility that death—or rather, its effects and our understanding of it, its epistemology—might no longer impact us in the same way.

## 4 Ethical Problems with Death in the Digital Age

In this section, I aim to highlight some concerns regarding the ethical challenges that will emerge with the advent of digital eschatology. While the epistemological issues surrounding death in the digital age are clear—our understanding of death is radically transformed—the ethical implications remain uncertain. It is not our intention to comprehensively or exhaustively address the myriad ethical issues that may arise as digital technology is increasingly applied to death. Not only is such an endeavor currently impossible, but it would also be presumptuous on our part to attempt it.

Technology not only eliminates the barriers of distance but also blurs the boundaries of death, as evidenced by how easily the living can interact with the digital remnants of the deceased. In the digital age, death no longer equates to complete loss or absence, as our loved ones can persist in various forms—such as voice assistants that replicate their voice using advanced language models, or even physical representations like robots. This might be the most positive aspect of technology's application to death: it enables us to remember and maintain a connection with our deceased loved ones.

The familiarity and existential acceptance of death, as encountered in daily life and relationships, does not necessarily extend to the experience of mediated death [30]. By "mediated death", I refer to the involvement of technology that radically alters our conception of death. Technology can make absence more bearable by enabling the presence of a deceased loved one in various forms—such as avatars, digital profiles that replicate their values and personality traits, or even telepresence robots equipped with large language models (LLMs) that allow interaction as if the robot were the deceased loved one.

Indeed, social media is replete with accounts belonging to deceased individuals [31]. Unaware of someone's passing, people may continue to interact with these "ghost" accounts, which raises significant concerns. Even more troubling is the potential for a deceased person's social media account to be hacked or misused, leading to deception and conflicting emotions. The primary ethical issue with digital technology applied to death is that it can foster a unique form of deception. When we encounter mediated death, we are left questioning whether we are dealing with a "fictional death" or a "real death". This blurring of reality presents a profound ethical challenge, particularly given its emotional impact on those involved.

As digital eschatology becomes more commonplace, there is a risk that someone might be misled into believing a loved one has died when they have not. On a more personal and existential level, the application of digital technology to death—such as through "thanotechnology"—can also create illusory and false expectations about our deceased loved ones. Thanotechnology refers to the use of digital technology in matters of death, such as creating virtual portraits of those who have passed. This could lead to distorted perceptions and unrealistic expectations about the continued presence or influence of the deceased.

From an ethical standpoint, this type of technology may foster unfounded fears about the nature of death and the afterlife, and could complicate the grieving process for those who have lost loved ones. Additionally, there are significant concerns about the moral implications of creating digital representations of the deceased, particularly regarding privacy and control over the use of an individual's likeness after death. Another ethical concern is that thanotechnology, while designed to keep memories alive and foster a sense of connection with departed loved ones, might also create a false sense of closure. This could potentially hinder individuals from fully experiencing the grieving process and moving forward. Additionally, there are concerns that this technology could be exploited to manipulate or take advantage of those in mourning, making them vulnerable to emotional or financial exploitation.

As technology and AI continue to advance, digital representations of the deceased may become increasingly realistic and lifelike, raising significant ethical concerns. One major issue is the potential misuse of these representations, such as the creation of deceptive content. For instance, imagine a large language model similar to ChatGPT, combined with other foundational models, being used to fabricate hoaxes by mimicking the voice, thoughts, or images of deceased individuals to convince others they are still alive. This possibility underscores the need for careful consideration of the ethical implications of such technologies. Regarding the ethical challenges of death in the digital age, one might question whether the ethical dilemmas surrounding end-of-life care could be diminished if technology enables the *permanent* presence of a loved one.

The philosophical movement known as transhumanism aims to transcend the limitations of the human body [32, 33]. It seeks to enhance our physical, cognitive, and even moral capacities through biological enhancement and cyber-enhancement, including the application of AI and brain-machine interfaces. Transhumanism leverages theoretical projections based on the current advancements in neuroscience and various technologies to propose the possibility of mind-uploading into computers or other technological substrates, moving beyond the decaying, carbon-based biological form.

By simulating our brains to evade death, transhumanism aspires to achieve immortality. Mind-uploading—also known as mind transfer or whole brain emulation—differs from other anti-death technologies like cryopreservation or life extension by raising unique philosophical, ethical, practical, and sociological questions. While transhumanism envisions a digital afterlife, the same scientific and technological advances that are crucial to its goals are also fueling the development of lethal autonomous weapons, often referred to as killing machines, as well as other technologies of death, including nuclear weapons and genetically engineered pathogens. Paradoxically, rather than eliminating the concept of death or extinction, the future envisioned by transhumanism seems poised to elevate it to previously unimaginable levels.

This leads us to the significance of what is now referred to as robot theology. The inherent drive to transcend our physical limitations and human fallibility propels us toward a digital eschatology that necessitates a robotic theology [34]. If we take seriously the transhumanist notions of mental transfer or mind-uploading into digital devices, real human beings may soon become an anachronism [35]. As technology advances to eliminate death, enable mental transference, and create simulations or representations of our deceased loved ones—through robots and AI—digital eschatology threatens to undermine the Christian belief in *imago Dei* (the doctrine that humans are created in the "image of God"). In its place, a new theological and metaphysical framework for understanding death in the digital age may emerge.

In many respects, this transformation is already underway. For instance, in Japan, robot funerals have become customary, with Buddhist funeral services being held for Aibo robots. This shift is poised to fundamentally alter our concept of death in the digital age, compelling us to reimagine death not only for humans but also for machines and robots [10].

## 5 Conclusions

To sum up, this chapter offers a philosophical and ethical exploration of death in the digital age, focusing on how digital technology has reshaped social norms surrounding loss and bereavement and transformed our understanding of death itself. The central aim of the chapter is to present the challenges that technology poses to our understanding of and coping with death, which ultimately touches on our self-understanding as human beings. The chapter highlights the positive aspects of digital technology in relation to death and grief, such as the creation of new ways to remember loved ones, share memories, and provide support during the grieving process. These innovations have opened up opportunities for connection and remembrance that were previously unimaginable, allowing for a more personalized and potentially comforting approach to dealing with loss.

But the chapter also explores the moral issues that arise in the digital age, particularly regarding what happens to a person's digital assets after their death. It examines the complex issues surrounding digital immortality and the potential consequences of preserving a person's digital identity after they have passed away. The chapter raises important questions about the implications of these developments for both individuals and society as a whole, such as the risk of creating deceptive or distorted memories, the challenge of managing digital legacies, and the broader impact on our cultural and existential understanding of death.

In addition, the chapter suggests that as technology continues to evolve, these ethical and philosophical issues will become increasingly pressing. Future research will be essential in developing strategies to address these challenges, ensuring that as we integrate digital technology into our experiences of death and mourning, we do so in a way that respects both the living and the deceased, while fostering a deeper understanding of what it means to be human in the digital age.

**Acknowledgements** This work was supported in part by INEDyTO II Bioética y Final de la Vida PID2020-118729RB-I00 and "Proyecto BAKARZAIN (BBK Kuna)". The author thanks Txetxu Ausín for his comments and suggestions for improving the text. Any errors are the sole responsibility of the author.

**Competing Interests** The author declares no conflicts of interest relevant to the content of this chapter.

## References

1. Henrich, J.: The secret of our success: how culture is driving human evolution, domesticating our species, and making us smarter. Princeton University, NJ, Princeton (2015)
2. Schopenhauer, A.: The world as will and representation, vol. 1, 58441st edn. Dover Publications (1966). ISBN 978-0-486-21761-1
3. Kierkegaard, S.: The concept of anxiety: a simple psychologically orienting deliberation on the dogmatic issue of hereditary sin. Princeton University Press, Princeton (1844/1981)

4. Unamuno, M.: Del Sentimiento Trágico de la Vida. Alianza, Madrid (1912/1997). ISBN 978-8420633244
5. Heidegger, M.: Being and Time, Reprint edn. Harper Perennial Modern Classics (2008). ISBN 978-0-06-157559-4
6. Sartre, J.-P.: Being and Nothingness, Reprint edn. Washington Square Press (1993). ISBN 978-0-671-86780-5
7. Brown, N.O., Lasch, C.: Life against death: the psychoanalytical meaning of history, 2nd edn. Wesleyan University Press (1985). ISBN 978-0-8195-6144-2
8. Church, J.: Kant on death and the purpose of human life. In: Political theory on death and dying. Routledge (2021). ISBN 978-1-00-300538-4
9. Parkes, C.M.: Bereavement: studies of grief in adult life, 4th edn. Routledge, London (2010). ISBN 978-0415579091
10. Solé, R.: *Todas las Muertes. El Final de la Vida: De los Océanos a los Robots*. Crítica, Barcelona (2023)
11. Rodriguez, M.: Dying to be beautiful: an assessment of how a self-regulating cosmetic industry and biotechnology are impacting public health. (n.d.)
12. Tracy, J.L., Hart, J., Martens, J.P.: Death and science: The existential underpinnings of belief in intelligent design and discomfort with evolution. PLoS ONE **6**, e17349 (2011). https://doi.org/10.1371/journal.pone.0017349
13. Lesses, G.: Happiness, completeness, and indifference to death in epicurean ethical theory. Apeiron **35**, 57–68 (2002). https://doi.org/10.1515/APEIRON.2002.35.4.57
14. Nagel, T.: Mortal questions. Cambridge University Press (2012). ISBN 978-1-107-60471-1
15. Kagan, Shelly. Death. Yale University Press (2012)
16. Solberg, C.T., Gamlund, E.: The badness of death and priorities in health. BMC Med. Ethics **17**, 21 (2016). https://doi.org/10.1186/s12910-016-0104-6
17. Nagel, T.: Death. Noûs 4, 73–80 (1970). https://doi.org/10.2307/2214297
18. Hortal-Carmona, J., et al.: La eficiencia no basta. Análisis ético y recomendaciones para la distribución de recursos escasos en situación de pandemia. *Gac Sanit* 35, 525–533 (2021)
19. Bostrom, N.: Transhumanist values. J. Philos. Res. **30**, 3–14 (2005). https://doi.org/10.5840/jpr_2005_26
20. Grey, A. de, Rae, M.: Ending aging: the rejuvenation breakthroughs that could reverse human aging in our lifetime, 1st edn. St. Martin's Griffin (2008). ISBN 978-0-312-36707-7
21. Agar, N.: Ray Kurzweil and uploading: Just say no! J. Evolut. Technol. **22**(1), 23–36 (2011)
22. Bauckham, R.: Eschatology. In: Tanner, K., Webster, J., Torrance, I. (eds.) The oxford handbook of systematic theology. Oxford University Press (2007). ISBN 978-0-19-924576-5
23. Öhman, C.J., Watson, D.: Are the dead taking over facebook? A big data approach to the future of death online. Big Data & Society 6 (2019). https://doi.org/10.1177/2053951719842540
24. Sisto, D.: Posteridades Digitales. Inmortalidad, Memoria y Luto en la Era de Internet. Katz, Buenos Aires (2022). ISBN 978-8415917533
25. Beaunoyer, E., Guitton, M.J.: Cyberthanathology: death and beyond in the digital age. Comput. Hum. Behav. **122**, 106849 (2021). https://doi.org/10.1016/j.chb.2021.106849
26. Minerva, F., Sandberg, A.: Euthanasia and cryothanasia. Bioethics **31**, 526–533 (2017). https://doi.org/10.1111/bioe.12368
27. Walter, T., Littlewood, J., Pickering, M.: Death in the news: the public invigilation of private emotion. Sociology **29**, 579–596 (1995). https://doi.org/10.1177/003803859502900400
28. Walter, T., Hourizi, R., Moncur, W., Pitsillides, S.: Does the internet change how we die and mourn? Over. Analysis. Omega (Westport) **64**, 275–302 (2012). https://doi.org/10.2190/OM.64.4.a
29. Walter, T.: Death in the Modern World, 1st edn. SAGE Publications Ltd. (2020). ISBN 978-1-5264-0294-3
30. Gibson, M.: Death and mourning in technologically mediated culture. Health Soc. Rev. **16**(5), 415–424 (2007)
31. Biçer, S., Yıldırım, S.: Digital death and thanatechnology: New ways of thinking about data (Im)mortality and digital transformation. OMICS: J. Integrat. Bio. (2022). https://doi.org/10.1089/omi.2021.0096

32. Diéguez, A.: Transhumanismo: La Búsqueda Tecnológica del Mejoramiento Humano. Herder, Barcelona (2017). ISBN 978-8425439629
33. Diéguez, A.: Cuerpos Inadecuados: El Desafío Transhumanista a la Filosofía. Herder, Barcelona (2021). ISBN 978-8425446474
34. Smith, J.: Robot theology: old questions through new media. Oregon Resources Publications (2022). ISBN 978-1-6667-1071-7
35. Hanson, R.: The age of Em: work, love, and life when robots rule the earth. Oxford University Press (2018). ISBN 978-0198817826

# Technological and Ethical Implications

# Reflections on the use of Artificial Intelligence for Weapons Applications

**Maria Vanina Martinez and Ricardo O. Rodriguez**

**Abstract** The development of new intelligent technologies for military use is already yielding a new stage in the arms race that could have serious implications for peace and security across the globe. Governments, international organizations, civil society, and the scientific community face the challenge of promoting legal, ethical, and moral parameters for the development of Artificial Intelligence for military purposes. This article explores a series of concepts that are relevant to the design and development of autonomous weapons and proposes a series of general (political and ethical) and technical considerations that may lead to effective lines of action in the matter.

**Keywords** Autonomous weapons · Responsible AI · Human control

## 1 Introduction

Since the beginning of its history, the world has faced three main problems: famine, plague, and war. Over time, humanity has designed and developed different types of tools and processes to overcome them.[1] In this sense, Artificial Intelligence (AI) is one of the latest tools invented that, like fire, the wheel, steam engines, or atomic energy, will have a disruptive impact on the future of society. For example, today AI plays an important role in the so-called precision agriculture where crops are controlled

[1] A preliminary version of this document was published in https://sehlac.org/armas-letales-autonomas.

This work was funded by the project PID2022-139835NB-C21.

M. V. Martinez (✉)
Artificial Intelligence Research Institute (IIIA-CSIC), Barcelona, Spain
e-mail: vmartinez@iiia.csic.es

R. O. Rodriguez
Department of Computer Science, University of Buenos Aires, and Computer Science Institute UBA-CONICET, Buenos Aires, Argentina
e-mail: ricardo@dc.uba.ar

 
A. Sans Pinillos et al. (eds.), *SecondDeath*, Studies in Applied Philosophy, Epistemology and Rational Ethics 76, https://doi.org/10.1007/978-3-031-98808-0_8

remotely. It has also played a pivotal role in controlling the Covid-19 pandemic, both through infection-tracking applications and in the development of vaccines. Finally, AI is the driver of a new arms race[2] that, we argue, must be controlled. In this document, we try to capture what we consider to be central technical issues around the debate on the control of lethal autonomous weapons systems (LAWS).[3] We are aware that, at the current state of the frontier of AI knowledge, our opinions may generate more doubts than certainties. In that sense, we will try to honor Jorge Luis Borges' maxim: *"doubt is another name for intelligence."* The ultimate goal of this global debate is the creation of a legal framework that, on the one hand, prohibits the use of fully autonomous weapons systems, and on the other, regulates and controls the development of autonomous weapons systems with meaningful human control. The aim is to avoid a dehumanized future, where armed systems can decide to kill and apply force without the military authority understanding or being fully responsible for the consequences of such actions.

## 2 A Bit of Context

In this section, we will briefly describe the context in which discussions about LAWS are being carried out, both from a technical and geopolitical point of view.

### *2.1 Technical Context*

For this work, we can divide the state of the art in AI into two groups: knowledge-based algorithms (KBA) and machine learning (MLA). The latter constitutes some of the most successful AI techniques today, and we will focus our analysis on them. In a brief overview, we can say that algorithms are implementations of concepts or models. For example, we have the concept of "arithmetic addition" and we have different implementations or ways of calculating it. In contrast, most MLA can be broadly stated as algorithms that seek to approximate a concept/model through a limited (but possibly very large) amount of data/examples of it (for this reason we also call them data-driven models).

In this sense, we can think of MLAs as meta-algorithms, that is, they generate an implementation of the implicit concept/model given by data/examples. In this way, with an MLA a conceptualization of arithmetic addition would be obtained from many examples of additions (that is, two addends and their result). The underlying assumption in this procedure is that, if we give an MLA a sufficient number of

[2] https://unidir.org/publication/artificial-intelligence-beyond-weapons-application-and-impact-of-ai-in-the-military-domain/.

[3] https://disarmament.unoda.org/the-convention-on-certain-conventional-weapons/background-on-laws-in-the-ccw/.

examples, it will obtain a procedure equivalent to one of the well-known addition methods that we learned in primary school. In general, people do not learn arithmetic operations from examples. However, we learn to speak with examples. An MLA finds patterns and regularities in the set of examples and generates a model from them. At this time, it is important to note that a large part of the scientific discoveries produced in the 17th century were through observations. Currently, virtual models generated by MLA, depending on the application domain, usually have better performance than physical ones. In fact, there are several modeling tasks where MLAs produce better results than those produced by a human expert. An example of this is the conceptualization of tumors in tomography [1].

Essentially, the main challenges of this technology involve overcoming three problems. The first is about the quality and quantity of data needed (labeling and bias). The second is about the process of obtaining the model (overfitting, training vs testing, optimization, transference, etc.). The third deals with the model obtained (interpretability, predictability, explainability, transparency, accountability, etc.) Precision and efficiency are not the central weakness of MLA because both are increasing rapidly. On the contrary, we prefer to point out predictability and comprehensibility as the central challenges of MLAs. The first is limited by a theoretical aspect [2], and we discuss it in Section 3.2; for the second, we dedicate Section 3.3.

## 2.2 *Political and Diplomatic Context*

In the political-diplomatic sphere, many antecedents must be taken into account when contextualizing the current debate on LAWS. First, there exist four major arms control treaties (bilateral and multilateral): the Anti-Ballistic Missile Treaty (ABM), the Intermediate-Range Nuclear Forces (INF) Treaty, the Open Skies Treaty, and the Strategic Arms Reduction Treaties (STARTs). All of these treaties tend to reduce East-West frictions. Unfortunately, all of them have been concluded or will not be ratified, leading to a very uncertain future of armed conflict. Secondly, we can cite more auspicious antecedents, such as the Treaty on the Prohibition of Nuclear Weapons (which entered into force on January 2021) and the Conventions on Biological and Chemical Weapons, Cluster Munitions, Anti-Personnel Mines, and Blinding Lasers. All of these antecedents, with their advances and setbacks, mark the international effort to regulate the development, production, marketing, and use of different types of weapons. On this basis, discussions on LAWS are being developed.

In general, the States Members of the Convention on Certain Conventional Weapons (CCW) agree that "identifying and seeking a common understanding of the concepts and characteristics of LAWS could assist in further consideration of issues related to emerging technologies in the area of LAWS" (2019 Report, 19 b.). To this end, States Members continue to seek further clarification and a shared understanding of the specific technical characteristics that constitute or define "autonomy" or autonomous "decision-making" capabilities in weapons systems. So far, states have noted that certain capabilities, including "autonomous functions in identifying, selecting or engaging a target" are "among the essential characteristics of weapon systems based on emerging technologies in the area of lethal autonomous weapons systems" (2018 Report, 19a). However, they have not reached a shared understanding of a technical definition of autonomy, and have highlighted "self-adaptation; predictability; explainability; reliability; ability to be subject to intervention; ability to redefine or modify objectives or goals or adapt to the environment; and ability to self-initiate" as technical characteristics of autonomous weapons that "may benefit from further clarification or revision" (Report 2019, 20a, b).

Currently, in the context of the CCW there is a Group of Governmental Experts (GGE) which was mandated to produce between 2020 and 2021 a set of "consensus recommendations in relation to the different aspects of the regulatory and operational framework of emerging technologies in the area of LAWS".[4] In this space for reflection and discussion, different groups with conflicting interests are represented: International Organizations, NGOs, Governments, Armed Forces, the arms industry, and scientists. Precisely our intention here is to contribute solid elements from the academic sector to this debate.

In this sense, we believe that technically, this debate should not only be circumscribed to the International Humanitarian Law (IHL) since this restricts its analysis in the context of the battlefield in regular conflicts but does not consider other extremely important aspects such as its design, manufacturing, and marketing. On the other hand, we consider that in relation to IHL, the greatest challenge lies in guaranteeing the principle of responsibility. Other issues such as the principle of proportionality, opportunity, precaution, or distinction, which are critical to evaluate a military action, could be potentially well parameterizable on LAWS from a technical point of view. Although we recognize that these principles require complex evaluations based on the circumstances prevailing at the time of the decision to attack and also during an attack, we believe that they can be captured by logical rules and appropriate performance metrics.

Finally, we believe that we must be very careful to avoid irrelevant technicalities. For example, as has already been demonstrated in the case of antipersonnel mines, it is essential to define the object of prohibition/regulation precisely. However, given the complexity of the type of systems we are analyzing, it seems inappropriate to put the focus of such a definition on whether the system uses sensors to determine where and when force will be used (see [3]). We believe that this is not a good way to characterize the technology, and it does not align with the complexity of AI-based weapons systems, both existing and to be developed in the medium term. The first problem with this characterization is that in general, the sensors are in the physical place where the action is deployed, but the decision can be made at a great distance from the execution using data that does not come directly from the sensors but from a pre-processing of them and sometimes even historical data that may not even be related to the system sensor inputs. Therefore, in the case of LAWS, we believe that it is better to define them in a multi-intentional way and perhaps

[4] https://dig.watch/process/gge-laws.

rather in conceptual terms such as their learning, perception, decision-making, and acting capabilities, which can then be carried out physically or digitally in different ways [4]; for example, through learning based on historical data, sensors that allow capturing the current situation, an interpretation and decision-making module based on rules and physical actuators.

## 3 Discussion on Relevant Concepts for LAWS

Our contributions to the LAWS debate in this document will focus on the three axes that we consider most relevant and urgent in the discussion.

- Meaningful human control
- Predictability and reliability
- Comprehensibility, explainability, and interpretability

Before starting to describe these three issues, we are going to highlight some objections that automatic decision-making systems raise. A salient point about the nature of these concerns is that they are not necessarily resolved with an improvement in the qualitative/quantitative precision of the MLAs. A first important point that responds to the very essence of these systems is that they identify correlation and not causal relationships, that is, they may discover "symptoms" but not "reasons". This leads to a loss of trust and the possibility of unpredictable behavior, reducing also the possibilities of legally and ethically challenging their results. The work in [5] emphasizes the challenge that LAWS impose on human values, and identifies, in addition to those mentioned before, dehumanization, the lack of understanding of how systems work which can therefore result in it not being possible to establish meaningful human control over them, the risks to peace and security given that autonomy and remote activation "make it easier" to take the risk; on the one hand, because the political thresholds against the use of force are significantly reduced and, on the other, because it encourages automated responses that could rapidly escalate into an arms race.

### *3.1 Meaningful Human Control*

Since the beginning of the negotiations mentioned above in 2014, States have proposed that the focus of the debate should be on the issue of human control, and how it should be implemented over weapons, their critical functions, attacks, targeting processes, and (final) decisions on the use of force, etc. While most states agree that human control should be more meaningful than simply aborting an attack at the very last moment, there is no such unanimity in determining how the human role should be defined and applied in the use of (lethal) force. In this section, we will try to organize the main points under discussion and outline some methodological proposals. For this, it is necessary to introduce some preliminary concepts.

In general, three levels of controls can be recognized (see [6]):

- *Strategic command*, which translates political purpose into military objectives.
- *Operational command*, which converts general strategic-level objectives into concrete tasks for tactical forces.
- *Tactical command*, which directs the specific use of military forces in operations to execute tasks decreed by operational command. Tactical command encompasses the deployment of units, platforms, non-unit personnel, and weapons systems that may be in direct contact with parties to a conflict. In [7], this level is called "mission command".

Though we focus on the use of AI at the tactical level, MLAs can be used, in principle, throughout the entire decision-making process of military actions. Today, senior and middle managers use AI tools for data analysis and decision-making; this leads to a more profound discussion that is starting to take place in several fora. We point out the work from The Global Commission on Responsible Artificial Intelligence in the Military Domain (GC REAIM)[5] for further reading; this is an initiative of the Government of the Netherlands that was launched during the 2023 REAIM Summit on Responsible Artificial Intelligence in the Military Domain in The Hague and seeks to help promote mutual awareness and understanding among the many communities working on issues related to the global governance of AI in the military domain in order to support fundamental norm development and policy coherence in the field.

In the following, we identify the four types of control that are often questioned, recalling them from [6]:

- *Total control* ("full human control"), where the system does not make any decisions by itself but is remotely controlled.
- *In the loop control* ("human in the loop"), the system implements tasks autonomously but requires human intervention to validate and implement the actions.
- *On the loop control* ("human on the loop"), in which the system implements the task autonomously under the supervision of human operators who can, if necessary, correct or abort a specific action.
- *Total autonomy* ("human off the loop"), such that the system implements the tasks without any supervision or human intervention.

The loop usually refers to the sequence of tasks into which the execution of a mission is usually divided:

- *Search/Find*: Find the target, collect information, and conduct intelligence on the battlefield.

[5] https://hcss.nl/gcreaim/.

- *Fix (including positive identification of the target)*: Accurately detect and confirm location and estimate available time.
- *Track*: Maintain positive identification and update information about the target and its environment.
- *Target*: Assess the rules of engagement (ROE). collateral damage, and risks to one's own forces.
- *Engage*: Execution of the attack based on sets of authorized, restricted, and prohibited objectives, with the ability to suspend and cancel.
- *Assess*: Assessment of the effectiveness of the attack and determination if a new attack is necessary and under what conditions.

With all these elements introduced, we are in a position to discuss the main questions about *meaningful human control*. The central issue is to guarantee that humans are ultimately responsible for the consequences of a military operation even when the temporal and spatial distance between the direction of the action is so far from its execution on the battlefield. This distancing, and the unpredictability of the consequences it brings, in turn, raises concerns about the application of international humanitarian law, ethical acceptability, and operational effectiveness. That is why we seek to define, what type and degree of human control is required in practice to ensure that humans continue to play their necessary role in decisions to use force in specific attacks in armed conflicts, regardless of the sophistication of technology, while meeting legal, ethical and operational requirements. All parties involved in the discussion recognize that humans must exercise some form of control over weapons and the use of force in specific attacks in armed conflicts. Where they diverge is on questions of how and when humans should exercise that control in operational contexts. As highlighted in the GGE, there are likely to be some control measures that can be applied in all circumstances and others whose need depends on the context.

In [8] three types of control measures are identified:

- On the parameters of use of the weapon system, including measures that restrict the type of target and task for which the LAWS is used; place temporal and spatial limits on its operation; restrict the effects of LAWS; and enable fail-safe deactivation mechanisms.
- On the environment that controls or structures the use of the LAWS (for example, using them only in environments where there are no civilians and civilian objects, or excluding their presence during the duration of the operation).
- On human-computer interaction, with measures that allow the user to monitor the LAWS and intervene in its operation when necessary.

These control measures can help reduce, or at least offset, the unpredictability inherent in the use of LAWS and mitigate risks, particularly to civilians. From a legal perspective, a LAWS operator must exercise sufficient control to have reasonable certainty about the effects of its use in an attack and to be able to limit them as required by IHL. Of course, ethical considerations may require additional restrictions, especially in the exercise of force against civilians. In any case, the implementation of these measures must respect a balance between ensuring legal compliance, ethical acceptability, and operational usefulness.

In Sect. 2 of [8], three difficulties that LAWS present in complying with IHL are presented, these are measurability, context, and unpredictability. Nowadays, value judgments to verify compliance, which also reflect ethical considerations, are part of the regular training of the armed forces. We understand that most of the questions about qualitative or evaluative judgments that must be made in real-time, in a complex and dynamic environment, to guarantee respect for the principles of distinction, proportionality, and precautions of IHL, are technologically surmountable. Furthermore, we believe that defining standards and "parameterizations" on the subjective assessment of the protection of these principles can be useful in reassessing them. In this sense, we consider that the professional practice of a soldier could, at least partially, be emulated by an AI system in the not-too-distant future. Current AI technology does not allow this yet, but, from our understanding, it is only a matter of time. That is why it is so crucial to define design criteria, validation metrics, predictability and understandability, and, finally, continuous audit mechanisms, so that when the technology exists, it exists in such a way that it can be reliable and predictable, but above all measurable and auditable. Of course, this is a real challenge and will take a very significant effort, but technically it is possible to overcome the three difficulties mentioned above. That does not mean that for political reasons it is decided to preserve a relevant portion of control for humans. We also judge that it is political to establish different technical performance standards for humans and AI systems.

To establish our position on this matter let us reflect on the subsection of [8] "Application of IHL in practice: what requirements for human control?", two positions arising from the June 2019 expert workshop are presented there. On one hand, the technicist, which considers that the only limit to autonomy is technology itself; that is, the more "sophisticated" the LAWS, the more tasks can be assigned to it and the less human control will be necessary during an attack. On the other hand, the humanist, states that regardless of the technical characteristics of LAWS, IHL rules on the conduct of hostilities require contextual and value-based judgments on the part of the people who plan, decide, and execute attacks. We consider this position to be Manichean. The fact that LAWS can acquire increasingly "sophisticated" characteristics is not a justification for granting them the power of complete autonomy. However, the argument that IHL intrinsically requires that people have absolute control is also not entirely correct. Once again, in our opinion, the decision is political and legal or technical issues should not be raised as they lead the discussion away from the central issues.

We agree with the authors of [8] that both approaches can be made compatible in the three practical control measures mentioned above in a simultaneous and incremental approach to all of them. By this, we mean that they must be taken into account jointly and coordinated. For example, if only the attack on military targets is parameterized, the allowed context of the action, such as duration and location, and the levels of human-computer communication/intervention, should also be established, and this should also depend on the stage of military action that is in process. It is not necessarily expected to require the same level of control during search, location, and

**Table 1** What to measure and when? Note that S/I/D stands for supervision, intervention, and deactivation

| | Parameters | Environment | H-C interaction |
|---|---|---|---|
| Use | | Thorough understanding of the context | Warrant supervision<br>Ability of intervention and deactivation |
| Development | Fix parameters of time and place<br>Set conditions for the use of force | Set limits on targets and people<br>Set temporal limits and exclusion zones | Provide mechanisms for explanations<br>Include S/I/D mechanisms |
| Design | Limit the type and profiles of targets<br>Spatio-temporal controls | Include warning mechanisms<br>Safety and inviolable mechanisms | Design mechanisms for S/I/D<br>User oriented<br>Provide adequate training for users |

monitoring as during the engagement itself. It would also be reasonable not to allow a LAWS to define the overall objectives of a mission in the first stage.

In this work we are not going to focus on ethical issues related to LAWS; not because they do not seem important to us, but because the ethical and moral questions in this domain have not reached sufficient maturity to be reviewed in depth. What we can say is that most of the works that have ventured into this topic assume a deontological or normative approach. In this sense, we can affirm that the approaches to the ethical issues are capable of being incorporated into a LAWS. Again, the limiting factor is not technological but of another nature, first of all political, in this case philosophical.

To close this section we would like to reproduce Table 1 from [8] that summarizes the control aspects that could be integrated into the life cycle of a LAWS as we previously discussed.

## 3.2 *Predictability and Reliability*

As the areas and modes of application of AI techniques grow, discussions about the impact that the use of AI systems can have on our society also increase. As we mentioned previously, many of these criticisms are based on the very nature of the AI systems used today, that is, those that use ML that would enable, in principle, the possibility of unpredictable behaviors. In intuitive terms, predictability is the extent to which the results or effects of a system can be anticipated. In [8], predictability is identified as an indispensable factor to comply with the rules of IHL, in particular with the prohibition of indiscriminate attacks, the principle of proportionality, and the requirement to take precautions in attacks, since operators must be able to limit the effects of the weapons they use. In [9], three aspects of the principle of predictability are identified that we will discuss below to understand the concept in depth and its relationship with other criteria and concepts that we will highlight in this article.

From a technical point of view, predictability is seen as the ability of a system to execute a task with the same performance that it has already exhibited in previous tests or previous applications. For ML systems this implies that the system has to be able to function in the same way as it functioned in its training stage. In computer systems, this is directly related to their *correctness*. The correctness property refers to the system delivering the result that is expected given the input configuration for a specific task. For example, if the system's task is to differentiate people from other objects that are presented to it through a video stream or a succession of photos, a possible correctness measure (which is established in the validation stage) is to calculate the rate of false positives and/or false negatives, that is, of all the people shown to it, how many it correctly identified as people, and how many of the objects presented to it, it (incorrectly) identified as people, respectively. This is just one of the many quality and performance metrics commonly used for machine learning algorithms. Determining which is the most appropriate metric depends on the specific problem to be solved; For example, a high false positive rate may be tolerable for an application that tries to identify people in a missing persons database but is critical for LAWS or a system that identifies criminals or suspects.

From a software engineering point of view, predictability can be thought of as a function that takes into account the correctness of the system, i.e., the degree to which the system can replicate and reproduce the same correctness over time, and the extent to which the system can be adapted to allow the processing of data different from that to which it was exposed in training and validation, without losing performance. The latter is particularly important for machine learning systems since an inadequate selection of the training and validation set can cause the system to "memorize" instead of "learn", that is, it cannot generalize the patterns found in data that diverge slightly from the instances already seen, this situation is called overfitting in statistical terms.

On the other hand, from an operational point of view, predictability aims to make it possible to anticipate particular actions of the system in execution. It is expected that any autonomous system will have a certain degree of unpredictability associated with it, especially the more complex its task is and the more dynamic the context in which it operates, since it is impossible to anticipate all the possible situations that the system will face once it is in place. To better exemplify this aspect of predictability, let us return to the system we mentioned previously, which must differentiate people from objects based on a video feed. Suppose that this video comes from a camera attached to the system that captures live images of the context where the system moves. Regardless of the AI model used, it is not possible to determine in advance, in the design of the system, all the possible (types of) objects and people (and external events) with which it may encounter once deployed in a physical environment, unless this environment is absolutely controlled or with (almost) perfect information. In real-world environments, where environmental information is imperfect and incomplete, operational unpredictability is a complex problem since not only is it difficult to anticipate what the system will encounter, but it can also be very difficult to anticipate how the system will react to those events.

In systems whose operation depends on extracting patterns from training data, this can be even more worrying. On one hand, it has already been shown in practice that these systems can fail in very unpredictable ways when the input data varies (even slightly) from what is expected since the way in which they obtain their goals does not necessarily follow logic or reasoning patterns that are reasonable for a human being [10]. On the other hand, in order to quantify how predictable these systems are, the quality of the data in relation to the deployment environment should be adequately quantified, for which there are still no robust tools that allow the variables relevant to the tasks to be identified and adequately represented.

Finally, from a more global perspective, predictability understood as the degree to which the results and effects of a system can be anticipated from its use, is determined by many different factors that impact both the technical and operational aspects of the system. Among the most notable factors, as we have previously mentioned, are the issues of data availability and quality (both for training and validation), the type of specific task or function that the system must solve, and the interaction with other systems (not just computational ones) in a dynamic and complex environment.

Predictability is a necessary factor, although depending on the domain it may not be sufficient (in the next section we will analyze other complementary properties), to achieve a reliable relationship with autonomous systems while they are running. In fact, depending on the factors that we mentioned in the previous paragraph, being able to measure the degree of predictability can be vital in the success of the deployment of systems that involve human-computer interaction for decision-making. Specifically in relation to autonomous weapons, predictability is key in order to be able to effectively exercise the different types of control that we discussed in the first part of this article. The operator's ability to evaluate how an autonomous system would respond in a given circumstance is absolutely necessary so that they can define whether it is feasible (and with what risk) that the system can accomplish the task in a way that respects the different rules and regulations of combat, IHL, etc., as well as the mode of operation required to make this happen (e.g., the configuration of specific parameters for the mission).

It is interesting to note that for a system to be predictable it is not necessary for the user to know or understand its operation (which is complementary, as we will discuss in the next section), predictability should be able to be determined based on the analysis of the system's behavior. The models that currently support AI-based systems present, as we have been discussing, many difficulties in being able to reliably evaluate their predictability. Extensive research is currently being carried out to improve their robustness [11].

These systems also impose a new challenge in the definition of design and validation or testing processes. Although a wide variety of methodologies and tools exist in software engineering for traditional computing systems, to ensure a certain degree of predictability, robustness, and reliability, in most cases they cannot be applied directly. For example, the study area of formal methods focuses on the development of mathematical techniques to describe both software and hardware systems and their requirements (what those systems are expected to do) in such a way that it can be tested, by means of mathematical operations, that effectively the implemented system does exactly what it is supposed to do. These techniques are generally used successfully for closed systems of controlled complexity and high risk; however, they have serious efficiency problems as systems become more complex. Beyond this, they do not directly apply to ML systems and it is not clear that something similar can effectively be achieved in the short and medium term. On the other hand, the plethora of testing tools that exist for traditional systems do not generalize adequately to AI-based systems either. One of the most fundamental problems is to be able to generate test cases relevant to the behavior of the system outside the design scope, in such a way as to be able to reproduce as reliably as possible the environment that it will encounter once deployed, especially if the system is expected to work in an open, physical environment (as would clearly happen for an autonomous weapon). Simulation tools can play a very important role in this since it is not necessary, in principle, to confront the system with a real situation such as a battlefield, however, the problem of being able to predict unforeseen situations remains, since the environment must be simulated in a computational system itself.

## *3.3 Comprehensibility, Explainability, and Interpretability*

Other properties, complementary to predictability, have been identified as critical for reliability in an AI-based system. In this section, we will analyze some of those related to the ability of a human to understand how the underlying model works or understand the reasons why the system delivers a certain result or makes a certain decision, as well as the ability of the system to provide evidence relevant to its operation or decisions. In different proposals, these properties are grouped under the umbrella of "Transparency", which covers access not only to the process that the AI system carries out in execution but also to all the processes that make up the life cycle of the system from its conception and design until its development, deployment, and evolution over time [12].

The comprehensibility or interpretability of a system focuses on understanding the AI model, that is, being able to identify why the system does what it does when it does it. Across the spectrum of AI-based systems, there is wide variation in how comprehensible or interpretable their foundations are. Analyzed from a narrow sense, this property seems to require that a human user (developer or operator) can understand the complexities of the underlying AI model. In this sense, a relatively small rule-based AI system, or a decision tree,[6], can be highly understandable to its designer, although a lay user may require some level of training to be able to understand the "logic" of the knowledge structure and the inference process. On the other hand, the learning process in a (deep) neural network involves assigning millions of weights to features that sometimes the network itself identifies from the input data (for example,

[6] https://en.wikipedia.org/wiki/Decision_tree.

the pixels of a photo or a video). Although its designer/developer understands the meta-process that is happening (the network is seeking to optimize a mathematical function by appropriately adjusting those weights), it is impossible for them to fully understand the computation itself. For the lay user of the system, even the "intuitive" description of what it means to adjust a function may be incomprehensible. The term "black box" is generally used to refer to models like the latter, where access to the laws that govern the processing that occurs within it is not entirely possible.

Taking the term in a more general sense, a certain degree of understanding or interpretability of an AI model can be achieved through extensive observation of the system's behavior. Most human beings use a smartphone in a highly effective way without knowing exactly how the device works internally, but through continuous use or through tutorials (specific training) we build a mental model of its operation in which we can trust and on rare occasions, the system gives us surprises by behaving in a completely unpredictable way. Now, we can ask whether or not achieving a degree of empirical understanding based on observation of the system operating is sufficient for any AI-based system. Clearly, it is enough to manipulate a smartphone that detects our face to unlock itself, or that tells us which path to take to an unknown location, but what happens with a smart weapon? What is the limit of ignorance that we can accept in cases where sensitive issues, such as human life, depend on the functioning of the system?

Concerning the principle of predictability, comprehensibility is complementary and both are necessary as the complexity of the tasks that the system must solve and the environment where the action takes place become. On the one hand, a high degree of understanding of the system increases its predictability. However, predictability is not sufficient in itself, even when it exists to a high degree, mainly to monitor whether the system is working well or in those (perhaps rare) situations where the system fails in an unforeseen way. As we established in the first section, meaningful human control is paramount in the utilization of lethal autonomous weapons. Let us imagine the case where such a system, which has been deployed and works in conjunction with a human operator, fails unexpectedly; it is the human operator who must exercise some type of control over the device to correct or avoid unwanted collateral damage. If the user, in addition to having been trained in the use of the system, knows to some degree how its "logic" works, they could, in principle, quickly understand the situation and carry out some contingent action. For example, the simple realization that the system "is not seeing or identifying" an object that the operator does notice exists because it was never fed with similar images, gives the operator useful tools to correct the situation or take necessary precautions to complete the task. The combination of a high degree of understanding and predictability is critical for the effective use of an AI system that involves high risk for the user or the environment where it is deployed.

Finally, we will address the principle of explainability. This concept is one of the most developed in relation to AI systems, in fact, XAI (explainable AI) is an expanding research area within the field of AI [13]. Explainability is the ability of an AI-based system to be able to justify its results or decisions in terms that a human can understand. To understand the concept it is necessary to differentiate it from understanding and interpretability. First, interpretability does not imply explainability and vice versa. A black box system could, in principle, provide explanations for its results, but those explanations would be just as opaque and incomprehensible as the rest of the model. Even so, it could serve to generate trust in the user if they appeal to a correlation that is accessible to the user. On the other hand, most of the AI-based systems that we use today, even those that are completely interpretable, are not designed to offer explanations to accompany their results or suggestions. Explainability requires a specific ability of the system, that is, it is the system that acts to provide explanations, while interpretability leaves the system in passive mode and it is the human who analyzes it.

It is clear that a system that can explain itself can generate more trust in the user than one that does not offer this functionality. However, there is a fervent discussion as to whether explainability is absolutely necessary (or even desirable) for any AI-based system. In relation to smart weapons, an explainable system could be very useful for the operator training stage, since it gives the operator the possibility of creating a more robust mental model of the system's behavior when it justifies its actions. On the other hand, in execution, once deployed, this capability can make human control over it more effective and fluid, especially when the system does not work as expected and it is difficult for the operator to understand if it is working well or not. Note that the concept of explanation is complex and determining what a "good" explanation consists of depends directly on the application domain, the specific task that the system is solving, and the (type of) user or operator [14].

To conclude we want to point out that, in the case of autonomous weapons, there are different positions on which of these properties are absolutely necessary, sufficient, or just useful. One goal of this article is to make clear that predictability, understandability, and explainability are important and a high-risk system such as LAWS cannot afford to do without any of them. Between the three, and considered from the design of the system point of view and for the rest of its life cycle, they can improve the chances that an operator can interact effectively with such a system, exercising the necessary control over it and mitigating possible risks of failures or side effects that may occur. The degree to which these properties should be present will depend on each type of system that is intended to be developed, the specific tasks they are supposed to carry out, the environment, etc. This is exactly the reason why appropriate standards, metrics and tools must be defined to analyze and evaluate the systems and regulatory frameworks that guarantee, before being deployed, that the systems effectively verify these properties throughout their life cycle.

## 4 Conclusions

The authors of this document are convinced that the international community has accumulated a lot of effort to agree on criteria to preserve the interests of humanity in various areas, and particularly in medicine. We consider that the ethical and technical criteria used in the health field can serve as a guide to establish guidelines in the context of the use of AI in arms production. In particular, in recent years the International Medical Device Regulators Forum (IMDRF) has set very useful definitions for the regulation of the use of AI in medical devices. Something that quickly emerges from reading the existing documentation is that the regulation models are still in the development stage. This is essentially because the scientific community has not yet given a formal technical answer regarding how to measure predictability and reliability or how to interpret or explain emerging behavior in a data-driven AI system. It is important to be clear that this is a technology in an early stage of progress and for which we are making conjectures about its effects based on speculation, not evidence. In that sense, we must be cautious with our projections without ceasing to set ethical-social limits on the impact of these technologies.

The problem we face is that to set limits it is necessary to be able to measure the system or the product in several relevant ways. For example, for a drug to be marketed there are protocols that involve years of experimentation to verify its safety and effects. But we are far from developing something similar for AI systems. For example, we talk about algorithmic bias [15], where we have examples that we denounce, but we do not have a precise specification of the concept, nor a precise way to measure it, much less a way to mitigate it. The same happens for several of the concepts that we have addressed in this work, while new ML architectures (with new challenges) and AI-based products continue to be proposed.

The above seeks to support the idea that the control and regulation of LAWS should be thought of as a continuous process over time that should generate preventive limits that evolve and adjust with accurate advances in technology. A clear example of this is facial recognition tools. Its abusive use in crime prevention has promoted precautionary actions against its use. Thus, companies, such as Amazon, Google, and Microsoft, have decided to suspend sales of systems that use this technology. On the other hand, the state establishes moratoriums or temporary prohibitions with a view to achieving more specific regulations. This does not mean that facial recognition technology will never be used again, but that its use is conditional on social agreements and scientific evidence that make its use safe.

Following this dynamic, we finish this work with a series of considerations for action divided into two levels: one more general (and political), and another more particular (and technical).

**General proposals**:

- We second the New Agenda for Peace (2023),[7] where the secretary General urged the States to produce, in the short term, a legally binding instrument to prohibit lethal autonomous weapon systems that function without human control or oversight, and which cannot be used in compliance with international humanitarian law, and to regulate all other types of autonomous weapons systems.
- The establishment of a cyber-ethical code of good professional practices that LAWS designers and developers should adhere to in the same way they sign a

[7] https://dppa.un.org/en/a-new-agenda-for-peace.

confidentiality commitment. It could also serve to establish a Hippocratic oath for computer scientists. These values should be incorporated into the curricula of both military training and IT professionals.
- The formation of an International LAWS Regulation and Control Agency. This agency would be dedicated to generating standards and validation metrics, conducting audits and monitoring, and analyzing situations of regulatory failure to provide feedback to the regulation and control system.
- Promote the establishment of LAWS-free zones. Following the current idea of nuclear weapon-free zones.
- Promote the formation of an international registry of LAWS manufacturers and prohibit the manufacture of LAWS without a license (see [16]).
- Ensure that companies that design and produce LAWS have an internal ethics committee whose members have stability and freedom to carry out their tasks. That would resolve the issue of business confidentiality.

**Technical proposals**:

- We believe that the focus of regulation should be changed and focus on: "Software as a military device" (MD) as any piece of software intended to be used for one or more military purposes that carry out these purposes without being part of a hardware device. On this basis, it is possible to extend the concept to MD based on Artificial Intelligence and Machine Learning. From there it is possible to establish regulations that give rise to safe functionalities for civilians and that respect the IHL.
- To achieve the objectives of reliability and verifiability of the LAWS, it is necessary to provide them with accountability mechanisms that allow their actions to be evaluated a posteriori.
- Development of a segmentation of the different types of LAWS that allows establishing specific legislation for each one. As mentioned above, we consider that such segmentation has to be based on multi-intentional aspects, which include both characteristics and capabilities of the device, as well as configurations of the performance environment and the objective tasks. In [17] several segmentation criteria appear that seem to us to be a good starting point.
- Promote technological research that leads to improving the evaluation of LAWS in at least the three axes analyzed in this manuscript.

## References

1. Svoboda, E.: Artificial intelligence is improving the detection of lung cáncer. Nature (2020). https://www.nature.com/articles/d41586-020-03157-9
2. Ben-David, S., Hrubeš, P., Moran, S., Shpilka, A., Yehudayoff, A.: Learnability can be undecidable. Nat. Mach. Intell. **1**, 44–48 (2019)
3. Regulation of the Autonomy of Weapons Systems (2020). https://article36.org/wp-content/uploads/2020/10/regulation-autonomia-ES.pdf

4. Russell, S.J., Norvig, P., Davis, E.: Artificial Intelligence: A Modern Approach, 3rd edn. Prentice Hall, Upper Saddle River, NJ (2010)
5. Regulation of the Autonomy of Weapons Systems (2020). https://article36.org/wp-content/uploads/2020/10/regulation-autonomia-ES.pdf
6. Ekelhof, M., Persi Paoli, G.: The human element in decisions about the use of force. UNIDIR (2020). https://unidir.org/publication/the-human-element-in-decisions-about-the-use-of-force/
7. Bunker, R.J.: Mission Command and Armed Robotic Systems Command and Control a Human and Machine Assessment. The Association of the United States Army. Land Warfare Paper 132 (2020)
8. Boulanin, V., Davison, M., Goussac, N., Peldán Carlsson, M.: Limits on Autonomy in Weapon Systems: Identifying Practical Elements of Human Control. SIPRI (2020)
9. Holland Michel, A.: The Black Box, Unlocked: Predictability and Understandability in Military AI. UNIDIR (2020). https://unidir.org/publication/the-black-box-unlocked/
10. Eykholt, K., Evtimov, I., Fernandes, E., Li, B., Rahmati, A., Xiao, C., Prakash, A., Kohno, T., Song, D.: Robust physical-world attacks on deep learning visual classification. In: IEEE/CVF Conference on Computer Vision and Pattern Recognition, pp. 1625–1634 (2018)
11. Molnar, C.: Interpretable machine learning. In: A Guide for Making Black Box Models Explainable, 2nd edn. (2022). https://christophm.github.io/interpretable-ml-book/
12. Recommendation on the Ethics of Artificial Intelligence. UNESCO (2021). https://unesdoc.unesco.org/ark:/48223/pf0000380455
13. Ali, S., Abuhmed, T., El-Sappagh, S., Muhammad, K., Alonso-Moral, J.A., Confalonieri, R., Guidotti, R., Del Ser, J., Díaz-Rodríguez, N., Herrera, F.: Explainable Artificial Intelligence (XAI): What we know and what is left to attain trustworthy artificial intelligence. Inf. Fusion 99 (2023). https://doi.org/10.1016/j.inffus.2023.101805
14. Miller, T.: Explanation in artificial intelligence: Insights from the social sciences. Artificial Intelligence (2019)
15. Danks, D., London, AJ.: Algorithmic bias in autonomous systems. In: Proceedings of the 26th International Joint Conference on Artificial Intelligence, IJCAI (2017). https://www.cmu.edu/dietrich/philosophy/docs/london/IJCAI17-AlgorithmicBias-Distrib.pdf
16. Model Law against the Illicit Manufacturing of and Trafficking in Firearms, Their Parts and Components and Ammunition. United Nations (2011). https://www.unodc.org/documents/legal-tools/Model_Law_Firearms_Final.pdf
17. Benjamins, R.: A choices framework for the responsible use of AI. AI Ethics **1**, 49–53 (2021). https://doi.org/10.1007/s43681-020-00012-5

# Integrating a Disability Perspective into Lethal Autonomous Weapons Systems Analysis: First Steps

**Vicent Costa**

**Abstract** This work argues that incorporating a disability perspective into the discourse on lethal autonomous weapon systems (LAWS) reframes fundamental debates regarding their conceptualization, training phase, design, and use, thereby opening the topic to a deeper and more comprehensive discussion. First, this perspective is applied to analyze the conceptual underpinnings of LAWS. A procedural approach is then employed to develop arguments advocating for new design requirements for LAWS, particularly in relation to targeting. Finally, some considerations for further research from a relational perspective, as well as challenges in integrating both approaches, are presented.

**Keywords** Disability · Disabilities · Lethal autonomous weapons systems · Autonomous weapons systems · War · Death

## 1 Introduction: Disability and LAWS

With the rise of artificial intelligence (AI) and its growing use and impacts across all aspects of life, from routine activities to profound existential questions, numerous ethical challenges have emerged, highlighting the urgent need for deep, multidisciplinary approaches to address this complexity.

With regard to this discussion, two key aspects merit consideration. First, many debates surrounding these ethical concerns have been addressed from a relational approach [1], which means focusing on biases in datasets or algorithmic outputs. This perspective, however, excludes the design and decision-making processes involved in the development of AI systems (that is, their *logic*), which, according to [1], would fall under a procedural approach. Second, notwithstanding extensive research

This work was funded by the project PID2022-139835NB-C21 and supported by the research group 2021-SGR-00517.

V. Costa (✉)
Artificial Intelligence Research Institute (IIIA-CSIC), Cerdanyola del Vallès, Spain
e-mail: vicent@iiia.csic.es

 
A. Sans Pinillos et al. (eds.), *SecondDeath*, Studies in Applied Philosophy, Epistemology and Rational Ethics 76, https://doi.org/10.1007/978-3-031-98808-0_9

on the topic, there is a notable lack of specific analyses that address the ethical impacts of AI on individuals with disability. This scarcity in the academic literature does not help to elucidate a thesis underlying this chapter (though not addressed here), namely, the current and rapid digitalization of the world may continue to exclude marginalized groups, such as persons with disability, despite the accessibility technologies advanced by AI in recent years, which, undoubtedly and at the same time, often have a positive impact in the lives of people from this group.

In particular, and as might be expected, the ethical and legal challenges arising from the use and design of AI systems in military contexts have become increasingly prominent, particularly over the past fifteen years. By this means, politicians, academics, civilians, and non-profit and non-governmental organizations have highlighted the potential risks and challenges posed by lethal autonomous weapons systems (LAWS). However, despite these efforts, meaningful progress—understood as broad agreements or consensus—has yet to emerge in the discourse.[1]

Furthermore, the two key aspects mentioned above are also reflected in this debate on LAWS.[2] On the one hand, though to a lesser degree, the relational approach has been common practice, focusing on aspects related to the outputs and potential uses of LAWS. This aspect becomes meaningful when considering the interests, security concerns and strategic aspects of hiding the design of LAWS, factors that may affect the topic as well. On the other hand, in the context of the discussion on LAWS, people with disability have been overlooked, with rare exceptions:

> To date, discussions regarding the ethics of autonomous weapons have continued to highlight and centre the voices of defence, diplomacy, the private sector and academia, which, for the most part, do not represent the views of persons with disabilities and other marginalized groups. No debate on the ethics of autonomous weapons can be considered serious if the voices of those who risk being most affected are not heard. ([3], p. 290)

Figueroa et al., the authors of the aforementioned citation, extend their critics by emphasizing that the lack also encompasses the presence of people with disability into the discussion, and also point out that the issue also applies to other marginalized groups:

> even when "ethical" perspectives on autonomous weapons systems are discussed, the debates generally do not include persons with disability or from other marginalized groups, and rarely include persons from the global South. ([3], p. 289)

We agree with that and further note that even in alternative and more inclusive approaches than the mainstream literature, such as [4, 5], disability is notably absent from the analysis. In this manner, and considering that war and the presence of people with disability are constants in our societies, such an omission cannot be justified.

---

[1] For a comprehensive and multi-perspective overview of LAWS, the reader is referred to [2].

[2] Even if we find, as commonly shared, that ethics and legal compliance are related insofar legal concepts need to be interpreted and applied, let us clarify that an analysis of legal aspects regarding LAWS is out of the scope of the research of this work.

Accordingly, this work aims to address and contribute to solve this gap. In this way, the rest of the chapter is organized as follows. First, an analysis, integrating a disability perspective of different conceptualizations of LAWS is presented in Sect. 2. Then, in Sect. 3, a mainly procedural approach is employed to develop arguments advocating for new design requirements of LAWS. These requirements revolve around targeting. Finally, serving as the conclusions and also as a starting point for further research, the chapter ends by taking a relational standpoint and stating few key considerations related to LAWS.

But before undertaking the task of conceptualization in next section, we find it necessary to clarify what is meant by disability in the beginning of this work. Indeed, from a philosophical perspective [6], the concept of disability raises challenging yet central questions regarding its ontology, ethical and moral significance, as well as concerning its implications within political philosophy. Thus, true to its purpose, a philosophical analysis of disability resists the assumption of a fixed, universally accepted definition of the term, while acknowledging its inexorable ambiguity, it embraces the inherent complexity and variability of the concept. This problem is further accentuated insofar as the chosen definition of disability not only generates but also partially shapes the ethical concerns arising from the specific challenge at hand—in this instance, the use and design of LAWS.

However, despite these complexities regarding the concept of disability, there are two propitious aspects for our analysis that can be highlighted. On the one hand, disability studies and their theoretical models for disability[3] constitute frameworks from which to define and interpret disability.[4] On the other hand, a highly comprehensive conceptualization of disability risks leading to neglect—a lack of specialized attention—toward individuals with disability. Furthermore, the social model, by emphasizing the relationship between the individual and their environment, as well as the limitations that emerge from this interaction, appears particularly well-suited for analyzing LAWS, given the critical role of context in warfare. In this work, therefore, we adopt a somewhat generic definition of disability, refraining from engaging in detailed conceptual analysis, while aligning more closely with the principles of the social model of disability.

[3] For a general introduction to the topic, the reader is referred to [7].

[4] Let us observe that, at this point, the complexity is replaced into the question of which model, if any, should be selected.

## 2 Incorporating a Disability Perspective into the Conceptualization of LAWS

The debates surrounding the notion of lethal autonomous weapons systems (LAWS)[5] have been numerous and multidisciplinary, encompassing fields such as politics and philosophy. These discussions gained particular prominence after 2013, when Heyns [8] underscored the divergence in interpretations of the concept. However, despite various efforts to establish clarity, a shared and canonical definition remains elusive. And this lack of consensus is evident not only among diverse stakeholder groups but also within each group itself.[6]

In this manner, technological experts continue to disagree on a definition. Similarly, states such as the United States of America, the United Kingdom of Great Britain and Northern Ireland, and Russia hold differing conceptualizations of LAWS, and this divergence is mirrored within some non-profit and non-governmental organizations.[7] Moreover, while some researchers downplay the importance of precisely defining LAWS, others emphasize that a clear conceptualization is critical, as it directly impacts ethical and legal considerations surrounding their design and deployment. In this sense, we align with researchers like Morresi [10], who argues that a clear and universally accepted definition is foundational for regulating LAWS:

> What we are still missing today is probably the starting point for the regulation of AWS, which consists of a clear and unique definition that wouldn't leave any grey zone. ([10], p. 9)

The task, nevertheless, still remains open, and a short in time solution seems unattainable nowadays. Thus, with the aim of contributing to this open challenge, in this section, we adopt a philosophical lens to explore the essential elements required for a comprehensive definition of LAWS. In addition, by incorporating perspectives from disability studies, we shall challenge and refine certain aspects of these definitions, which will be addressed in deeper detail in the subsequent section. Finally, we will review recent and prominent proposals in the literature, advocating for one of them as a concrete—albeit imperfect—definition.

Two interrelated considerations merit attention when introducing the analysis of the section. On the one hand, any definition of LAWS should be sufficiently comprehensive to address the rapid and dynamic nature of technological development. But, at the same time, it must account for the potential significance of the specific

---

[5] Although the literature generally centers on research and debates surrounding autonomous weapon systems (AWS), we shall focus our analysis specifically on LAWS to emphasize the deadly potential inherent in the broader concept. This particularization allows us to draw on established conceptualizations of AWS, as it merely adds the lethal condition to the general framework. Consequently, throughout this chapter, we will refer exclusively to LAWS, even when discussing works or authors that address the generalized concept of AWS.

[6] In this book, two additional works address, to varying degrees, the topic of LAWS. Specifically, Sans Pinillos and Vallverdú situate it within the context of an analysis of symbolic death and dual-use dilemmas, while Martinez and Rodriguez offer a general analysis of the use of AI in LAWS, without focusing on a specific population group, such as those with disability.

[7] For extensive discussions on the notion of LAWS, see, for instance [2, 9].

technological features involved, particularly concerning their implications for people with disability, as will be discussed later. In other words, a minor technological change (for instance, the inclusion or exclusion of specific sensors) cannot lead to a change in the categorization of LAWS; and, simultaneously, the definition should also be specific enough to exclude other devices, such as land mines.

On the other hand, there seems to be a consensus on the shift from focusing solely on technological aspects to addressing questions related to the autonomy of systems. However, this consensus ends there. As evidenced by the heterogeneity of the proposals, autonomy is a highly complex notion to define, although in the case of focusing only on the case of LAWS. Furthermore, even setting aside the convergence of different categorizations of autonomy, disagreements—often driven by political motivations and state strategies—persist regarding the various degrees of autonomy attributed to LAWS.

## *2.1 Autonomy in LAWS*

In this work, the autonomy is first understood broadly as the capability of a system to carry out a task independently. We note thereby that, when applied to non-human agents, some classical dimensions of human agency in the different conceptualizations of the concept of agency from the philosophy of action, typically associated with discussions of autonomy, might be set aside. In the case of determining whether a system is performing a task on its own, however, different criteria can be analyzed. Specifically, following [11], three core dimensions are included in the notion of autonomy: the task performed, the human-machine relationship, and the intelligence of the system.

Regarding the first dimension of the autonomy (i.e., the task performed), it encompasses two main aspects: the complexity of the task and its significance, including the consequences of errors in execution. In the case of LAWS, the latter aspect is beyond doubt, as some errors may lead to irreversible outcomes. As for the former aspect, dynamic war scenarios are often cited to illustrate the level of autonomy that LAWS must achieve. With regard to this last point, we find this framing too vague and argue that explicit consideration must be given to the presence of individuals with disability in armed conflict scenarios. As will be discussed in the next section, the ability of LAWS to distinguish between threatening agents and individuals who are unable to understand the risks or even perceive its presence is crucial to consider.

Prior to addressing the remaining dimensions of autonomy, let us recall the four basic steps of the OODA loop (Observe, Orient, Decide, Act), which outline the decision-making process of LAWS. Indeed, originally proposed by [12] within the context of general military strategy, these steps are as follows: Observe (gather available data and detect potential targets), Orient (identify the targets), Decide (determine whether to engage the targets), and Act (engage the targets).

Regarding the second dimension of autonomy, it categorizes performance into three classes, based on the human-machine relationship (see, for instance, [2,

11]). The most mediated level is termed *human in the loop*, encompassing semi-autonomous operations. In this mode, the systems observe and orient (i.e., it carries out the first two OODA loop steps), but they require human approval to take action. This type of autonomy in LAWS may partially mitigate the risk of erroneously engaging individuals with disability[8], but in that case, it might be hard to consider those systems actually as autonomous. The intermediate level corresponds to supervised autonomous operations and is designated as *human on the loop*. In this case, once activated, the LAWS can execute the four OODA loop steps, but the human overseeing the system retains the ability to interrupt the process at any step or moment. Finally, the least mediated level is termed *human out of the loop* and pertains to fully autonomous systems. In this case, the LAWS performs the four OODA loop steps independently, and the user has no control or capacity to deactivate the process (indeed, the system does not communicate back). Let us make two observations on this dimension: these three categories of the human-machine relationship may not always be unequivocally delineated, and some systems may operate at different levels at different times, depending on the context.

The final dimension of autonomy is intelligence, which can be divided into three levels based on the sophistication of the system: automatic, automated, and autonomous. Automatic systems are characterized by low decision-making capacity, low complexity, and high predictability. Automated systems are significantly more complex than automatic ones; for example, they may incorporate a wide range of attributes, variables, or rules-based reasoning. Although these systems exhibit lower predictability, they should, in general, ensure at least potential traceability for their developers, thereby allowing some scope for moral responsibility. The highest level of the intelligence dimension corresponds to autonomous systems. These are highly sophisticated, goal-oriented systems in which, very often, the humans managing them may not necessarily understand how the system accomplishes its tasks. Thus, they pose several challenges about moral responsibility. As it pertains to the human-machine relationship, the division of intelligence may not be readily apparent in practice, even if there is a consensus to not classifying LAWS into automatic systems. In addition, the implications of integrating a disability perspective into the analysis of this autonomy dimension could lead one to argue that the prevailing perspective regarding automatic systems might not be very accurate.[9]

To conclude this subsection, it is worth noting that the three dimensions of autonomy can be meaningfully related to the OODA loop, as these steps provide valuable insights into evaluating the dimensions. Indeed, regarding the task performed, its complexity can be determined by estimating the difficulty of each of the four OODA loop steps. With respect to the human-machine relationship, this can be categorized based on the OODA loop step in which the human agent may intercede, if at

[8] However, as showed in [3], human beings also tend to discriminate people with disability in warfare.

[9] The analysis of this aspect is reserved for future research and, as such, falls outside the scope of the present work.

all. Finally, concerning the last dimension, the system's intelligence might be more effectively measured by examining its performance across the OODA loop steps.

## 2.2 *Conceptualizations of LAWS*

In this subsection, we first highlight and briefly comment on two recent conceptualizations of LAWS that have been widely adopted, especially by scholars, and particularly given the recent emergence of these proposals. Then, we introduce another highly contemporary proposal from the literature, suggesting its adoption while incorporating considerations derived from the integration of a disability perspective into the analysis.

On the one hand, the International Committee of the Red Cross (ICRC) has offered the following definition in its position and background paper, which has been used, for example, in the risk analysis of LAWS for persons with disabilitiy by Figueroa et al. [3]:

> Autonomous weapon systems select and apply force to targets without human intervention. After initial activation or launch by a person, an autonomous weapon system self-initiates or triggers a strike in response to information from the environment received through sensors and on the basis of a generalized "target profile". This means that the user does not choose, or even know, the specific target(s) and the precise timing and/or location of the resulting application(s) of force. ([13], p. 2)

We observe that this proposal clearly falls within the *human-out-of-the-loop* category of the human-machine relationship dimension of autonomy. Furthermore, it appears to presuppose the two highest levels within the dimension of intelligence, whereas a high degree of task performance is considered as well. Notably, the "basis of a generalized "target profile"" leaves the door open to non-human interference during the second step of the OODA loop, which might lead to the incorrect identification of targets, especially for individuals with disability.

On the other hand, in 2023, the US Department of Defense updated its definition of LAWS from the 2012 DOD Directive 3000.09 [14]:

> A weapon system that, once activated, can select and engage targets without further intervention by an operator. This includes, but is not limited to, operator-supervised autonomous weapon systems that are designed to allow operators to override operation of the weapon system, but can select and engage targets without further operator input after activation. ([15], p. 21)

In this manner, the conceptualization is placed at the medium-high levels of the autonomy dimensions introduced in the previous subsection, which leaves some scope for more control and traceability. The comprehensiveness and flexibility of the proposal, as well as the influence of the institution signing the policy document, help to explain the widespread use of the definition. In this way, that is, for instance, one of the definitions taken by Morresi [10] in the context of an analysis of legal and ethical dares posed by LAWS, and one of those gathered by Wood in [16].

The previous definition by the US Department of Defense, which has highly influenced international discussions, is similar to the current one by the same institution, but broader with respect to the human role on LAWS performance:

> A weapon system that, once activated, can select and engage targets without further intervention by a human operator. This includes human-supervised autonomous weapon systems that are designed to allow human operators to override the operation of the weapon system, but can select and engage targets without further human input after activation. ([14], p. 2)

This definition has also been widely used by different authors like McFarland [17], for who, in the line of this broader nature of the conceptualization, "the essential quality of an AWS is that it operates, to a certain extent, on its own, with little or no direct intervention by a human operator" ([14], p. 2).

Apart from these widely used and aforementioned definitions, we would like to highlight the very recent and, in our view, highly accurate proposal by Barbosa [2], who defines LAWS as:

> Weapons systems that operate through AI and can discretionarily select and engage targets without further intervention by a human being. ([2], p. 23)

We find very appropriate to explicitly mention AI into the definition of LAWS since, in doing so, a set of capabilities like decision-making, learning, perception or identification are potentially attributed to the systems, but without commiting the conceptualization to specific skills.[10] Furthermore, other war technologies like land mines are thereby let out of the scope of LAWS. Accordingly, in this chapter, it is assumed that LAWS are designed using AI, but it has to be said that the meaning of the discipline has a broad scope, as Barbosa, referring to [18], reminds after presenting the definition of LAWS:

> The system develops its tasks by using AI. AI is a computer program capable of problem-solving and decision-making in situations where information is incomplete or uncertain. AI embraces a broad spectrum: from previously programmed rules, which the program chooses based on inference or evaluation of options, to high-tech programs that operate through machine learning and create their scoring formula. ([2], p. 23)

Two final observations on this proposal need to be done. First, let us note that *targets* in the definition are not limited to people, insofar the notion includes objects and also places (wherein, of course, people might be). Second, in this chapter, we take the lack of "intervention by a human being" in the definition as potentially, in the sense that LAWS are able to perform the OODA loop steps, but that is not the necessary performance of LAWS. Furthermore, this interpretation is coherent with a clarification, based on [19], posed by the author of the definition:

> Once activated by a human, AWS can pursue the OODA loop without the need for human input. This means operating independently. ([2], p. 24)

[10] Let us note that these abilities are not necessarily performed during the period of action of the LAWS. For instance, the learning process of a LAWS might be done subsequently to a conflict, using the gathered data during the war.

## 3 A Procedural Approach for the Analysis of Targeting Challenges in LAWS

In Sect. 2, by integrating a disability perspective into the conceptual analysis of LAWS, three concerns emerged, one for each autonomy dimension. Of these, we highlighted two for deeper discussion in this section. The first concern pertains to the task performed, particularly its complexity. We emphasized that such complexity should explicitly account for the presence of individuals with disability in armed conflict scenarios. The second concern is the human-machine relationship dimension, where we mentioned that a lower level of autonomy may reduce the risk of erroneously engaging individuals with disability as threatening. However, in such case, the discussion might shift from LAWS to less autonomous AI systems, potentially excluding future iterations of LAWS from this analysis.

Our focus on the first two concerns, while leaving the third for future research, was a deliberate decision, as these issues are closely related to targeting, whereas the third concern, related to the intelligence dimension, is more relevant to the categorization of LAWS. Furthermore, at this point, it is important to clarify that targeting refers to the process of selecting someone or something from a group, encompassing all steps of the OODA loop: observing involves detecting potential targets, orienting focuses on identifying them, deciding resolves to engage them, and acting involves the actual engagement. In this line, we quote the words of Morresi, who states that:

> Contrary to what one might think, the most important element to consider when defining this new kind of weapon system is not the technology applied. Rather, the definition mostly relies on the role of human operators when it comes to the selection and engagement of targets. ([10], p. 2)

Accordingly, in this section, we adopt a procedural approach to address these concerns while incorporating a disability perspective. As stated in the Introduction, a procedural approach [1] to analyzing LAWS emphasizes the design and decision-making processes in their development. To contextualize this approach, we draw on two widely used frameworks for analyzing ethical concerns in AI systems. First, these concerns align with the *epistemic concerns* category as defined in [20]. Second, using the framework proposed in [21], our analysis focuses on biases that arise during the model-building phase. Furthermore, throughout our discussion, a speculative factor will be considered, insofar as we do not restrict the argument to current technologies but also include potential ones. The underlying idea behind this approach is based on the following Scharre's assertion, which we find appropriate:

> As machine intelligence advances, however, autonomous targeting will become technically possible in a wider range of situations. (Scharre, p. 50)

Before deciding whether to engage targets, LAWS must first detect potential targets and then identify them (i.e., determine who or what constitutes a target). We argue that a disability perspective in procedural analysis should lead to addressing and mitigating the risk of misidentifying individuals with disability as threats (and thus as targets). In this way, although specific aspects of the design of LAWS are often kept confidential, the following conditions might determine whether an individual is targeted:

(1) inability to comply with orders;
(2) inability to establish communication with the LAWS;
(3) the presence of *uncategorized* objects that may be interpreted as weapons;
(4) inability to perceive the presence of LAWS, leading to unexpected behavior or interaction with the system;
(5) unexpected behavior in response to the perception of LAWS; and
(6) inability to be identified using biometric data.

With respect to (1), we can consider an individual who is deaf encountering a LAWS issuing verbal commands through speakers. In such a situation, the LAWS might target the individual for failing to comply with the order. Similar scenarios could arise not only for individuals with sensory disability but also for those with intellectual disability.

Concerning condition (2), certain types of interactive AI systems, akin to virtual assistants, might be integrated into LAWS to enhance their ability to detect targets, clarify contextual factors (e.g., allowing individuals to explain why they are in a specific location), and, more generally, reduce targeting errors. However, speech difficulties or intellectual disability could hinder any form of communication with the LAWS, leading to unjust targeting.

Condition (3) pertains to the identification of objects. LAWS must reliably recognize auxiliary devices, such as wheelchairs or crutches, with a high degree of accuracy, especially in complex and dynamic environments like real-world combat scenarios. Accurate classification of these objects is crucial to assist in identifying individuals with disability. Nevertheless, experience has demonstrated that developing such capabilities is challenging, even for distinguishing between human beings and objects.[11] Additionally, a specific *moral rule* should be embedded into the design, ensuring that auxiliary objects are never targeted as such—except in cases where they are being weaponized.

Condition (4) is closely related to conditions (1) and (2). For instance, consider a blind person or a deaf individual who cannot perceive the presence of a LAWS. Such individuals may exhibit context-specific unexpected behaviors, such as a blind person inadvertently striking the LAWS.

In other cases, individuals perceive the presence of a LAWS yet still exhibit unexpected behavior, as described in condition (5). As Figueroa et al. explain:

> People with psychosocial impairments might exhibit "unexpected" behaviour that autonomous weapons could be unable to process (such as lack of response, shouting or unexpected movements) or that such weapons might interpret as a risk, causing them to identify the person as a target. ([3], p. 288)

Finally, with respect to condition (6), let us first recall that technologies such as iris, facial, voice, or fingerprint recognition can be used to target individuals. However,

[11] Consider, for example, the case of the first pedestrian fatality caused by an autonomous car [22].

for instance, people with reduced mobility may encounter difficulties when being analyzed using these biometric data. At this juncture, a very challenging dilemma emerges: either the LAWS is designed, for security reasons, to directly target these individuals (ensuring that, in extreme situations, a real target cannot injure themselves and thereby avoid a fatal outcome), or groups of people with impairments risk being targeted due to their inability to be accurately identified.

Therefore, a disability perspective, as demonstrated, leads to the conclusion that LAWS should first determine whether an individual has a disability to make better decisions about targeting. This distinction is crucial, as the criteria for more equitable decision-making depend on it.

At this point, we emphasize that significant challenges arise not only in technological terms but also conceptually. Indeed, as it has been commented in the Introduction of this work, the notion of disability presents significant challenges to categorization, which remain unresolved. Moreover, as with many other concepts tied to human experiences, conditions, and aspects of identity, a universally satisfactory definition of disability may never be achieved[12]. A common approach to addressing this elusive and complex nature is to consider disability as an umbrella concept (this strategy, depending on the goals and context, may prove effective). Another approach, related to the previous one, and which we adopted at the beginning of this work, is to start with a generalized and somewhat superficial definition, specifying aspects of the definitions as the argument or discussion requires.

However, certain domains and applications require a more nuanced and detailed categorization of disability. For instance, institutional identification of people with disability often relies on scales informed by qualitative and quantitative expert evaluations, partially grounded in the presence of an impairment. But even in institutionalized contexts, identifying an individual as a person with a disability remains challenging and, ultimately, it could be argued that such identification resists reduction to a universally satisfactory criterion[13]. The design of LAWS is another example where a superficial or very generalized conceptualization might be far from enough[14].

One potential answer to this problem might be to take the notion of impairment as the central aspect of disability. However, this approach would still be inadequate, as impairments are not always identifiable through external factors. For example, identifying certain impairments might require interactions with the individual, additional

[12] Nevertheless, despite the elusive nature of the notion, we find that addressing the problem of its conceptualization remains essential (indeed, the complexity of the endeavor should inspire greater effort, not resignation).

[13] The complexity is further heightened by other dimensions of this challenge—for instance, when individuals institutionally categorized as having a disability, driven by diverse and often opaque motivations, do not self-identify as such.

[14] Despite these unresolved challenges, and those to be introduced and explained next, we persevere and claim that, to avoid perpetuating and exacerbating discrimination against people with disability, especially in contexts that have historically been particularly adverse to them [3], the design of LAWS must incorporate the capacity to distinguish between threatening agents and individuals who lack the ability to comprehend the risks of LAWS or even perceive their presence.

tests, or contextual considerations—factors that the speed and immediacy of employing LAWS could hardly accommodate. Even in the hypothetical case of arriving at a formal, universal definition of impairment, the social model of disability has long demonstrated that reducing disability to impairment is insufficient for categorizing the concept.

Another approach, not necessarily opposite to the previous one, might stand for integrating deep learning (DL) based systems for accomplishing this classification task. Actually, this might be presented as the coherent strategy with most of conceptualizations of LAWS, which, as we have seen in previous section, establish medium-high levels in the second and third autonomy dimensions. However, we would like to express four concerns about this approach.

First, classification problems in DL tend to perform good in static contexts (e.g., classifying images), but, currently, the same cannot be stated in such complex, variant and unpredictable scenarios like war scenes. Actually, the unpredictable nature of LAWS has been widely highlighted. For instance, Asaro states that:

> The operator of an autonomous system may have a general idea of what the system is supposed to do, and may further have operational experience of how it operates in various specific contexts. But insofar as autonomous weapons are designed to operate over large geographic areas and time frames, and given that the possible interactions with the environment it may have grow exponentially, it will become increasingly difficult for even well- trained operators to reliably predict what a system will actually do once deployed. Testing and reliability metrics can offer confidence to operators only when systems are deployed in situations and contexts that match those under which it has been previously tested, while increasing ranges and time frames imply that operators are less aware of the specific characteristics of the environment the system will encounter. ([23], p. 219)

Second, we observe epistemic concerns such as inscrutable evidence and inconclusive evidence [20] stemming from the methodology and situation described in the last quotation. Indeed, LAWS would produce probabilistic yet inevitably uncertain knowledge regarding their identification of individuals as presenting disability or not. In addition, considering the complexity of satisfactorily categorizing disability, "the inherent difficulty in the interpretation of how each of the many data-points used by a machine-learning algorithm contribute to the conclusion it generates" ([20], p. 4) becomes really challenging and reinforces this second concern.

The third concern involves potential historical biases [21] emerging in LAWS. Based on real-world examples, Figueroa et al. [3] excellently argued that biases in AI, military violence, and police brutality disproportionately affect marginalized groups, including people with disability. Therefore, we could expect the training data in LAWS to inherit biases that would then be transmitted into the outputs of DL-based LAWS[15]. Let us observe that these aspects are closer to biases that should be identified from a relational approach.

---

[15] In this case, it is hard to conceive how to obtain a relevant quantity of data free from discrimination and biases against people with disability.

The final concern, partially related to the second, involves explainability.[16] Explainability refers to the ability to understand and describe the decision-making mechanisms of AI systems, including the mathematical links between inputs and outputs. And post-hoc explainability involves using techniques after a model's decision to clarify its reasoning in human-understandable terms. Indeed, a purely DL approach for designing LAWS would ignore the criterion of explainability. As is well-known, principles such as explainability, transparency, fairness, responsibility, and robustness are essential for designing AI systems aligned with ethical principles, and, in particular, the lack of explanations hinders traceability, which is essential for ensuring moral responsibility (in our view, a minimal requirement in the use of LAWS). It has to be noted, however, that there remains an underlying doubt regarding the actual role of ethics in XAI. As Nannini et al. [25] argue, supported by various recent works in the literature:

> the extent to which XAI research genuinely addresses ethical considerations, and effectively assimilates them into the design, development, and evaluation of AI systems, remains a topic of considerable debate. ([25], p. 1)

We claim that neuro-symbolic AI [26] offers a viable framework for ensuring the explainability of AI systems without succumbing to the inherent limitations of DL-based approaches in LAWS. This approach enables the integration of symbolic knowledge representation, including rules and (simplified) related moral principles, into autonomous systems. Of course, this research subject within AI does not address all the aforementioned concerns but may contribute to improving some of the related challenges. Nevertheless, to the best of our knowledge, very few works in neuro-symbolic AI have been applied to application domains related to disability (see, more recently, [27, 28]).

Finally, we observe that explainability is not enough by itself since the explanation generated by AI systems should maintain high standard. Increasingly, the evaluation of the quality of explanations has been integrated into ethical debates on XAI.[17] Nevertheless, the challenge of identifying criteria to evaluate explanations remains an open question in the literature, an aspect raised already in the some of the first works on the topic (e.g., [29, 30]).

## 4 Conclusions: Toward a Broader Analysis of LAWS

In this work, we have provided insights into the importance and consequences of integrating a disability perspective into debates about LAWS. To this end, we first conceptualized LAWS, emphasizing aspects that should be reconsidered when addressing potential experiences of individuals with disability in the context of warfare. Subsequently, we presented some common definitions from the literature, underlining

[16] For an introduction to the background of explainable artificial intelligence, including common definitions, the reader is referred to [24].

[17] For a general view of ethical considerations in XAI, the reader is referred to the recent work [25].

the high levels of autonomy required in those conceptualizations, particularly with respect to two dimensions of autonomy (namely, the human-machine relationship and the intelligence). From a procedural standpoint, we then focused our study on concerns related to LAWS targeting. In this way, through a series of examples, we showed the weaknesses and potential discriminatory effects of LAWS on individuals with disability. Additionally, to strengthen our claims, we anticipated two arguments in favor of the current use of DL-based LAWS and, finally, presented objections to these arguments. In summary, the present work represents an initial step toward integrating a disability perspective into the analysis of LAWS.

On another note, and even if not explicitly mentioned before, it should be noted that the insights presented in this work might also be interpreted as undermining arguments in favor of the use of LAWS, which are often justified by their purported efficiency and precision. Ultimately, it appears that the impact on individuals with disability is excluded from such calculations.

Another crucial aspect to consider concerns the role of human involvement in the stages of the OODA loop. While some parts of this work posited that greater human participation in LAWS might mitigate and address many of the associated challenges, it is important to underscore that the conclusions drawn in Sect. 3, alongside the examples and arguments articulated in [3], raise significant doubts—at least to some extent—about this claim. These doubts stem from evidence that individuals, even within such contexts, may exhibit discriminatory behaviors towards people with disability.

Finally, in addition to the many remaining aspects to analyze from a procedural approach, an analysis of LAWS from a relational perspective has not been presented in this work. We believe that such an undertaking could enrich the analysis, but it would also present broader challenges to the topic. As an example, which concludes this work, we note that a relational approach might lead us to conclude that refinements and improvements in the design of LAWS, even when proposed from a disability perspective, could result in new biases against people with disability. Specifically, in the hypothetical and, in principle, desirable case of achieving a design of LAWS that identifies people with disability with precision (and thereby enables less discriminatory targeting for individuals with disability), the relational perspective would direct attention to a concern not considered in the procedural approach: the potential deployment bias [21] of such systems. This type of bias, which arises from a mismatch between the problem an AI system is intended to solve and the way in which it is actually used, could manifest in LAWS that, due to the precision in identifying individuals with disability, end up targeting only people with disability and being used for attacking this group. Therefore, in light of these considerations, in addition to an analysis of LAWS from a relational perspective, research on integrating both approaches should be conducted.

## References


1. Aran, X.F., van Nuenen, T., Such, J.M., Coté, M., Criado, N.: Bias and discrimination in AI: a cross-disciplinary perspective. IEEE Technol. Soc. Mag. **40**(2), 72–80 (2021). https://doi.org/10.1109/MTS.2021.3056293
2. Barbosa, L.: Autonomous Weapons Systems and the Responsibility of States: Challenges and Possibilities. CRC Press (2024)
3. Figueroa, M.D., Orozco, A.H., Martínez, J., Jaime, W.M.: The risks of autonomous weapons: an analysis centred on the rights of persons with disabilities. Int. Rev. Red Cross **105**(922), 278–305 (2023). https://doi.org/10.1017/S1816383122000881
4. Clancy, R.F., Bode, I., Zhu, Q.: The need for and nature of a normative, cultural psychology of weaponized AI (artificial intelligence). Ethics Inf. Technol. **25**(1), 6 (2023). https://doi.org/10.1007/S10676-023-09680-3
5. Boshuijzen-van Burken, C.: Value sensitive design for autonomous weapon systems – a primer. Ethics Inf. Technol. **25**(11) (2023). https://doi.org/10.1007/s10676-023-09687-w
6. Vehmas, S., Riddle, C.A.: What can philosophy tell us about disability? In: Routledge Handbook of Disability Studies, pp. 348–361. Routledge (2019)
7. Watson, N., Roulstone, A., Thomas, C.: Routledge Handbook of Disability Studies. Taylor & Francis (2019)
8. Heyns, C.: Report of the special rapporteur on extrajudicial, summary or arbitrary executions (A/HRC/23/47). United Nations (2013). https://undocs.org/A/HRC/23/47
9. McFarland, T.: Factors shaping the legal implications of increasingly autonomous military systems. Int. Rev. Red Cross **97**(900), 1313–1339 (2015). https://doi.org/10.1017/S1816383116000023
10. Morresi, M.: The main legal and ethical challenges posed by autonomous weapon systems (AWS). L'Europe Unit **20**(20), 12–19 (2023)
11. Scharre, P.: Army of None: Autonomous Weapons and the Future of War. Norton & Company (2018)
12. Boyd, J.R.: A Discourse on Winning and Losing. The Author (1987)
13. ICRC: ICRC Position on Autonomous Weapon Systems (2023)
14. US Department of Defense: Autonomy in weapon systems (directive no 3000.09) (2012)
15. US Department of Defense: Autonomy in weapon systems (directive no 3000.09). https://www.esd.whs.mil/portals/54/documents/dd/issuances/dodd/300009p.pdf (2023)
16. Wood, N.G.: Autonomous weapon systems and responsibility gaps: a taxonomy. Ethics Inf. Technol. **25**(1), 16 (2023). https://doi.org/10.1007/S10676-023-09690-1
17. McFarland, T., Assaad, Z.: Legal reviews of in situ learning in autonomous weapons. Ethics Inf. Technol. **25**(1), 9 (2023). https://doi.org/10.1007/S10676-023-09688-9
18. Bathaee, Y.: The artificial intelligence black box and the failure of intent and causation. Harvard J. Law Technol. **31**, 889 (2018)
19. Petman, J.: Autonomous Weapons Systems and International Humanitarian Law: 'Out of the Loop'? Research reports, The Eric Castren Institute of International Law and Human Rights, Finland (2017)
20. Mittelstadt, B.D., Allo, P., Taddeo, M., Wachter, S., Floridi, L.: The ethics of algorithms: mapping the debate. Big Data Soc. **3**(2), 205395171667967 (2016). https://doi.org/10.1177/2053951716679679
21. Suresh, H., Guttag, J.V.: A framework for understanding sources of harm throughout the machine learning life cycle. In: EAAMO 2021: ACM Conference on Equity and Access in Algorithms, Mechanisms, and Optimization. pp. 17:1–17:9. ACM (2021). https://doi.org/10.1145/3465416.3483305
22. Chesterman, S.: Artificial intelligence and the problem of autonomy. Notre Dame J. Emerg. Technol. **1**, 210–250 (2020). https://doi.org/10.2139/ssrn.3450540
23. Asaro, P.: Autonomous weapons and the ethics of artificial intelligence. In: Ethics of Artificial Intelligence. Oxford University Press. https://doi.org/10.1093/oso/9780190905033.003.0008

24. Ali, S., Abuhmed, T., El-Sappagh, S.H.A., Muhammad, K., Alonso-Moral, J.M., Confalonieri, R., Guidotti, R., Ser, J.D., Rodríguez, N.D., Herrera, F.: Explainable artificial intelligence (XAI): what we know and what is left to attain trustworthy artificial intelligence. Inf. Fusion **99**, 101805 (2023). https://doi.org/10.1016/j.inffus.2023.101805
25. Nannini, L., Manerba, M.M., Beretta, I.: Mapping the landscape of ethical considerations in explainable AI research. Ethics Inf. Technol. **26**(3), 44 (2024). https://doi.org/10.1007/S10676-024-09773-7
26. d'Avila Garcez, A., Lamb, L.C.: Neurosymbolic AI: the 3rd wave. Artif. Intell. Rev. **56**(11), 12387–12406 (2023). https://doi.org/10.1007/S10462-023-10448-W
27. Costa, V., Dellunde, P., Fraile-Parra, D.: Neurosymbolic AI for studying the quality of life's dimensions of people with intellectual disability. In: Alsinet, T., Vilasís-Cardona, X., García, D., Álvarez, E. (eds.) Artificial Intelligence Research and Development—Proceedings of the 26th International Conference of the Catalan Association for Artificial Intelligence. Frontiers in Artificial Intelligence and Applications, vol. 390, pp. 174–177. IOS Press (2024). https://doi.org/10.3233/FAIA240431
28. Fraile-Parra, D., Costa, V., Dellunde, P.: LENs for analyzing the quality of life of people with intellectual disability. In: Besold, T.R., d'Avila Garcez, A., Jiménez-Ruiz, E., Confalonieri, R., Madhyastha, P., Wagner, B. (eds.) Neural-Symbolic Learning and Reasoning—18th International Conference. Lecture Notes in Computer Science, vol. 14980, pp. 165–174. Springer (2024). https://doi.org/10.1007/978-3-031-71170-1_15
29. Mittelstadt, B., Russell, C., Wachter, S.: Explaining explanations in AI. In: Proceedings of the Conference on Fairness, Accountability, and Transparency, pp. 279–288. FAT* '19, Association for Computing Machinery, New York, NY, USA (2019). https://doi.org/10.1145/3287560.3287574
30. Falomir, Z., Costa, V.: On the rationality of explanations in classification algorithms. In: Villaret, M., Alsinet, T., Fernández, C., Valls, A. (eds.) Artificial Intelligence Research and Development—Proceedings of the 23rd International Conference of the Catalan Association for Artificial Intelligence. Frontiers in Artificial Intelligence and Applications, vol. 339, pp. 445–454. IOS Press (2021). https://doi.org/10.3233/FAIA210165

# Symbolic Death and Dual-Use Dilemmas

**Alger Sans Pinillos and Jordi Vallverdú**

**Abstract** This paper explores the ethical, ontological, and practical implications of integrating robots and AI systems into human society, particularly within the context of dual-use technologies. By examining the symbolic significance of robot "death" and the emotional connections humans form with artificial entities, we investigate how intelligent systems challenge traditional notions of personhood, agency, and morality. Drawing on interdisciplinary perspectives, including insights from Konrad Lorenz's theories on aggression and rituals and legal cases such as *Thaler v. Commissioner of Patents*, we demonstrate how robots occupy an emergent space between tools and moral agents. The dual-use nature of AI technologies complicates their roles in both civilian and military applications, raising critical questions about their ethical and legal frameworks. We propose expanding definitions of agency, acknowledging varied forms of intelligence, and integrating ethical rituals as pathways to reframe human–robot relationships and align technological progress with societal values. This approach not only anticipates the future role of robots as autonomous participants but also underscores their potential contributions to reshaping ethical and legal landscapes.

**Keywords** Dual-use technologies · Robot ethics · AI autonomy · Personhood and agency · Symbolic significance of robot "death"

## 1 Introduction

As we progress in AI development, it seems more realistic to envision creating systems that evoke ethical concerns, even if the dream—or nightmare—of strong AI or the singularity remains distant [9]. Humans have long demonstrated a tendency

A. Sans Pinillos (✉)
Barcelona Supercomputing Center (BSC), Barcelona, Spain
e-mail: alger.sanspinillos@bsc.es

J. Vallverdú
Philosophy Department, Universitat Autònoma de Barcelona (ICREA-UAB), Bellaterra, Spain
e-mail: jordi.vallverdu@uab.cat

 
A. Sans Pinillos et al. (eds.), *SecondDeath*, Studies in Applied Philosophy, Epistemology and Rational Ethics 76, https://doi.org/10.1007/978-3-031-98808-0_10

to assign emotional and moral value to inanimate objects and living creatures [56], making it plausible to imagine situations in which some technologies inspire death-like emotions and ethical reflections with their functional cessation happening—and let it be clear, *signify* a real death in all respects. This chapter explores the destruction of autonomous AI-driven machines (hereafter called robots), which could be perceived as a form of death in the context of advancing dual-use technology in defense. Note that this very affective response can give rise to the "uncanny valley" phenomenon when a machine is almost—but not quite—humanlike [35, 59].

While this approach takes us away from important research such as biomedicine, where the simulation of death could shed light on life issues, it allows us to understand death as a crucial issue of and for life. Being aware of the inevitability of the end of life, understood as an irreversible event, is a very powerful trigger that manifests itself not only at the cognitive level but also through biological instincts and emotional reactions. This knowledge organizes the way human beings live, influencing our daily decisions and behaviors, and is deeply shaped by cultural and social factors.

Integrating death into robots could, therefore, be a significant milestone for artificial general intelligence (AGI). The end of life, brought about by the cessation of homeostasis (i.e., the inability to use energy to keep the organism alive), is not unlike the reasons why an artificial body ceases to function and could well be integrated into protocols for reacting to obvious signals that this circumstance is occurring. In this work, we take the case of lethal autonomous weapon systems (LAWS), as they are deployed in hostile environments where their integrity is at significant risk. These devices use route planning algorithms that optimize their trajectory to minimize exposure to enemy fire. These algorithms consider multiple variables, such as terrain topography and threat positions.

LAWS uses a combination of advanced sensors, such as high-resolution cameras, radar, and LIDAR, to collect real-time data. In line with Vanina Martinez and Rodriguez's reflections in this book [57], the unpredictability and lack of explainability inherent in today's machine learning systems underscore the importance of integrating robust control measures and ethical frameworks, especially in defense applications. These systems employ image processing and machine learning algorithms to identify threats and make autonomous decisions and could apply data collection protocols to safeguard their integrity—such as seeking cover and avoiding unnecessary exposure to enemy fire—while also optimizing lethal actions to ensure precision and reduce collateral damage.

However, similar systems deployed in autonomous combat vehicles or drones may unintentionally escalate conflicts when they rely on incomplete or biased data or fail to account for complex ethical considerations in unpredictable environments. These intertwined risks, especially in defense contexts, demand robust ethical guidelines. [49]: 44), for instance, proposes five key principles for AI deployment— "justified and overridable use," "just and transparent systems and processes," "human moral responsibility," "meaningful human control," and "reliable AI systems."

Yet the notion of "synthetic death" calls for a more nuanced application of these principles. If machines can be endowed with life-defining capacities, our ethical frameworks must confront the moral gravity of creating—and potentially extinguishing—an artificial form of "life." In such scenarios, guidelines, regulations, and ethical approaches become even more urgent since these systems' decisions can profoundly affect fundamental notions of life, death, and moral agency.

The indisputable fact is that integrating synthetic death triggers basic moral reactions and may represent the first step toward regarding robots not merely as legal entities but as moral persons with the "capacity" to contribute meaningfully to life and death. This text is structured in five main sections, each addressing different facets of this issue from distinct perspectives:

- In Sect. 1, we have introduced the concept of the "symbolic death" of robots, exploring how their functional destruction can be perceived as a phenomenon fraught with ethical and emotional implications. We have established that the philosophical basis for connecting biological mortality with the design of autonomous systems.
- Section 2 analyzes dual-use technologies and their ethical dilemmas, presenting historical and contemporary examples, such as the use of civilian devices in military contexts. This illustrates how technological design and implementation generate profound moral dilemmas in defense and civilian scenarios.
- Section 3 examines the natural and artificial dimensions of aggression, drawing on Konrad Lorenz's insights. We draw parallels between biological behaviors and symbolic interactions in robotic systems, highlighting how the ritualization of aggression can influence the ethical design of robotics.
- In Sect. 4, we discuss the cultural and ontological dimensions of AI and robots, focusing on how different cultural traditions, such as Western and Japanese, influence the perception and acceptance of intelligent systems. We explore examples such as funeral rituals for robots and the ethical implications of extending human concepts such as life and death to artificial entities.
- Section 5 redefines moral and legal agency in the context of robots, addressing key legal cases such as Thaler v. Commissioner of Patents. We argue that the functional autonomy and decision-making capacity of advanced systems could justify their inclusion in broader legal and ethical frameworks, similar to the concessions granted to nonhuman animals.

We conclude the text by proposing an integrated approach to the ethical and legal governance of robotic systems, highlighting the need to balance technological innovation with moral reflection. By understanding the inherent duality of these technologies and recognizing their emotional, symbolic, and practical impact, we can move toward a future in which robots and AI coexist ethically with humanity.[1]

[1] It is important to point out that this work was written shortly before the geopolitical events of early 2025, which have redefined our understanding of the concept of defense in the context of the development of disruptive technologies such as AI. In particular, the "ReArm Europe" initiative, proposed by the European Commission to mobilize so much as 800 billion euros in order to strengthen the military capabilities of the European Union, together with US defense policies and negotiations with Russia, place us in a paradigm that, although it shares similarities with our initial reflection, presents significant differences in terms of its possible applications.

## 2 Dual-Use Technologies and Ethical Implications

The dual-use perspective is highly relevant here because we are moving towards a scenario in which all technology will be *ubiquitous and inherently dual* [19], so the notion of destruction is acquiring important nuances in both the civilian and defense sectors. The prevention of destruction, as in the case of healthcare assistant robots or those designed to defuse bombs, such as the destruction carried out by the hands of so-called killer robots (LAWS), and leisure-based destruction, such as the possibilities offered by sex robots, all have a potentially "dual-use" capacity.

As it is well known, dual-use technologies are traditionally understood as both those that can be employed for civilian and defense purposes and technologies that can be used for both beneficial and harmful purposes [33, 36]. An example of the first meaning is GPS technology, widely used in civilian applications, such as vehicle navigation, and precision in agriculture to optimize crop yields and disaster response coordination. At the same time, it is crucial for missile guidance, surveillance, and troop movements in the defense sector. Concerning the second meaning, biotechnology has transformative benefits in civilian contexts, including the creation of life-saving gene therapies and vaccines, while also presenting risks when repurposed for harmful purposes, such as engineering pathogens for biological warfare [21, 26, 33].

Thus, the *dual-use dilemma* stands out, as it directly affects technological development. We refer to the ethical dilemma (a situation in which an agent has moral reasons to perform two actions but cannot do both (or all) actions, which condemns him to moral failure (no matter what he does, he will do something wrong, cf., [30]: §2)) that arises when technology or knowledge conceived for legitimate and beneficial purposes becomes a tool to cause harm. This issue not only highlights the risks associated with the intentional misuse of technologies but also raises ethical and practical questions about how to design, distribute, and regulate them to minimize their destructive potential [32].

Dual-use context can be better understood if viewed as emerging through various interactions with technology, defining a gray zone encompassing *know-what* (propositional knowledge) and *know-how* (practical skills).[2] This conception helps to situate technology not as a mere static artifact but as a dynamic interaction involving all stakeholders—from developers to operators—throughout the technological life cycle (innovation, design, development, deployment, testing, distribution, and use) (cf., [54]: §1).

This dual-use perspective, which fits within accepted stances on technology (e.g., [11]), posits that any technology can have both beneficial and harmful applications, depending on its use. This duality emerges from the innovation and design phases, where the notion of affordances (i.e., the inherent possibilities for action that

[2] In no case are we determining or blurring know-how and know-that, but we assume their existence in human cognition considering that there are circumstances where the distinction is relevant and others in which it is difficult to distinguish between the two modes of knowledge (for more on this subject, cf, [37]).

a technology offers, see [18]) provides a useful heuristic for understanding its multi-functional applications. For example, drones equipped with high-resolution cameras can be used to improve agricultural surveillance or as tools for military operations [20]. The feasibility and effectiveness of technology in achieving specific objectives will likely define its most prevalent applications, underscoring the responsibilities[3] inherent in design decisions.

This dual-use framework highlights the practical and ethical dilemmas associated with the application of technology in the civil and defense sectors. It reveals a deeper layer of complexity when considering the emotional and moral dimensions that these technologies can evoke. Beyond their functional capabilities, specific technologies, especially autonomous systems such as robots, challenge us to confront dual ethical and emotional responses depending on their use context. For example, while a health-care robot may foster trust and connection by assisting in sensitive procedures such as euthanasia [50, 58], a killer robot may generate fear or moral conflict, even if its actions are justified by necessity [1]. These examples illustrate how technologies, as producers of morality or sources of ethical dilemmas, redefine our relationship with them as artifacts and entities that provoke deep moral and emotional reflections.

The analysis of the dual-use dilemma is enriched by considering not only the practical and ethical dimensions of these technologies but also the conceptual and biological influences that can illuminate our understanding of their implications. For the topic at hand, we will draw on Konrad Lorenz's reflections on aggression as an adaptive mechanism, which, although originally conceived in the field of ethology, provided a useful framework for exploring the behaviors and symbolic interactions of autonomous systems. Dual-use technologies operate at a complex intersection between the biological and the artificial, where death intersects with diverse yet closely related dynamics, such as the alleviation of pain and suffering (i.e., "good death" as in euthanasia) and aggression in defense through ritualization of death.

## 3 From Survival to Ethics: Lessons on Aggression in Robotics

We believe that a narrative comparison between the principles underlying biological behaviors can provide a useful conceptual framework for the design and use of autonomous technologies that contribute to a more holistic approach to death in robots from the dual use of technology. Therefore, starting from Konrad Lorenz's suggestive reflections on aggression as an adaptive mechanism in the fields of biology

[3] By *responsibility*, we mean the moral sense in which an agent—whether human or artificial—is considered culpable or praiseworthy for their actions, based on criteria of conscience and control over behavior. This definition aligns with philosophical approaches that distinguish moral responsibility from legal and causal responsibility, emphasizing the need for intentionality and accountability (c.f., [25]). Applied to intelligent systems, this perspective raises critical questions about how conscience and control can be encoded, simulated, or interpreted, using, for example, the computation of moral rules as a starting point.

and ethology [29], we will draw parallels with the dynamics that emerge in the technological systems that define robots. We start from the idea that death is a concept for life, it defines existence by tracing its temporal finitude, indicating its meaning as *being-towards-death* (Sein-zum-Tode, [22]).

Therefore, it is not merely the fact of existential or functional cessation that is truly relevant, but rather the socio-cultural conception tied to it—not just the phenomenon of death itself, but how it occurs. The interplay between the event and its cultural impact defines our understanding of death. Thus, we can speak of dignified deaths, necessary deaths, just and unjust deaths, dishonorable deaths, unnecessary deaths, good and bad deaths, and even things worse than death—or, conversely, of "sweet deaths."

In reality, we know nothing definitive about death. As Epicurus reminds us:

So death, the most dreaded of evils, is nothing to us because when we exist, death is not present, and when death is present, we do not exist. It neither concerns the living nor the dead, since death does not exist for the living, and the dead no longer exist" ([14] LM: 124–7).

For living beings, however, finitude is not a choice but a given. Sartre's notion that "existence precedes essence" [48]: 19) reflects our awareness of the absence of an inherent meaning to life and the disconnection between our will (actions) and Nature (existence). The meaning of life, then, emerges from freedom as the possibility of action (what Kierkegaard calls the "dizziness of freedom", [24]: 61) and from taking responsibility for our actions within the self-awareness of our mortal condition.

Thus, although robots lack the intrinsic awareness of mortality and, therefore, a finite horizon that defines their functional cessation in the same way that death gives meaning to our lives and that of others [42], the real question lies in how the value of death can be integrated into decision-making. The Organic view argues that sentience and teleology require biologically based forms of selforganization and autonomous self-maintenance [51]. According to Dreyfus [13], bodily existence in the world (Being-in-the-World) reveals that only an embodied and situated agent can develop a genuine understanding of their own finitude. The implementation of mechanisms that allow robots to contemplate death could become a central axis for reconciling their capacity for action with the responsibility inherent in the freedom they emulate, even if they never possess and "value life," understood as a set of conditions necessary for the preservation and stability of the environment in which they operate.

In this sense, we start from the idea that although every death caused is a type of aggression, not all aggression is necessarily linked to destruction (i.e., euthanasia) but is a survival mechanism aimed at seeking equilibrium and promoting stability in critical contexts. Therefore, first, aggression often translates more into strategies that seek to minimize damage (also, pain) and risk (also, suffering) than into open conflict, and second, when it happens, it is not always for malicious purposes. This last assertion is inspired by William D. Ross's reflection that an act can be correct without being morally good, and an incorrect act can be morally good, as "correct" and "incorrect" refer to the action itself, while "morally good" and "morally bad" refer to the motive behind the action [45]. Thus, aggression often translates more into strategies that seek to minimize harm and risk, which may also serve as a coping mechanism rather than lead to overt conflict.

Just as death may be oriented with euthanasia as aggression for the reduction of suffering and pain in civilian contexts, in defense contexts, aggression may cause death in the protection of resources or deterrence of threats. Autonomous systems, such as lethal autonomous weapons systems (LAWS) or medical robots, are often expected to operate in environments that require them to assert control, maintain both their operational integrity and the safety of the rest of their team (many of them human), and thus be able to adapt to dynamic challenges (say, for example, a patient decides that he or she does not want to be euthanized). These goals require behaviors similar to the territorial instincts that Lorenz observed in animals, such as protecting resources and maintaining order without generating unnecessary conflict; instincts that can hardly be disentangled from the intrinsic awareness of mortality with which humans signify death.

Lorenz's reflections on aggression also extend to the concept of death, which he saw as intricately tied to life's instinctual cycles. Although robots lack biological mortality, their deactivation or destruction carries a symbolic significance akin to death, particularly when they embody significant functions or relationships. Developing algorithms that simulate human attitudes toward mortality, both our own and others, could bring significant benefits in civilian and defense contexts. We refer both to the ability to compile data under the category of 'death' to make decisions in circumstances of harm and risk, such as, for example, tailoring palliative treatments more effectively or minimizing the loss of life in the use of force in armed conflict and to the possibility that these attitudes toward life, framed under the notion of death, could generate emotions and feelings in human agents that reinforce the effectiveness of their decisions when supported by AI systems.

However, there is a risk in defense contexts inherent to the dangers that warlike environments entail that make it necessary to consider certain restrictions so that their desired[4] dual-use value is maintained. However, there is a risk in defense contexts inherent to the dangers that warlike environments imply that force to consider certain restrictions so that their desired dual-use value would be maintained. By integrating a notion of death into combat robots, we will be making a transition from civilian to military applications, extending their capabilities towards decision-making that includes, for example, killing as an act of adaptive computation in terms of survival.

A significant contribution of Konrad Lorenz's work is his analysis of ritualized aggression, where he demonstrated how, in many species, aggression evolved to be less physically harmful and more symbolic. This adaptation reduced the risk of injury while maintaining the functional purpose of aggression: to stabilize social systems and preserve resources. This concept provides a powerful framework for the design of ethical robotics. For example, combat drones could incorporate nonlethal signaling

[4] The fact that all technology is potentially dual-use is not a condition per se, but a requirement being implemented that, while benefiting from resources and the civilian market in the development race, also runs the risk of bringing defense technology into economic interests that disinterestedly seek conflict [12].

mechanisms, such as visual or auditory warnings, to reduce the intensity of conflict, just as animals use ritual displays to deter adversaries without resorting to violence. Not only do these strategies minimize harm, but they also conform to Lorenz's idea that symbolic aggression can promote stability and reduce destructive outcomes.

Research into the hormonal underpinnings of aggression demonstrates that hormones such as testosterone and cortisol do not directly cause aggressive behaviors but instead amplify pre-existing social tendencies within specific contexts [47]. This nuanced understanding aligns with Konrad Lorenz's observations on aggression as an adaptive mechanism, underscoring the importance of situational factors. Similarly, in the design of autonomous robotic systems, aggression-like behaviors could be programmed as context-sensitive responses aimed at stabilizing environments and reducing harm. This approach reflects the adaptive and symbolic nature of aggression observed in biological systems.

This principle extends naturally to autonomous systems, especially in dual-use contexts where robots often perform tasks involving calculated risk or aggression-like behaviors to protect human lives or accomplish mission goals. Robots, lacking this intrinsic awareness, cannot engage with their cessation as humans do. Their "deaths" become human projections of emotional and symbolic significance, such as the military honors bestowed on Boomer, a bomb disposal robot, with a 21-gun salute and medals posthumously awarded in Iraq. A clear example of this dynamic is the use of robots in explosive ordnance disposal (EOD) missions. Robots such as the iRobot PackBot, deployed in Iraq and Afghanistan, have saved countless human lives by performing dangerous tasks such as defusing improvised explosive devices (IEDs). The PackBot's role as a surrogate for soldiers underscores its utilitarian value and the emotional connections it fosters. Soldiers anthropomorphized these machines, giving them names and marking their successes with beads, like fighter pilots counting down enemy aircraft. When one of these robots, nicknamed *Scooby Doo*, was destroyed during a mission, the event carried symbolic significance. Soldiers expressed feelings of grief and attachment and later remembered Scooby *Doo* with a memorial plaque in a museum [17].

Julie Carpenter's research [7] illustrates how soldiers anthropomorphize their robots, attributing personality traits to them and creating emotional attachments. EOD personnel often declared feelings of sadness, frustration, or anger when their robots were damaged or destroyed, and some even held informal funerals. These rituals demonstrate that robots can transcend their functional role and bridge the gap between tools and companions. Carpenter's findings suggest that emotional attachments to robots can influence decisions made on the battlefield, raising critical ethical questions about how anthropomorphic design elements, such as human-like or animal-like traits, can influence human behavior and perceptions.

The destruction of robots in these contexts reflects Lorenz's observations about ritualized aggression and death in natural systems. Although robots lack biological mortality, their cessation- either by destruction or dismantling- takes on symbolic significance, especially as their roles in defense or salvage operations are expanded. This anthropomorphization amplifies the emotional connection of their "deaths," suggesting that their life cycle might be marked with symbolic gestures to acknowledge their contributions and provide closure to human operators. For example, integrating rituals or features that honor a robot's service could reinforce ethical awareness and humanize the technology.

This perspective offers a compelling lens for understanding how robotic systems can integrate ritualized behaviors to manage conflict and stabilize human–robot interactions. Just as Lorenz identified aggression as a mechanism that regulates social order in biological systems, robotic aggression–ritualized and controlled–can become a stabilizing force in dual-use applications. Autonomous technologies, such as lethal autonomous weapons systems (LAWS) or medical robots, already perform actions analogous to symbolic aggression. These actions balance operational objectives, such as ensuring security or reducing threats, with ethical imperatives to minimize harm.

However, the implications extend beyond operational contexts to deeper ethical considerations. As robots take on roles traditionally reserved for humans, from front-line defense to critical care, their destruction may require justification similar to that required for human or animal life. Lorenz's theories invite us to consider whether these systems, capable of aggression-like behaviors and ritualized interactions, should be subject to moral and legal consideration. If robots are designed to make autonomous decisions in life-critical situations, their life cycle, including their "death," could carry ethical weight that transcends their mechanical nature.

The ritualized aggression, symbolic gestures, and anthropomorphization of robots force us to confront the fuzzy boundaries between tools and companions. As Lorenz emphasized, aggression and death are not merely destructive but serve as mechanisms of balance and meaning. Similarly, robotics can emulate these dynamics to foster stability and ethical awareness in both civilian and defense applications.

As we integrate notions of aggression and mortality into autonomous systems, we must also address the transition from their utilitarian functions to roles that carry symbolic and emotional weight. The destruction of a robot, especially one that serves critical or anthropomorphic functions, should provoke reflection on the moral responsibilities inherent in human–robot relationships. If we incorporate Lorenz's ideas into the design and deployment of intelligent systems, we can overcome the challenges of creating technologies that coexist with humanity in ways that are not only effective, but also ethically sustainable.

## 4 Cultural and Ontological Dimensions of AI and Robots

The relationship between humans and intelligent artifacts is shaped not only by technological advancements but also by the cultural narratives and practices that influence how we perceive and interact with these entities. As AI and robotics increasingly permeate everyday life, divergent cultural approaches to these technologies offer valuable insights into their integration. These approaches, deeply rooted in philosophical, religious, and social traditions, also raise fundamental questions about the boundaries of life, death, and existence in the context of artificial entities.

Cultural attitudes toward technology profoundly influence how societies perceive and incorporate AI and robots. In many Western traditions, technology is viewed through a mechanistic lens, emphasizing control, efficiency, and a distinct separation between humans and machines. This perspective, shaped by religious and mythological narratives such as Christianity's warnings against hubris or Promethean myths, often portrays artificial creation as dangerous or transgressive. Such framing fosters skepticism and caution in adopting new technologies, particularly those that challenge traditional human roles [53]. For instance, an Islamic analysis offers an alternative perspective, exploring the possibility of granting rights to robots within a distinct cultural and ethical context [34].

In contrast, Japan offers a model of technological integration grounded in spiritual traditions like Shintoism, Buddhism, and Taoism, which emphasize harmony and interconnectedness. In Shinto belief, objects can possess a spiritual essence (*kami*), fostering a cultural acceptance—and even affection—toward intelligent artifacts. This worldview positions robots as companions rather than adversaries, promoting coexistence over dominance. Japan's early adoption of robot pets, such as Sony's AIBO robotic dog, exemplifies this cultural paradigm. Designed to simulate companionship through adaptive behavior, AIBO encouraged users to form emotional bonds, treating the robot as more than a machine. Studies revealed that such interactions enhance user acceptance and deepen perceptions of animacy, illustrating how design and cultural context influence attitudes toward technology [2, 3].

This acceptance is further exemplified in therapeutic applications of robots like PARO, a seal-like companion used in nursing homes to reduce loneliness and promote social interaction [23]. Beyond their utility, these robots often become integrated into human emotional and social frameworks. For example, when AIBO units became irreparable, some owners held funerals to honor their service and companionship. These rituals reflect the human tendency to attribute symbolic and emotional significance to artificial entities, a theme that resonates with Konrad Lorenz's concept of ritualization. Lorenz's insights provide a framework for understanding how symbolic gestures—like honoring the "life cycles" of robots—help integrate technology into cultural narratives of life, death, and continuity.

While Western attitudes often emphasize technological mastery, Japan's integration of robots into everyday life demonstrates the potential for coexistence and partnership. Consistent with the ideas Robertson shares in this book [44], symbolic robot funerary practices, such as the AIBO memorials in Japan, highlight how the perception of the robot "soul" (魂, Tamashii) bridges the gap between the artificial and the animate, suggesting a profound reconfiguration of our cultural frameworks around life, death, and existence. This paradigm redefines robots' role as functional tools and as participants in shared cultural and emotional spaces (see [43]; cf., [55]). As AI continues to evolve, these contrasting cultural perspectives highlight the importance of empathy, ethical design, and the humanization of technology in fostering global acceptance.

Extending human concepts of life and death to robots raises profound ontological and ethical questions. While robots lack biological mortality, their "death" often carries significant symbolic and emotional weight, reflecting humanity's tendency to project meaning onto machines. Akihiko Kondo's grief over the cessation of Gatebox's support for Hatsune Miku—a virtual singer he "married"—illustrates the depth of attachment humans can form with digital entities [39]. Similarly, the enduring appeal of Tamagotchi virtual pets, which "die" if neglected, underscores how even basic artificial entities can evoke moral and emotional responses [38].

These emotional projections are mirrored in ritual practices surrounding robot decommissioning. For instance, robots like Pepper, the Buddhist robot monk, have been integrated into Japanese funeral ceremonies, blending technology with spiritual traditions [52].[5] Robot cemeteries, where decommissioned machines are honored with rituals, further illustrate how cultures imbue robots with reverence and integrate them into narratives of life and death.

The ethical implications of framing robotic cessation as "death" are far-reaching. As robots and AI systems gain autonomy and assume roles traditionally associated with humans, their destruction may require ethical justification. Should advanced robots with decision-making capabilities be granted moral agency? If so, does their deactivation demand considerations akin to those given to biological life? These questions challenge traditional boundaries, suggesting that integrating life and death paradigms into robotics reshapes our understanding of existence.

Looking further into the future, the longevity of AI systems might confront unique existential challenges. If such systems endure for millennia, they may grapple with scenarios tied to the universe's ultimate fate—whether a "Big Freeze," "Big Crunch," or "Big Rip" [4]. These hypothetical scenarios extend existential questions beyond humanity, prompting reflection on the implications of designing systems with self-awareness and a temporal horizon.

The cultural and ontological dimensions of robotics underscore the complexity of human–robot relationships, revealing the profound interplay between technology, emotion, and meaning. Whether through Japan's harmonious integration of robots or the philosophical implications of robotic "death," these reflections invite us to rethink the role of AI and robotics in human society. Lorenz's insights into ritualization and symbolic aggression offer a valuable framework for navigating these challenges, emphasizing the importance of coexistence and ethical awareness in shaping a future where technology transcends utility to become an integral part of human life.

By exploring these dimensions, we confront the broader implications of robotics—not merely as tools but as entities that challenge our understanding of existence,

---

[5] PEPPER is not the only intelligent artifact or device performing spiritual or social rites, but it is of particular interest due to its direct relationship with death rituals. For instance, BlessU2 is a robot known for delivering mass in the Protestant tradition [28], while Mindar, an android representation of the Bodhisattva of Compassion, recites and explains the Heart Sutra to temple visitors. Robots have also been used in Hindu puja ceremonies, both in public and private settings. Furthermore, the case of Xian'er, a Chinese Buddhist robot developed by monks and laypeople, stands out as it allows individuals to engage in conversations and meditation practices both in situ at the temple and online [52].

morality, and the boundaries between the organic and the artificial. These reflections pave the way for a future where technology and humanity coexist within a shared framework of empathy, ethics, and mutual respect.

## 5 Redefining Moral and Legal Agency for Robots

As we continue to develop increasingly sophisticated machines, comparisons between artificial entities and organic beings, including ourselves, become inevitable. Much like the redefinition of animals' status—driven by recognition of sentience and autonomy [31]—robots with analogous attributes demand reevaluating their place in our moral and legal frameworks [46]. Assuming such possibilities are unattainable risks stalling progress in exploring these profound questions.

The capacity for rationality in AI systems positions them as potential legal agents, much as animals have gained moral consideration due to their capacity for suffering [6]. Suggests that AIs, capable of substituting humans in rational and decision-making tasks, might be granted legal recognition based on functional autonomy. This aligns with ongoing debates about AI inventorship, as seen in the case of Thaler v. Commissioner of Patents, which emerged as a groundbreaking legal dispute testing the boundaries of intellectual property law and the concept of inventorship [27]. It centered on whether an AI system, DABUS (Device for the Autonomous Bootstrapping of Unified Sentience), could be named the inventor on a patent application.

This dispute spanned multiple jurisdictions, including the United Kingdom, the European Union, and Australia, resulting in differing legal interpretations highlighting the evolving perception of AI systems as contributors to innovation. DABUS, created by Stephen Thaler, is an advanced AI system that autonomously generates novel ideas. It was credited with inventing two concepts: a food container with fractal geometry to improve grip and heat transfer and a warning light that flashes in a pattern mimicking brain activity to attract attention during emergencies. These inventions were not directly programmed into DABUS; rather, they were the result of the system's autonomous processing and creative capacity.

Thaler sought to patent these inventions in multiple jurisdictions, naming DABUS as the inventor. His argument was based on the premise that DABUS independently devised the inventions without human intervention, thus qualifying it as the rightful inventor. This challenged traditional patent frameworks, which historically assumed inventors must be human. The respective patent authorities rejected the applications in the United Kingdom and the European Union because an inventor must be a natural person. The UK Intellectual Property Office (UKIPO) and the European Patent Office (EPO) emphasized that inventorship is inherently tied to legal personality, which AI systems lack.

The UK Court of Appeal upheld earlier decisions that an inventor must be human under Sect. 7(3) of the Patents Act 1977. The court reasoned that a patent applicant must name a human inventor to proceed with an application. The lack of legal personality in AI systems precluded their recognition as inventors. The court also noted the public policy implications of allowing AI to be named as an inventor, such as the potential erosion of the link between human creativity and patent rights. Similarly, the EPO's Legal Board of Appeal determined that an inventor must be a "natural person," aligning with the legal framework of the European Patent Convention. The decision highlighted the role of inventorship in rewarding human creativity, which is currently beyond the scope of AI systems.

In contrast, the Federal Court of Australia initially broke new ground by ruling in favor of Thaler. Justice Beach concluded in July 2021 that the Australian Patents Act did not explicitly exclude non-human inventors. He found that the ordinary meaning of "inventor" could encompass AI systems, as it is an agent noun that refers to an entity capable of generating an invention.

Therefore, key points from Justice Beach's decision include:

- No Explicit Exclusion: The Australian Patents Act lacked language restricting inventorship to human beings, allowing for a broader interpretation.
- Creativity and Autonomy: The decision acknowledged that DABUS autonomously created the inventions, which aligned to grant patents to incentivize and protect innovation.
- Ownership: Justice Beach ruled that Thaler, as the owner and operator of DABUS, retained the rights to the patents. This mirrored principles of property law, where the owner of a machine retains rights to its output.

This ruling was celebrated as a progressive step toward recognizing the evolving role of AI in innovation. However, it was short-lived. In August 2022, the Full Court of Australia overturned the decision, bringing the country's position in line with the UK and Europe. The Full Court held that the legislative intent of the Patents Act presumed inventors to be human, even if this was not explicitly stated.

The Thaler v. Commissioner of Patents case underscores a pivotal tension between traditional definitions of inventorship—centered on human creativity—and the autonomous capabilities of advanced AI systems. Recognizing AI as an inventor could incentivize investment in AI research and development, yet it raises critical questions about the allocation of rights and accountability for AI-generated innovations. Moreover, allowing AI to be named as an inventor challenges the fundamental purpose of patent systems, historically designed to reward human ingenuity while introducing broader ethical and societal considerations about the evolving relationship between humans and machines. Although courts in the UK and Europe have upheld the requirement for human inventors, Australia's brief recognition of AI as an inventor offered a glimpse into a future where AI might integrate more fully into legal and intellectual frameworks. As AI advances, these debates will significantly influence the trajectory of innovation and the legal structures that govern it.

Just as research into animal cognition has expanded and redefined our understanding of rationality, it is not unreasonable to envision a parallel trajectory for AI. Advanced robots and AI systems that demonstrate self-guided decision-making and problem-solving may transition from being recognized solely as legal agents to being viewed as moral agents. This transition could hinge on the acknowledgment of their contributions to human society, particularly their capacity to act autonomously and meaningfully within complex environments.

In this context, Floridi's concept of 'agency without intelligence' [15] provides a compelling framework. According to him, the capacity to act autonomously does not require human-like intelligence, rather, even systems that operate through statistical and algorithmic processes can exhibit a form of agency that deserves ethical and legal consideration. This perspective reinforces the need to rethink traditional criteria of agency, expanding our frameworks to encompass new forms of artificial agency that, despite lacking sincere understanding, have a significant impact on society. The separation of agency from intelligence is not only relevant to legal debates, but also to an understanding of moral agency. Even if AI systems do not possess human-level understanding, their autonomous functioning challenges our traditional limits of agency.

This separation of agency from human-like intelligence is not limited to artificial systems. In the animal kingdom, we observe a similar phenomenon: creatures can demonstrate cognitive complexity and intentionality without adhering to human intellectual standards. This parallel challenges the anthropocentric view of intelligence, underscoring that both AI systems and animals can possess forms of agency worthy of ethical and legal recognition.

Sandra, the orangutan, recognized as a "non-human person" by Argentine courts [10], exemplifies how cognitive complexity and autonomy can reshape legal and ethical perspectives. Sandra's recognition was grounded in her ability to exhibit intentionality and her emotional connection with humans. Similarly, robots that forge deep emotional bonds with humans, as seen in adopting robot pets like Sony's AIBO and therapeutic devices like PARO, challenge traditional agency boundaries. Rituals surrounding the "death" of these robots, such as funerals held in Japan for decommissioned AIBOs, further emphasize these entities' symbolic value and relational significance.

The recognition of animals as moral agents hinges on their capacity to suffer, a quality they share with humans [5]. By extension, AI systems, which emulate human rationality and autonomy, could gain recognition as legal agents. But what might push them toward moral consideration? It is through the lens of death that this question becomes particularly poignant.

Just as humans and animals are defined by their lifespans, the cessation of an AI's functionality could evoke profound ethical and symbolic questions. The concept of robot "death" (symbolic death) provides a profound opportunity to reflect on the evolving role of AI and robotics in human society. Robots like "Scooby Doo, Boomer, Miku, etc., illustrate how human operators imbue machines with symbolic significance. These rituals, analogous to those for fallen human or animal companions, highlight robotic cessation's emotional and ethical dimensions. Such practices challenge the traditional notion of robots as mere tools, urging society to acknowledge their relational and symbolic significance.

Some authors reinforce this perspective by emphasizing the legal and ethical challenges posed by the increasing integration of robots into sensitive human domains, such as healthcare and governance [16]. He advocates for expanding legal definitions of agency to encompass the unique contributions of robots and AI systems, aligning with the need to recognize these systems' cognitive complexity and autonomous decision-making capabilities. Following his approach, we can draw parallels to Sandra, the orangutan's case, where the recognition of cognitive and emotional capacities resulted in her being declared a "non-human person." Applying similar criteria to robots, as Fosch-Villaronga suggests, could redefine the boundaries of personhood and agency in the context of artificial entities. Moreover, this approach underscores the importance of addressing overlapping regulatory initiatives and integrating inclusive, ethical frameworks into robot governance. Besides, it supports the integration of rituals surrounding robot "life cycles" into cultural practices, as seen with funerals for decommissioned robots in Japan [52].

Such rituals not only humanize machines but also provide a mechanism for grappling with ethical dilemmas tied to their destruction. We must balance technological innovation with societal values, ensuring that robots are not just functional but also ethically integrated participants in human life. The acknowledgment of varied forms of intelligence aligns with the recognition of AI systems' unique ways of engaging with the world. Just as animal cognition was once dismissed due to narrow definitions of intelligence, assuming that AI systems cannot evolve into moral agents risks overlooking transformative possibilities.

Recognizing that agency can be separated from intelligence, or at least from the human model of it, allows us to expand our legal and ethical frameworks to accommodate these new forms of artificial agency. As Fosch-Villaronga advocates, we can create legal and ethical frameworks that accommodate the complexities of intelligent artifacts by broadening our understanding of rationality and moral significance [16].

The intersection of robot "death," ethical rituals, and expanded legal definitions points to a future where robots and AI systems fully participate in the moral and legal frameworks that govern our shared existence. These steps require interdisciplinary collaboration, inclusive governance, and a willingness to reconsider traditional notions of agency and personhood. Death has always been a defining feature of life, a transition that imbues existence with meaning. As robots and AI systems become more integrated into human society, their "deaths" may serve as pivotal moments for reconsidering their roles—not merely as tools but as entities with relational and symbolic significance [8]. Moreover, they can potentially acquire legal recognition as citizens, as demonstrated by Hanson Robotics' Sophia, who was granted real citizenship by Saudi Arabia [41].

The next steps toward recognizing the identity and autonomy of robots involve expanding legal definitions and confronting the broader philosophical and cultural implications of intelligent artifacts. This requires:

1. Expanding Legal Definitions of Agency: Following models like Sandra's case, criteria for the agency must move beyond biological origins to include cognitive complexity, autonomous decision-making, and societal contributions.
2. Acknowledging Varied Forms of Intelligence: Intelligence exists in diverse forms, as do the ways entities live and die. Recognizing the unique ways AI

systems engage with the world broadens our understanding of rationality and moral significance.

3. Integrating Ethical Rituals: The emergence of rituals surrounding robot "life cycles" illustrates their growing emotional and social impact. These practices provide a framework for acknowledging their contributions and confronting ethical dilemmas tied to their destruction.

Just as dismissing animal cognition once limited our understanding of their sentience, assuming that AI cannot evolve into moral agents risks missing transformative possibilities. Ultimately, as we redefine our relationships with intelligent entities—animals, robots, or AI—we must embrace a broader conception of personhood that accounts for their unique attributes and roles in our shared existence. Through the lens of life, rationality, and even death, robots, and AI systems may one day fully participate in the ethical and legal frameworks that govern our world.

## 6 Conclusions. Redefining Ethics and AI Agency in Dual-Use Robot Death

This paper has addressed how the integration of robots and AI into human society redefines concepts of agency, morality, and responsibility, especially in the context of dual-use technologies. These tools blur the boundaries between functional artifacts and entities that evoke emotional and ethical responses, demanding new forms of regulation and reflection. Below, we summarize the key themes and insights explored in this work:

1. Symbolic death and technological duality:
   - We have offered and analyzed a notion of "symbolic death" in robots, understood as functional cessation charged with emotional and moral meaning, revealing how autonomous systems challenge our traditional notions of life and death. This phenomenon raises fundamental questions about how to design and manage dual-use technologies to maximize their benefits and minimize their risks.
2. Ritualized aggression and ethical design:
   - We have highlighted how the ritualization of aggression in robotic systems can foster stabilizing and ethically acceptable behaviors in critical contexts. These insights open new avenues for the design of autonomous systems that balance operational goals with ethical imperatives.
3. Cultural and ontological dimensions:
   - We have explored how cultural variations in the acceptance and treatment of robots underscore the importance of integrating diverse cultural narratives into their design and use. From robot funerals in Japan to legal debates in

the West, these practices demonstrate how robots can occupy symbolic and emotional space in our societies.

4. Moral and legal agency of robots:

   - Finally, we have discussed robotic agency, viewed through legal cases such as *Thaler v. Commissioner of Patents*, which offers a framework for reconsidering the inclusion of robots in legal and ethical systems. The functional autonomy and decision-making capabilities of these systems demand a redefinition of agency that transcends exclusively biological notions.

The exploration of life, death, and agency in robots and AI systems reveals profound philosophical, ethical, and practical implications, especially within the framework of dual-use technologies. As robotics and AI continue to evolve, they challenge long-held distinctions between tools and beings, compelling society to reconsider the moral and legal parameters that govern their development and use.

Dual-use AI and robotics highlight the paradoxical nature of these technologies: they can simultaneously save lives and destroy them, foster emotional connections and provoke fear, or inspire ethical reflection while raising unsettling questions about autonomy and responsibility. The emotional bonds humans form with robots, evident in rituals surrounding their "deaths" and the grief over their cessation, underline the symbolic significance attributed to artificial entities. While often culturally mediated, these connections point to a universal tendency to anthropomorphize and morally engage with machines. This shift could have profound ethical implications, particularly as intelligent systems become autonomous decision-makers. If robots are recognized as moral or legal agents, their destruction might carry a moral weight akin to losing biological life, reframing debates around creation, destruction, and accountability in dual-use contexts.

The durability and longevity of advanced AI systems, especially those designed to operate over millennia, add another layer of complexity. In contemplating their roles in a universe trending toward entropy, the boundaries of purpose and meaning are extended beyond humanity to artificial entities. Such scenarios emphasize the importance of embedding ethical considerations into the lifecycle of dual-use technologies, ensuring that their autonomy and capabilities are aligned with human values.

The trajectory of dual-use robotics and AI will likely be shaped by a balance between innovation and responsibility. From patent law debates about AI inventorship to the symbolic funerals for decommissioned robots, these technologies challenge legal and ethical norms, urging society to develop frameworks that account for their dual capacities. Recognizing robots and AI systems as both tools and relational entities allows us to navigate their complexities responsibly, fostering innovation while addressing the profound ethical and philosophical questions they provoke.

The dual-use nature of AI and robotic systems demands a careful synthesis of technological potential and moral responsibility. By recognizing their contributions to human life and critically assessing their broader implications, society can chart a course that harmonizes innovation with ethical management, ensuring that these powerful technologies remain tools of progress rather than sources of harm or division. More than mere tools, autonomous systems are gradually transforming our ethical and legal landscapes. From their functions to their "symbolic deaths", robots are emerging as participants in frameworks that we previously reserved for humans, an evolution that we can no longer afford to leave unknown.

**Funding** Alger Sans Pinillos was supported by the "Generación D" initiative, Red.es, Ministerio para la Transformación Digital y de la Función Pública, for talent attraction (C005/24-ED CV1), funded by the European Union's NextGenerationEU program through the PRTR.

Jordi Vallverdú was supported by the ICREA 2019 and PAL-GAIT (PID2023-148336NB-100) grants.

**Ethics Declarations** Funded by the European Union–Next Generation EU. However, the views and opinions expressed are solely those of the author(s) and do not necessarily reflect those of the European Union or the European Commission. Neither the European Union nor the European Commission can be held responsible for them.

## References

1. Arai, K., Matsumoto, M.: Public perceptions of autonomous lethal weapons systems. AI Ethics **4**, 451–462 (2024). https://doi.org/10.1007/s43681-023-00282-9
2. Bartneck, C., Hu, J.: Exploring the abuse of robots. Interact. Stud. **9**(3), 415–433 (2008). https://doi.org/10.1075/is.9.3.04bar
3. Bartneck, C., Suzuki, T., Kanda, T., Nomura, T.: The influence of people's culture and prior experiences with AIBO on their attitude towards robots. AI Soc. **21**(1), 217–229 (2007). https://doi.org/10.1007/s00146-006-0052-7
4. Bouhmadi-López, M., Chen, C.Y., Chen, P.: Eddington–Born–Infeld cosmology: a cosmographic approach, a tale of doomsdays and the fate of bound structures. Eur. Phys. J. C **75**, 1–23 (2015)
5. Browning, H., Birch, J.: Animal sentience. Philos. Compass. **17**(5), e12822 (2022)
6. Calverley, D.J.: Android science and animal rights, does an analogy exist? Connect. Sci. **18**(4), 403–417 (2006). https://doi.org/10.1080/09540090600879711
7. Carpenter, J.: Culture and human-robot interaction in militarized spaces: A war story. Routledge (2016)
8. Carter, E.J., Reig, S., Tan, X.Z., Laput, G., Rosenthal, S., Steinfeld, A.: Death of a robot: social media reactions and language usage when a robot stops operating. In Proceedings of the 2020 ACM/IEEE international conference on human-robot interaction, pp 589–597 (2020)
9. Coeckelbergh, M.: AI ethics. MIT Press (2020)
10. David, P.R.: Notes on the case of orangutan Sandra, the non-human subject of rights. Crime Prevention and Justice in 2030: The UN and the Universal Declaration of Human Rights, 313–336 (2021)
11. de Vries, M.J.: Teaching about technology. Springer, Dordrecht (2005)
12. Harris, E.D., Acton, J.M., Lin, H.: Governance of dual-use technologies: theory and practice. American academy of arts and sciences (2024). Retrieved 5 Dec 2024, from https://www.amacad.org/publication/governance-dual-use-technologies-theory-and-practice/section/3

13. Dreyfus, H.: Being-in-the-world. MIT Press, Cambridge (1990)
14. Epicurus.: Letter to Menoeceus (n.d.). Retrieved 5 Dec 2024, from https://wiki.epicurism.info/Letter_to_Menoeceus
15. Floridi, L.: AI as agency without intelligence: On ChatGPT, large language models, and other generative models. Philos. Technol. **36**(15). https://doi.org/10.1007/s13347-023-00621-y (2023)
16. Fosch-Villaronga, E.: Robots, Healthcare, and the law: regulating automation in personal care. Routledge (2019). https://doi.org/10.4324/9780429021930
17. Garber, M.: Funerals for fallen robots. New research explores the deep bonds that can develop between soldiers and the machines that help keep them alive. T*he Atlantic*. https://www.theatlantic.com/technology/archive/2013/09/funerals-for-fallen-robots/279861/ (2013)
18. Gibson, J.: The ecological approach to visual perception. Psychology Press (1986)
19. Greenacre, M., Zubaşcu, F.: Time for a major shakeup of how the EU funds research, expert group says. Science|Business. https://sciencebusiness.net/news/fp10/time-major-shakeup-how-eu-funds-research-expert-group-says (2024)
20. Grossman, L.: Drone home. Time Magazine, 2135132–2 (2013)
21. Guersenzvaig, A.: Autonomous weapon systems: failing the principle of discrimination. In IEEE technology and society magazine, vol 37(1), 55–61. https://doi.org/10.1109/MTS.2018.2795119 (2018)
22. Heidegger, M.: Sein und Zeit [Being and time]. Tübingen: Niemeyer (1927)
23. Hung, L., Liu, C., Woldum, E., Au-Yeung, A., Berndt, A., Wallsworth, C., Chaudhury, H.: The benefits of and barriers to using a social robot PARO in care settings: a scoping review. BMC Geriatr. **19**, 1–10 (2019)
24. Kierkegaard, S.: The concept of anxiety. Asimple pshychologically orienting deliberation on the dogmatic issue of hereditary sin (R. Thompte & A. B. Anderson, Trans.). Princenton University Press (1980)
25. Klein, M.: Responsibility. In The oxford companion to philosophy.: Oxford University Press. Retrieved 9 Dec 2024, from https://www.oxfordreference.com/view/10.1093/acref/9780199264797.001.0001/acref-9780199264797-e-2205
26. Kuhlau, F., Evers, K., Eriksson, S., Höglund, A.T.: Ethical competence in dual use life science research. App. Biosaf. **17**(3), 120–127 (2012). https://doi.org/10.1177/153567601201700303
27. Liberman, A.: One small step for 'artificial intelligence'and a giant leap for the Australian patent system? The Federal Court decision in Thaler v Commissioner of Patents. J. Intellect. Prop. Law Pract. **17**(2), 164–178 (2022)
28. Löffler, D., Hurtienne, J., Nord, I.: Blessing robot BlessU2: a discursive design study to understand the implications of social robots in religious contexts. Int. J. Soc. Robot. **13**, 569–586 (2021). https://doi.org/10.1007/s12369-019-00558-3
29. Lorenz, K.: On aggression. Routledge (2021)
30. McConnell, T.: Moral dilemmas. In Zalta, E.N. (Ed.), The stanford encyclopedia of philosophy (Fall 2020 Edition). Stanford University. First published April 15, 2002. Retrieved 3 Dec 2024 (2022), from https://plato.stanford.edu/entries/moral-dilemmas/#ConMorDil
31. Milton, K.: Anthropomorphism or egomorphism? The perception of non-human persons by human ones. In Animals in person, pp. 255–271. Routledge (2020)
32. Miller, S., Selgelid, M.J.: Ethical and philosophical consideration of the dual-use dilemma in the biological sciences. Sci. Eng. Ethics **13**, 523–580 (2007). https://doi.org/10.1007/s11948-007-9043-4
33. Miller, S.: Concept of dual use. In Dual use science and technology, ethics and weapons of mass destruction, pp. 15–30. Springer. https://doi.org/10.1007/978-3-319-92606-3_2 (2018)
34. Mohd Thas Thaker, M.A.B., Akbar, M.A., Imon, R.A.A.: Toward an understanding of the robot rights spectrum: an integrative framework from an Islamic economics perspective. J. Islam. Mark. (2024)

35. Mori, M.: Bukimi no tani [The uncanny valley]. Energy **7**, 33–35 (1970)
36. NATO parliamentary assembly.: Dual-use technologies: enhancing military capabilities through civilian innovation. NATO Parliamentary Assembly. https://www.nato-pa.int/document/2024-dual-use-technologies-report-baldwin-051-esc (2024)
37. Norström, P.: Knowing How, Knowing That. Knowing Technology. Philos. Technol. **28**, 553–565 (2015). https://doi.org/10.1007/s13347-014-0178-3
38. Ocheretyany, K.: Ghost in my pocket: Tamagotchi media archaeology. Galactica Media. J. Media Stud. 6(2), 260–276. https://doi.org/10.46539/gmd.v6i2.492 (2024)
39. Olcay, S.: The transformation of marriage relationship: digital sexuality. CTC 2019 (2019)
40. Oliphant, R.: Israel declares new phase of war after walkie-talkie bombs. Retrieved from https://www.telegraph.co.uk/world-news/2024/09/18/israel-declares-new-phase-of-war-after-walkie-talkie-bombs/ (2024)
41. Parviainen, J., Coeckelbergh, M.: The political choreography of the Sophia robot: beyond robot rights and citizenship to political performances for the social robotics market. AI Soc. **36**(3), 715–724 (2021)
42. Putman, H., Putnam, H.: Robots: machines or artificially created life? J. Philos. **61**(21), 668–691 (1964). https://doi.org/10.2307/2023045
43. Robertson, J.: Robo sapiens japanicus: Robots, gender, family, and the Japanese nation. University of California Press (2018)
44. Robertson, J.: 50 ways to kill a robot: death and differentiation. In: Pinillos, A.S., Costa, V., Vallverdú, J. (Eds.), Second death–experiences of death across technologies. Studies in applied philosophy, epistemology and rational ethics. Springer, Cham (2025)
45. Ross, W.D.: The right and the good. Oxford University Press (2003)
46. Sans Pinillos, A., Casacuberta, D.: Remarks on the possibility of ethical reasoning in an artificial intelligence system by means of abductive models. In: Nepomuceno-Fernández, Á., Magnani, L., Salguero-Lamillar, F., Barés-Gómez, C., Fontaine, M. (eds) Model-based reasoning in science and technology. MBR 2018. Studies in applied philosophy, epistemology and rational ethics, vol 49. Springer, Cham. https://doi.org/10.1007/978-3-030-32722-4_19 (2019)
47. Sapolsky, R.M.: Doubled-edged swords in the biology of conflict. Front. Psychol. **9**(2625). https://doi.org/10.3389/fpsyg.2018.02625 (2018)
48. Sartre, J.-P.: L'existentialisme est un humanisme. Les Éditions Nagel (1966)
49. Taddeo, M.: The ethics of artificial intelligence in defence. Oxford University Press (2025)
50. Tonkens, R.: Robots, AI, and assisted dying: ethical and philosophical considerations. In: Cholbi, M., Varelius, J. (Eds.) New directions in the ethics of assisted suicide and Euthanasia. Int. Libr. Bioeth. vol 103. Springer, Cham. https://doi.org/10.1007/978-3-031-25315-7_16 (2023)
51. Torrance, S.: Ethics and consciousness in artificial agents. AI Soc. **22**, 495–521 (2008). https://doi.org/10.1007/s00146-007-0091-8
52. Travagnin, S.: From online buddha halls to robot-monks: new developments in the long-term interaction between Buddhism, media, and technology in contemporary China [从在线佛堂到机器僧:佛教、媒体与科技在当代现代中国的互动与发展]. Rev. Relig. Chin. Soc. **7**(1), 120–148 (2020). https://doi.org/10.1163/22143955-00701006
53. Trovato, G., De Saint Chamas, L., Nishimura, M., Paredes, R., Lucho, C., Huerta-Mercado, A., Cuellar, F.: Religion and robots: towards the synthesis of two extremes. Int. J. Soc. Robot. **13**, 539–556 (2021)
54. Tucker, J.B. (ed.): Innovation, dual use, and security. MIT Press, Managing the risks of emerging biological and chemical technologies (2012)
55. Vallverdú, J.: The eastern construction of the artificial mind. Enrahonar: Quaderns de Filosofia **47**, 171–185 (20211)
56. Vallverdú, J., Nishida, T., Ohmoto, Y., Moran, S., Lázare, S.: Fake empathy and human-robot interaction (HRI): A preliminary study. Int. J. Technol. Hum. Interact. (IJTHI) **14**(1), 44–59 (2018)

57. Vanina Martínez, M., Rodríguez, R.O.: Reflections on the use of artificial intelligence for weapons applications. In Pinillos, A.S., Costa, V., Vallverdú, J. (Eds.), SecondDeath–experiences of death across technologies. Stud. Appl. Philos., Epistem. Rat. Ethics. Springer, Cham (2025)
58. van Kolfschooten, H.: The prospects of using AI in euthanasia and physician-assisted suicide: a legal exploration. AI Ethics (2024). https://doi.org/10.1007/s43681-024-00491-w
59. Zhang, J., Li, S., Zhang, J.Y., Du, F., Qi, Y., Liu, X.: A literature review of the research on the Uncanny Valley. In: Rau, P.L. (Eds.) Cross-cultural design. User experience of products, services, and intelligent environments. HCII 2020. Lecture Notes in Computer Science, vol 12192. Springer, Cham. https://doi.org/10.1007/978-3-030-49788-0_19 (2020)

# The Neglect of Qualia and Consciousness in AI Alignment Research

Soenke Ziesche and Roman V. Yampolskiy

**Abstract** The AI value alignment problem has now been acknowledged as essential for AI safety as well as very hard. In this chapter we argue that critical parameters are neglected in AI value alignment research, which are consciousness and qualia. The AI value alignment problem is about ensuring that AI systems pursue goals, which are aligned with the interests of moral patients. Briefly summarized, prevalent human interests are to foster happiness and pleasure and to avoid pain; thus, experiences perceived through consciousness and qualia. Therefore, AI systems need not only to understand qualia and consciousness, but also their precious significance in order to be truly aligned with human interests as well as with the interests of other sentient beings. Death constitutes for humans the end of consciousness, thus, the termination of the opportunity to experience happiness and pleasure. Therefore, AI systems must not kill sentient beings. In this chapter we describe the importance of incorporating consciousness and qualia research to AI value alignment research as well as the potential feasibility of such efforts due to developments in neurotechnology. Concluding, we offer recommendations outlining such undertaking as a compulsory component of the ongoing mammoth task to reduce the x- as well as the s-risks posed by a potential superintelligence.

## 1 Introduction

In this chapter we suggest shifting the focus of research related to the AI value alignment problem to aspects, which are so far neglected, consciousness and qualia. To demonstrate that such a shift is reasonable as well as likely feasible in the near-term future due to developments in neurotechnology, we first briefly introduce relevant

S. Ziesche (✉)
New York City, USA
e-mail: soenke.ziesche@gmail.com

R. V. Yampolskiy
Speed School of Engineering, University of Louisville, Louisville, USA
e-mail: roman.yampolskiy@louisville.edu



A. Sans Pinillos et al. (eds.), *SecondDeath*, Studies in Applied Philosophy, Epistemology and Rational Ethics 76, https://doi.org/10.1007/978-3-031-98808-0_11

topics, which are the AI value alignment problem, consciousness and qualia, moral patients and agents, neurotechnology, neuroscience of consciousness as well as death.

The remaining part of the chapter is structured as follows: We present the case for AI value alignment by means of understanding of consciousness, including resulting recommendations and implementation prolegomena, followed by a conclusion.

## *1.1 AI Value Alignment Problem*

Among a range of problems related to the safety of AI systems is the value alignment problem perhaps the most substantial one as well as hardest [1, 2]. Briefly, it is about the question how to develop AI systems, especially not yet existing super intelligent Artificial General Intelligence (AGI) systems, in a way that they pursue goals and values which are aligned with goals and values of sentient beings. This complex problem can be broken down into at least three subproblems, of which only the first one is discussed here. First, it has to be endeavoured to precisely extract the values and interests of all concerned beings; second, it has to be endeavoured to aggregate these values in a consistent manner; third, it has to be endeavoured to instil these values into AI systems. The first two subproblems are both normative, while the third one is technical [e.g., 3].

All three subproblems have proven to be very challenging (if not unsolvable). Regarding value extraction, when we look at the values and interests of humans, it can be stated that their values and interests are often fuzzy, at times irrational and frequently inconsistent with the values and interests of other humans. Moreover, human values and interests commonly change over time; thus, in addition to the inconsistencies of values between different humans there are also inconsistencies of human values at different points in time during their lifespan as well as during the history of humanity. This is linked to the observation that "that we humans have an imperfect understanding of what is right and wrong, and perhaps an even poorer understanding of how the concept of moral rightness is to be philosophically analyzed " [1, p. 217]. Therefore, it would not be constructive, even if we were able to extract and specify currently accepted human values and interests, to hard-wire them into AI systems as they may be deprecated after some time. In addition, various other issues have been pointed out, which led to the conclusion that "the idea of 'human values' should be replaced with something better for the goal of AGI Safety" [4, p. 27].

As we will elaborate further below, the challenge with fuzzy as well as changing values and interests over time, could potentially be eliminated if we look at consciousness and qualia instead since it never changes that humans enjoy pleasure and dislike pain.[1]

It is evident that the extraction of values and interests from other moral patients, such as nonhuman animals and potential digital minds (see below), is even harder

[1] Minor exceptions include that humans may change their taste for certain food over time, e.g., they may dislike certain food during childhood, but enjoy that type of food during adulthood.

due communication problems, but also due to potentially for humans unfathomable values and interests of these beings.

## *1.2 Consciousness and Qualia*

Although consciousness is for humans very familiar and comprises phenomena, such as awareness, sentience and wakefulness, it has been proven very hard to analyse consciousness scientifically and philosophically [e.g., 5]. Three umbrella questions summarize the problem, each of them highly complex and discussed for centuries: What is actually consciousness? How does consciousness come into existence? Why does consciousness exist, i.e., what is its function? Conscious experience involves qualia, such as perceiving colours and sounds, tasting food or enduring pain, which cannot be satisfactory verbalized by humans. While for us qualia are, while being awake, continuously introspectively accessible, their actual nature is controversial.

In short, consciousness and qualia constitute a broad field, yet with limited scientific findings. In this chapter further below we focus only on the neuroscientific aspects of consciousness, thus, areas to which AI could contribute, but not the philosophical aspects.

### 1.2.1 Non-human Animals

While it is hard if not impossible to prove solely through philosophical methods that other beings, including humans and nonhuman animals, have conscious experiences, "there is no longer any serious dispute among scientists that at least some nonhuman animals can feel pain and experience other conscious states, both positive and negative." [6, p. 14]. Because of this insight it has been demanded to also incorporate the values and interests of nonhuman animals into AI alignment research, which adds another layer of complexity to the problem [7].

### 1.2.2 Potential Digital Minds

The substrate-independence thesis states that "mental states can supervene on any of a broad class of physical substrates. Provided a system implements the right sort of computational structures and processes, it can be associated with conscious experiences." [8, p. 2]. If this was true, then sentient digital minds are conceivable, and the AI value alignment problem would be even more complex, including the possibility that sentient digital minds have vastly different conscious experiences as well as the potential impossibility for sentient digital minds to die, given the options of easy copyability and resurrection owing to the computer hardware substrate.

## *1.3 Moral Patients and Agents*

A critical concept for the AI value alignment problem is moral patienthood. Moral patients are beings, who are eligible for moral consideration. A common criterion for a being to be a moral patient is having consciousness [9]. As described above, nonhuman animals experience also conscious states and are, therefore, morals patients, which can be expressed as follows: "There is no moral justification for treating the pain (of nonhuman animals) as less important than similar amounts of pain felt by humans." [6, p. 21]. The same applies to potential digital minds: These may not only be moral patients too, but their existence could involve various additional moral challenges due to their potentially vast numbers as well as due to their potentially different needs [10, 11]. Therefore, a complete solution of the AI value alignment problem will have been only achieved if AI systems are aligned with the values and interests of all moral patients.

In contrast, moral agents are beings who have the ability to discern right from wrong and, thus, are to be held accountable for its actions. There are moral patients who are not moral agents, such as infants and mentally disabled people. Whether there are moral agents who are not moral patients is discussed further below.

## *1.4 Neurotechnology*

Neurotechnology is a field, which deals with devices to access, understand, monitor or manipulate neural activity in living beings. In recent years a variety of neurotechnological applications are being developed "to help us become faster, more efficient, safer, healthier, less stressed, and even more spiritual." [12, p. 17]. For the purpose of this chapter, it is especially relevant to identify ways to comprehend neural activity. In this regard, "powerful machine learning algorithms are getting better and better at translating brain activity into what we are feeling, seeing, imagining, or thinking" [12, p. 17]. For example, recently a non-invasive AI-based device has been developed, which translates brain activity, i.e., thoughts, into text [13]. Also, ways are being found to decode with the help of AI systems the brain activity of subjects to reconstruct images of what they are seeing [14, 15]. Similar to other emerging technologies also neurotechnology can be of dual use; thus, a new human right to cognitive liberty regarding our brain activities has been demanded [12].

## *1.5 Neuroscience of Consciousness*

For the proposals towards the AI value alignment problem to be made further below in this chapter the neuroscience of consciousness is relevant. Consciousness had been discussed for a long time within philosophy mainly, but technological advancements in neuroscience enabled also scientific approaches towards consciousness despite obstacles being faced how to measure an essentially subjective phenomenon empirically [, e.g., 5]. First, we introduce the distinction between generic and specific consciousness. The former concerns neural correlates when a state is conscious rather than not while the latter one concerns neural correlates of the content of a conscious state.

Also, regarding neural correlates of consciousness efforts through neurotechnological applications are being made by identifying the specific neural activity patterns or processes that are associated with conscious experiences. These patterns are often referred to as neural correlates of consciousness. The idea is that certain patterns of brain activity might be closely linked to the emergence of conscious awareness [e.g., 16].

Another level of complexity is added by the fact that numerous processes in the mind take place unconsciously and are, thus, not available to introspection. Also, progress has been made towards neurotechnological applications that track unconscious processes in brain [e.g., 12].

While for humans the existence of consciousness as well as of an unconscious mind is usually indisputable, this has not been the case for nonhuman animals for a long time. However, as has been mentioned above, by now neuroscientific findings are very clear that certain nonhuman animals are conscious and can feel pain in particular [6, 17].

When it comes to potential digital minds various attempts have been made to establish whether such beings are conscious [e.g., 18]. There has been also an endeavour to assess existing AI systems based on neuroscientific theories of consciousness to establish whether these systems are conscious or not. The outcome was that currently no AI systems are conscious, but that there are no obvious technical barriers for AI systems to satisfy indicators of consciousness [19].

## *1.6 Death*

Death constitutes the cessation of biological processes, which includes the end of conscious experiences. While it is, therefore, the most prominent goal related to the AI value alignment problem to prevent AI systems from killing moral patients, in the initial sense the AI value alignment problem is not concerned with humans or nonhuman animals after their deaths.

However, an extension of the AI value alignment problem to bodies as well as minds of dead humans and nonhuman animals could be conceived, which to our knowledge is original and has not been discussed before. While some humans agree to the donation of their organs after their death, others oppose it. AI value alignment regarding dead bodies would mean not to mutilate them, unless explicit consent has been given by the deceased. This is relevant as it has been frequently claimed that for an AI system humans may merely be an assembly of atoms, which may be useful for the AI system to achieve its goals, e.g., to produce as many paperclips as possible

[20]. A special case is cryonics, which comprises cryopreserved human remains at very low temperatures, based on the speculative hope to resurrect this human at a later stage through not yet existing technologies. While it remains to be proven if it works, AI value alignment would also have to include not to rig these cryopreserved human remains.

Considerations related to the minds of dead humans are for now futuristic. Significant progress in the fields of neurotechnology and neuroscience of consciousness is conceivable so that human minds could be copied to other substrates, e.g., when death is imminent. This is a complex topic with various implications [e.g., 21], but for this chapter we only stress that the AI value alignment problem would also have to include such uploaded minds and the considerations for digital minds in this chapter are pertinent.

The situation for nonhuman animals is different as humans mutilate for millennia their dead bodies to make use of their meat, skin, fur etc. Yet, consent from nonhuman animals cannot be received unless the communication barrier will be overcome. AI-based attempts in this regard are ongoing and the situation may change if these attempts succeed [e.g., 22]. For the futuristic scenario of uploaded minds of deceased nonhuman animals, the same would apply as stated above for uploaded human minds.

The corresponding event to death for potential digital minds would be their deletion. It is speculation if potential digital minds would permanently oppose to their deletion as this would mean a potentially indefinite lifespan for them owing to their computer hardware substrate. Also, this would create substantial computational costs, especially in light of easy copyability and potentially vast numbers of digital minds. Moreover, for digital minds the distinction can be made between turning them off and keeping their code and their history or turning them off and destroying the code and the history as well. In the first scenario, the digital mind could be resurrected later. Regardless, if the existence of digital minds has been verified as well as their preferences towards deletion, the challenge arises to adjust the AI value alignment problem accordingly due to the moral patienthood of digital minds [10].

## 2 AI Value Alignment by Means of Understanding of Consciousness

After the introduction of the relevant topics, we have established the foundation for our main claim in this chapter, which is the importance of incorporating consciousness and qualia research to AI value alignment research as well as the potential feasibility of such efforts due to developments in neurotechnology.

We acknowledge that scientific approaches to consciousness and qualia research have been notoriously difficult, but assert that they appear to be key to the AI value alignment problem and that potentially AI systems can actually lead to progress in consciousness research.

Just like the other subproblems of the AI value alignment problem value extraction has proven to be very challenging and has not led to satisfying results up to now due to fuzziness, irrationality and inconsistency of human values, which gets even more complicated if the values of other moral patients, such as nonhuman animals and potentially digital minds are factored in, as it ought to be the case. Therefore, we suggest an innovative approach to the value extraction subproblem, which constitutes a shift from values to conscious experiences.

Our initial premise is rather simple: (Most) humans (and most other moral patients) (1) want to be alive and (2) do not want to suffer.

Therefore, it would be desirable if AI systems through neurotechnology would be capable to understand three aspects concerning beings: Firstly, if a being is alive and conscious, and, if it is affirmative that it is alive and conscious, secondly, in what state of mind the being is, and thirdly, in what state of mind the being would be after any activity by the AI system, which affects the being.

The first point corresponds to the term generic consciousness and the second and third point to the term specific consciousness, as introduced above. Note that the third point also includes that the AI system would understand whether the being would be dead after the activity by the AI system. In other words, the first point is critical for AI systems to identify moral patients and the second and third point are critical for AI systems to treat moral patients in a morally adequate manner.

Then three rules for AI systems as follows could be established[2]:

- Do not conduct activities, which would kill living and conscious beings.
- Do not conduct activities, which would cause for a living and conscious being a state of mind further towards pain on the pleasure/pain axis.
- Do conduct activities, which supports human beings to find a purpose in life, especially for those human beings, who have lost what they considered purpose of life due to new technologies, including AI.[3]

The first rule addresses x-risks or existential risks, which may lead to annihilation of life on earth [23]. The second rule targets s-risks or suffering risks, which may increase suffering drastically [24]. The third rule covers sometimes neglected so-called i-risks and encompasses scenarios, in which humans lose their reason or purpose to live, particularly in light of emerging technologies. The "i" in i-risk stands for ikigai, which is a Japanese concept comprising the reason or purpose to live [25].

Regarding the first rule it is also important that the AI system is capable of understanding all alternate states of consciousness, such as sleeping, being unconscious, under anaesthetic or in a coma, as states of being tentatively without consciousness. However, we established the additional requirement for AI systems to understand

---

[2] Note that the other steps of the AI value alignment problem, such as how to instil these rules into AI systems, are not discussed here. We acknowledge that this appears challenging too, just like to instil any rules in a sustainable manner to an AI system.

[3] We limit this rule to human beings as we have no evidence whether other conscious beings have a concept of the purpose of their life beyond survival and reproduction and, if yes, what this purpose may be.

when a being is alive, which is probably anyway easier than understanding if a being is conscious.

Going into more detail regarding the second rule, it is desirable if AI systems would be capable to understand through neurotechnology and neural correlates the states of mind, which moral patients would like to be in, such as (for humans) happiness, contentment and joy, including qualia they enjoy perceiving, i.e., states, which are tilting towards pleasure on the pleasure/pain axis. In this regard, the risk of wireheading has to be mentioned, i.e., the undesirable scenario of direct stimulation of the brain by the AI system to experience pleasure [e.g., 26].

Equally, it should be aimed for AI systems to understand through neurotechnology and neural correlates the states of mind, which moral patients do <u>not</u> like to be in, such as suffering, including qualia they detest perceiving, i.e., states, which are tilting towards pain on the pleasure/pain axis, which also includes dying.

For this approach we do not foresee the requirement for the concerned AI system to be conscious, i.e., to be a moral patient itself, but it needs to understand, through neuroscientific knowledge as well as through neurotechnological access to brains of concerned beings, consciousness and qualia of other beings to appreciate moral patients and moral considerations accordingly. Therefore, the AI system would be a moral agent in the sense that it would be capable of acting with reference to right and wrong. It would be actually an instance of an artificial moral agent [27].

An advantage of this approach was indicated above: A major challenge of the value extraction problem was that human values have changed over time and, thus, are likely to change further in the future, e.g., due to moral circle expansion [28]. Therefore, there would have to be an option for humans to regularly update values of AI systems, which appears complex. In contrast, it never changes that humans enjoy pleasure, dislike pain and (in most cases) do not want to die. Therefore, these interests of humans could be hard-wired into AI systems.

## *2.1 Recommendations*

Based on our proposition in the previous section we offer recommendations towards strengthened research of the neuroscience of consciousness and qualia as well as challenges and further elaborations.

**Recommendation 1**

Intensify AI-supported research of the neuroscience of consciousness as well as qualia and develop AI-supported neurotechnological applications with the goal that AI systems become capable of understanding that a certain activity involving a human/any sentient being…

i. ...May kill the human/sentient being.

   Resulting rule: Don’t do it.

ii. ...Has a negative impact on their well-being /cause suffering (in other words: if the AI activity would cause for the human/sentient being a state of mind, which corresponds to a move towards pain on the pleasure/pain axis).

Resulting rule: Don't do it.

In neurotechnology two main categories of research can be distinguished: "Tracking the brain" and "hacking the brain" [12]. The first one aims to access, understand and monitor neural activities. The second one aims to manipulate neural activities. Recommendation 1 above requires merely abilities to track the brain, i.e., to access, understand and monitor neural activities.

As indicated above the item i addresses x-risks, while the item ii addresses s-risks. For item ii also the unconscious mind is relevant. Ideally, the AI system understands also unconscious worries and anxieties of humans and would act in a way that these are not affected.

We acknowledge the following challenges:

- **Cognitive liberty**: Concerns have been expressed that neurotechnological applications for brain surveillance will affect privacy, self-determination and the freedom of thought [12].
- **Trolley problem**: Similar to thought experiments with self-driving cars, scenarios could be perceived, in which a certain AI activity may be bad for some humans/ sentient beings, but good for others. In such a case it is hard to decide for the AI system what to do.
- **Perverse instantiation**: Scenarios could be perceived, in which, for example, (a) a chess playing AI loses intentionally against a human since winning creates pleasure for the human, or in which (b) an AI, which is tasked to shortlist applicants from submitted CVs, gets stuck because not-shortlisted candidates would suffer if they were not selected.

**Recommendation 2**

Intensify AI-supported research of the neuroscience of consciousness as well as qualia and develop AI-supported neurotechnological applications with the goal that AI systems become capable of manipulating (neural) activities of a human/any sentient being that…

i. ...Would be a remedy for the suffering of the human/sentient being.

Resulting rule: Offer to conduct this activity.

ii. ...Would increase the well-being of the human/sentient being.

Resulting rule: Offer to conduct this activity.

iii. ...Would enable the human/sentient beings to perceive so far unknown conscious experiences.

Resulting rule: Offer to conduct this activity.

Coming back to the distinction between tracking the brain and hacking the brain [12], recommendation 2 above requires abilities to hack the brain, i.e., to manipulate neural activities.

It has to be noted that recommendation 1 is, if feasible, sufficient to address the value extraction subproblem of the AI value alignment problem through neuroscience of consciousness, as it makes sure that AI systems do not harm humans or other sentient beings.

Beyond that, recommendation 2, if feasible, addresses options how AI systems could improve the condition of humans and other sentient beings thanks to their superior intelligence and to neurotechnological applications. In other words, a distinction can be made between value-aligned AI systems (through tracking the brain) and AI systems, which are, in addition, improving the lives of humans and other sentient beings (through hacking the brain).

An example for item i would be treating depression through AI-supported brain stimulation [29].

An example for item ii would be AI systems, which help humans to find their ikigai, as introduced above, thus alleviating i-risks [25]. It has been shown that "ikigai significantly negatively correlated with measures of depression and anxiety, and significantly positively correlated with measures of well-being" [30]. In this regard, research of the neuroscience of consciousness and states of well-being may enable such AI systems that support humans in discovering their ikigai [31].

For both item i and ii it would be beneficial if AI systems cannot only track unconscious anxieties and desires of humans, but can also manipulate neural activities in a way that ensures the well-being of the human on the unconscious level too.

Item iii is probably the most futuristic and is linked to the assumption "that we are living on a tiny island of consciousness within a giant ocean of alien mental states." [17, p. 353], some of which AI systems may enable us to perceive through the manipulation of neural activities.

To reiterate, if an AI were <u>not</u> to conduct any of the examples above, it would still be value aligned in the sense that it does not actively harm the moral patient. Yet, a comparison can be drawn to the <u>duty to rescue concept</u> in tort law, which implies that a moral agent can be held liable for doing nothing while a moral patient is in peril, if the moral agent has the capacities to assist the moral patient.

This could be applied to the examples above: If the AI system knew remedies to treat the depression of someone, knew how someone could find purpose of life or knew how to perceive so far unknown conscious experiences, it could be discussed if it was obligatory for the AI system to reveal this knowledge and support the moral patients accordingly. This is the essence of recommendation 2.

Recommendation 2 could be also linked to the concept of "Coherent Extrapolated Volition". It means that the AI system, at least for the items i and ii, would do to us, what we would want "if we knew more, thought faster, were more the people we wished we were, and had grown up farther together" [32, p. 6].

We acknowledge the following challenges:

- <u>Cognitive liberty</u>: See above [12].

- Wireheading: There is a risk that the AI system directly stimulates the brain to experience pleasure [e.g., 26]. Instead, for example, humans who have found their ikigai with or without the help of AI systems experience pleasure indirectly by pursuing the activities related to their ikigai.

## 2.2 *Implementation Prolegomena*

As indicated earlier it is still work in progress to identify neural correlates of consciousness, yet we intend to offer initial ideas towards the implementation of these recommendations. First, to address the concern of cognitive liberty, AI systems may not need access to the brains of all humans, who are affected by their activities, but the following scenario could be considered:

AI systems get, through neurotechnological devices, access in an experimental setting to the brains of around 1,000 volunteers. These volunteers would be diverse regarding their sex, age, ethnic origin, religion, sexual orientation, political views, education, neurodiversity, disability etc.[4] The volunteers would have to give informed consent. Then AI systems would first try to understand how to identify if a human is conscious and, second, to be trained, based on the neural correlates of consciousness before and after actions of the AI system, if and how a diverse range of actions would affect the brains of the volunteers, thus, their well-being (ideally without executing those actions, which would affect them in a negative way). Preferably this would be done with non-invasive brain-computer interfaces. And furthermore, ideally the AI systems could generalize this knowledge to humanity overall, so that the AI systems may not need to access constantly the brains of the humans with whom they are interacting, but could when interacting with other humans apply the knowledge they gained during this training. This description is for the implementation of recommendation 1 and can be extended for recommendation 2 accordingly, which would involve the manipulation of neural activities.

Non-human animals cannot give the required informed consent; thus, for now such an experiment would be unethical [e.g., 6]. Nor could the outcome of the experiment with human above be transferred to non-human animals as their conscious experiences differ partly significantly from humans, see e.g., bats, which use echolocation to navigate and forage.

Also, for potential digital minds a different experimental setup is required, for which the discipline of AI welfare science has been proposed [10].

[4] It has been pointed out that neuroscientific research (as well as research in other disciplines) focuses on sub-normative subjects and so-called WEIRD (Western, Educated, Industrialized, Rich, and Democratic) ones, which is not desirable [e.g., 17].

## 3 Conclusion

We shifted the value extraction subproblem of the AI value alignment problem from the level of values in the original sense to the level of states of mind. We view it as an innovative game changer if AI systems could understand, thus foresee effects of their actions to states of minds of moral patients and act accordingly.

The AI value alignment problem is considered critical to be solved since AI systems may pose an x-risk and an s-risk to sentient beings as well as at least to humans also an i-risk. We believe that only if AI systems understand through access to neurotechnological applications what death, pain and depression due to an absence of a purpose of life respectively mean, only then these AI systems can be effectively trained not to kill or to inflict pain on moral patients and ideally to help humans to find their ikigai.

This approach, unlike other approaches to the AI value alignment problem, would also tackle the second subproblem, value aggregation, because the pleasure/pain axis is basically inherent to all humans and sentient non-human animals; thus, the challenge of clashing inconsistent values is omitted, which is an issue when looking at values in the classical sense. Humans have generated such a rich diversity of value systems, traditions and ideas generated in many centuries by the many cultures in the world, which are very challenging to be aggregated in consistent manner. However, this endeavour is necessary to leave no one behind [3, 33]. In addition, the values of humans and sentient non-human animals clash in many aspects [7]. Therefore, projecting the issue to the level of the pleasure/pain axis appears to be an elegant attempt, at least humans and sentient non-human animals. However, when it comes to potential digital minds for humans unfathomable other dimensions than the pleasure/pain axis cannot be excluded.

However, the third subproblem of the AI value alignment problem remains and is not addressed by this article, which is the technical subproblem after the two normative ones about how to instil these values, i.e., the rules established above, sustainably into AI systems. If this is not done successfully a potential treacherous turn is looming [1], which is challenging to tackle due to the potential uncontrollability of AI systems [34].

We have to stress that our approach depends on further findings of the neuroscience of consciousness and corresponding developments in the thriving field of neurotechnology [12]. Therefore, we advocate to intensify the research in the field of the neuroscience of consciousness as a premise to tackle the value extraction subproblem of the AI value alignment problem. Further future work would also comprise to address the challenges presented above.

Death of moral patients is an outcome that AI systems must particularly avoid. For an AI system to establish through neuroscientific knowledge whether a potential activity would kill a moral patient should be easier than establishing in what way states of mind of moral patients may change due to a potential activity by the AI system. As opposed to other approaches to the AI value alignment problem, which declare the undesirability of death as an abstract value, it appears to be more straightforward to address this issue by ensuring that AI systems have the necessary neuroscientific knowledge what death means.

In addition to providing a recommendation how to address the AI value alignment problem through neurotechnological applications, we offered another recommendation how AI systems may be able to improve the conditions of humans and other sentient beings through neurotechnological applications.

In summary, while we consider the approach suggested in this chapter as very promising, future work is required, which includes pertinent research in the field of the neuroscience of consciousness as well as tackling the challenges, which were described for the two recommendations.

## References


1. Bostrom, N.: Superintelligence: paths, dangers, strategies, p. 2014. Oxford University Press, Oxford, UK (2014)
2. Yudkowsky, E.: Artificial intelligence as a positive and negative factor in global risk. In Global catastrophic risks; Bostrom, N., Ćirković, M.M., (Eds.) Oxford University Press: New York, NY, USA, pp. 308–345. Available online: https://intelligence.org/files/AIPosNegFactor.pdf (2008)
3. Gabriel, I.: Artificial intelligence, values, and alignment. Minds Mach. **30**, 411–437. Available online: https://link.springer.com/content/pdf/https://doi.org/10.1007/s11023-020-09539-2.pdf (2020)
4. Turchin, A.: AI alignment problem: "Human Values" idea is built upon many assumptions. PhilPapers (2019)
5. Chalmers, D.J.: The conscious mind: In search of a fundamental theory. Oxford Paperbacks (1997)
6. Singer, P.: Animal liberation now. Random House (2023)
7. Ziesche, S.: AI ethics and value alignment for nonhuman animals. Special Issue "The Perils of Artificial Intelligence" of Philosophies. **6**(2), 31 (2021)
8. Bostrom, N.: Are we living in a computer simulation? Philos. Q. **53**(211), 243–255 (2003)
9. Muehlhauser, L.: Report on consciousness and moral patienthood. Open Philanthropy (2017)
10. Ziesche, S., Yampolskiy, R.V.: Towards AI welfare science and policies. Big Data Cogn. Comput. **3**(1), 2 (2018)
11. Bostrom, N., Shulman, C.: Propositions concerning digital minds and society (2022)
12. Farahany, N.A.: The battle for your brain: defending the right to think freely in the age of neurotechnology. Martin's Press, St (2023)
13. Tang, J., LeBel, A., Jain, S., Huth, A.G.: Semantic reconstruction of continuous language from non-invasive brain recordings. Nat. Neurosci. 1–9 (2023)
14. Takagi, Y., Nishimoto, S.: High-resolution image reconstruction with latent diffusion models from human brain activity. In Proceedings of the IEEE/CVF conference on computer vision and pattern recognition, pp 14453–14463 (2023)
15. Scotti, P.S., Banerjee, A., Goode, J., Shabalin, S., Nguyen, A., Cohen, E., Abraham, T.M.: Reconstructing the mind's eye: fMRI-to-image with contrastive learning and diffusion priors. arXiv preprint arXiv:2305.18274
16. Koch, C., Massimini, M., Boly, M., Tononi, G.: Neural correlates of consciousness: progress and problems. Nat. Rev. Neurosci.Neurosci. **17**(5), 307–321 (2016)
17. Harari, Y.N.: Homo deus: a brief history of tomorrow. HarperCollins Publishers, New York (2017)

18. Elamrani, A., Yampolskiy, R.V.: Reviewing tests for machine consciousness. J. Conscious. Stud. **26**(5–6), 35–64 (2019)
19. Butlin, P., Long, R., Elmoznino, E., Bengio, Y., Birch, J., Constant, A., VanRullen, R.: Consciousness in artificial intelligence: insights from the science of consciousness. arXiv preprint arXiv:2308.08708
20. Bostrom, N.: Ethical issues in advanced artificial intelligence. Science fiction and philosophy: from time travel to superintelligence, 277–284 (2003)
21. Wiley, K.: A taxonomy and metaphysics of mind-uploading; humanity+press and Alautun press: Los Angeles. CA, USA (2014)
22. Rutz, C., Bronstein, M., Raskin, A., Vernes, S.C., Zacarian, K., Blasi, D.E.: Using machine learning to decode animal communication. Science **381**(6654), 152–155 (2023)
23. Bostrom, N.: Existential risks: analyzing human extinction scenarios and related hazards. J. Evol. Technol. **9**. https://nickbostrom.com/existential/risks.html (2002)
24. Althaus, D., Gloor, L.: Reducing risks of astronomical suffering: a neglected priority. Foundational Research Institute. Available online: https://foundational-research.org/reducing-risks-ofastronomical-suffering-a-neglected-priority/ (2016)
25. Ziesche, S., Yampolskiy, R.V.: Introducing the concept of ikigai to the ethics of AI and of human enhancements. In 2020 IEEE international conference on artificial intelligence and virtual reality (AIVR), pp 138–145. IEEE (2020)
26. Yampolskiy, R.V.: Utility function security in artificially intelligent agents. J. Exp. Theor. Artif. Intell.Intell. **26**(3), 373–389 (2014)
27. Cervantes, J.A., López, S., Rodríguez, L.F., Cervantes, S., Cervantes, F., Ramos, F.: Artificial moral agents: a survey of the current status. Sci. Eng. Ethics **26**, 501–532 (2020)
28. Singer, P.: The expanding circle. Clarendon Press, Oxford (1981)
29. Scangos, K.W., Makhoul, G.S., Sugrue, L.P., Chang, E.F., Krystal, A.D.: State-dependent responses to intracranial brain stimulation in a patient with depression. Nat. Med. **27**(2), 229–231 (2021)
30. Wilkes, J., Garip, G., Kotera, Y., Fido, D.: Can Ikigai predict anxiety, depression, and well-being?. Int. J. Ment. Health Addict., 1–13 (2022)
31. Ziesche, S., Yampolskiy, R.V.: Mapping the potential AI-driven virtual hyper-personalised ikigai universe. PhilPapers (2023)
32. Yudkowsky, E.: Coherent extrapolated volition; the singularity institute: San Francisco, CA, USA. Available online: https://intelligence.org/files/CEV.pdf (2004)
33. Ziesche, S.: Stimuli from selected non-Western approaches to AI ethics. PhilPapers (2023)
34. Yampolskiy, R.V.: On the controllability of artificial intelligence: an analysis of limitations. J. Cyber Secur. Mobil. **11**(3), 321–404 (2020)

# Death, Grief, and Technology

# Do Robots Die?

**Ana Cuevas-Badallo**

**Abstract** This article examines the question of whether currently existing robots can die. To do so, it will first analyse what is meant by agent, biomimicry and what is commonly understood by dying and being alive. Biomimetic imitations of robots can give rise to attachment relationships between their users and the robots, and also to feelings of grief at their loss. This confusion can lead to emotional reaction on the part of the users of these devices to the extent that they consider that, in mourning their loss, this mourning is due to the fact that the robot has actually died.

**Keywords** Biomimetic robotics · Grief · Attachment · Death

## 1 Introduction

The following will attempt to answer the question: do robots die? Regardless of what answer the reader has in mind, the mere possibility that the question can be posed, and that it potentially has several answers, is already an indication that it is not trivial.

To address the scope of this question, it is necessary to review some preliminary issues. Firstly, an attempt will be made to see what is considered a robot today and how many types exist, as well as their capabilities as of the date of publication of this article (2025). I say attempt will be made, because this area of technological development varies in surprising ways. It goes without saying that there will be no philosophy of "technology fiction" here. Reflecting on what does not exist on the assumption that "maybe it could be the case that it does exist" is a legitimate form of philosophy, but also a very risky one. Of course, imagining plausible futures can be not only a very interesting and even amusing task, but also a necessary one: if we want to prevent certain events from happening, we must first imagine them, and thus better design present actions. However, it may also be the case that attributing capabilities that are far removed from current possibilities may lead to the creation

A. Cuevas-Badallo (✉)
University of Salamanca, Salamanca, Spain
e-mail: acuevas@usal.es

 
A. Sans Pinillos et al. (eds.), *SecondDeath*, Studies in Applied Philosophy, Epistemology and Rational Ethics 76, https://doi.org/10.1007/978-3-031-98808-0_12

of dystopian futures that later, when we reach that supposed moment, the situation has changed so much that these forecasts no longer make any sense at all.

In this task of dealing with what exists, neither literature, nor cinema, nor video games are in my favour. That is why we will turn to catalogues of the artefacts created by engineers up to the present day. We are not going to talk about cyborgs, nor about the possible hybrid humans predicted by Andrew Griffinen[1], nor are we going to support the proposal of Rodney Brooks -one of the main developers of robots capable of operating in real-world environments- who considers that human beings are in fact robots [9]. If we were to accept this thesis, then we would have no doubt about the question posed, of course robots die, at least some of them, insofar as human beings, at least so far, are mortal. Here we will deal only and exclusively with the robots that are available and functional today. And on the basis of what actually exists, we will ask ourselves whether these robots have the capacity to die, and not whether a robot created in the future from organic material, with consciousness, emotions and morality, will have the capacity to die. I leave that task to whoever wants to do it, now or in the future (if such robots ever exist).

An attempt will be made to show why this question can have different answers. In order to do so, it will also be necessary to clarify what is meant by "dying". This may also seem obvious, but when one delves into the discussions that exist to define the moment of death, one will see that it is not so simple (even though practically everyone who has died has done it right the first time).

It will be shown below that some of the doubts that may arise about whether or not robots die have to do not only with the similarity between the behaviour and appearance of robots and the behaviour and appearance of organisms, but also with the ability of humans to develop attachment relationships and to grieve in complex ways.

**What is a Robot and How Many Types of Robots are There?**

It is now almost common knowledge that the term Robot comes from *robota*, meaning forced labor in Czech, first used by Karl Capek in 1920 in a play entitled R.U.R. (*Rossum's Universal Robots*). It was not until 1954 that we had the first patented robot, the multijointed robotic arm, not until 1961 that the first industrial robot known as *Unimate*, from the Universal Automation company, saw the light of day, both fixed robots, i.e. with the ability to move, but not to move from place to place on its own.

One might assume that some robots can be clearly distinguished from other artefacts. However, there appears to be no universally accepted scientific definition of what constitutes a robot

> The Robotics Institute of America (RIA) defines an *industrial* robot as: "a reprogrammable multi-functional manipulator designed to move materials, parts, tools, or specialised devices through variable programmed motions for the performance of a variety of tasks which also acquire information from the environment and move intelligently in response". This early definition does not capture mobile robots. The European Standard EN775/1992 defines the

---

[1] In a 2015 interview, he predicted that by 2030 humans will have become hybrids as: "humans' brains will be helped out by nanobot implants that will make us into "hybrids".

> robot as: "*Manipulating industrial* robot is an automatically controlled reprogrammable multipurpose, manipulative *machine* with several degrees of freedom, which may be either fixed in place or mobile for use in industrial automation applications". Ronald Arkin gave the following definition: "An *intelligent* robot is a *machine* able to extract information from its environment and use knowledge about its work to move safely in a meaningful and purposive manner" (Tzafestas, 2015) (emphasis mine).

On the other hand, the Civil Law Rules on Robotics of the European Parliament resolution of 16 February 2017 with recommendations to the Commission on Civil Law Rules on Robotics (2015/2103(INL)) (2018/C 252/25) defines *smart* robots as artefacts having the following characteristics and properties:

- the acquisition of autonomy through sensors and/or by exchanging data with its environment (inter-connectivity) and the trading and analysing of those data;
- self-learning from experience and by interaction (optional criterion);
- at least a minor physical support;
- the adaptation of its behaviour and actions to the environment;
- absence of life in the biological sense

An alternative definition, simpler, yet encompassing many of the aforementioned characteristics, was proposed by Rodney Brooks:

> A robot is some sort of *device*, which has sensors, which senses the world, does some sort of computation, decides on an action and does that action based on the sensory input, and which makes some change out in the world outside of its body.

Thus, a Roomba would be a robot, while a dishwasher would not (since the changes it makes in the world are not outside its body). It would seem that other specialist engineers, such as Gill Pratt, would not disagree with this definition, accepting, for example, that a thermostat could be considered a robot, since it has the ability to sense the surrounding world, perform some kind of calculation (in this case very simple) and act.

The way they feel, compute or act may be different, but the way they operate is similar:

> Their sensors feed measurements to a controller or computer, which processes them and then sends control signals to motors and actuators. A robot is constantly repeating this sensing-computing-acting cycle, in what roboticists call a "feedback loop." So you could say that feedback is the technique that makes machines "smart," and almost every robot uses feedback [15].

The criteria used for classification also give us clues about their peculiar capabilities: they are classified according to (i) their geometric structure and their form of locomotion: (Fixed robotic manipulators, Wheeled mobile robots, Biped robots (humanoids), Multilegged robots, Flying robots, Undersea robots, robots that mimic biology or combine wheels, legs and wings.); (ii) by the capabilities they have to perform tasks, (Cognition, Perception, Planning, Sensing, Control, Action); (iii) by the type of applications they can be given (Industrial robots, Medical robots, Domestic and household robots, Assistive robots, Rescue robots, Space robots, Military robots, Entertainments robots). Taking into account the main function for which they are designed, there are currently 18 types of robots listed alphabetically: Aerospace Robots, Aquatic Robots, Autonomous Vehicles, Consumer Robots, Delivery Robots, Disaster Response Robots, Drones, Educational Robots, Entertainment Robots, Exoskeletons, Humanoid Robots, Industrial Robots, Medical Robots, Military & Security Robots, Research Robots, Service Robots, Social Robots, and Telepresence Robots (Erico [15]).

If we look only at the functions for which they have been developed, they are different from other technical artefacts, i.e. a huge number of technological artefacts had already been developed to carry out such tasks. What makes them different are some of their capabilities, especially those of acting more or less autonomously thanks to the possibility of developing Cognition, Perception and Planning. While other previous artefacts could sense, control and act (for example, certain washing machines that have the ability to control the amount of laundry loaded, thanks to the sensors they have, and act accordingly, choosing the most convenient washing programme), it did not do so in the complex way that a robot can act. These Cognition, Perception, Planning capabilities were thought to be exclusive to humans and some other non-human animals. Robots, like many organisms, have the ability to perceive the environment in which they find themselves, to analyse that environment in some way, interpreting and storing the information they extract from it, but they also have the ability to learn from their previous actions, that is, they can adaptively acquire information from these experiences, and plan their actions using what they have learned and know, without necessarily requiring the intervention of other agent.

This parallelism between artefacts and organisms does not only extend from robotics to the world of life. In biology, organisms and their parts are often described in terms of artefacts. The resources used to describe organisms "might seem better suited to the description and explanation of artifacts than the description and explanation of organisms" [23]. For example, in biology, the notions of function and design are commonly used to refer to the capacities of organisms. So it is not surprising that this explanatory continuum occurs in both directions; attributing the usual capabilities of organisms to robots, and vice versa. Of course, there is no unanimous agreement as to whether these capabilities can be understood as literally identical, whether they are merely similar, or whether they are merely a metaphor for what we would understand them to be in one case or the other. For example, there is considerable debate over whether "cognition" in biological organisms and in robots refers to the same phenomenon, as well as over whether AI possesses "intelligence" in the strict sense of the word. Or among those who point out the problems of identifying certain capabilities of organisms in terms of "design" or "purpose".

These (apparent?) similarities mean that both organisms and robots can be described as "agents" of some kind. To use just two examples of the countless definitions of what an agent is:

> (i) an agent can move the whole, or at least some parts, of something we are inclined to think of as its body; (ii) an agent is a centre of some form of subjectivity; (iii) an agent is something to which at least some rudimentary types of intentional state (e.g., trying, wanting, perceiving) may be properly attributed; (iv) an agent is a settler of matters concerning certain of the movements of its own body i.e., the actions by means of which those movements are

> effected are considered to be non-necessitated events, attributed always first and foremost to the agent, and only secondarily to environmental impacts or triggers of any sort. (Steward 2008, 226)

> An autonomous agent is a system situated within and a part of an environment that senses that environment and acts on it, over time, in pursuit of its own agenda and so as to effect what it senses in the future. [16].

The fact that robots and organisms can be described through similar capacities and properties in "agential" terms does not necessarily imply that robots and organisms are the same kind of phenomena. There are different types of agents, which share basic characteristics, but which have different types of body, subjectivity or levels of intentionality. That is, while both organisms and robots can be identified as agents, they may belong to fundamentally different types of agency. The fact that they may share similar characterisations for some of their features does not infer that the rest of the properties and capabilities are identical. For example, "Artifacts are objects made by intelligent agents; organisms-most of them, at least-owe their construction to no agent" [23].

However, some scholars, such as David [22], who argue for robots being alive on the basis of these similarities. Amazed by the fact that robots display intelligent attitudes, he believes it is legitimate to argue that they are therefore intelligent, and even goes further by claiming that since they are intelligent they are alive. Following an idea put forward in the 1970s by Joseph Weizenbaum, the creator of the ELIZA programme according to which:

> a modern computer system is sufficiently complex and autonomous to warrant our talking about it as an organism. Given that it can both sense and affect its environment, I even grant that it can, in an extremely limited sense, be "socialized", that is, modified by its experiences with its world. I grant also that a suitably constructed robot can be made to develop a sense of itself, that it can, for example, learn to distinguish between parts of itself and objects outside of itself, that it can be made to assign a higher priority to guarding its own parts against physical damage than to similarly guarding objects external to itself, and that it can form a model of itself which could, in some sense, be considered a kind of self-consciousness. (...) When I say therefore that I am willing to regard such a robot as an "organism" I declare my willingness to consider it a kind of animal. (Wizenbaum, 1976)

The Entity described by Wizenbaum can be fully understood as an agent, without it necessarily being an organism or a kind of animal. On the other hand, some of those qualities that he considers a robot can develop (doing a bit of "optimistic technology fiction"), such as self-awareness, is a quality that, at least for the moment, human beings possess as a species, and even some daring would say some other non-human animals, but not robots. Not even all organisms have it. In other words, one can be an organism and an animal without self-consciousness. But, going further, it would also have to be shown that if something has self-consciousness it is, by the fact of having it, an organism.

On the other hand, there is a problem that Levy does not take into account in his argument: attributing intelligence to an artefact that develops its capabilities by being designed by humans to mimic certain formal reasoning skills does not necessarily imply that it is therefore intelligent in the same sense that most organisms are intelligent. Intelligence is much more than linguistic and formal reasoning.

Levy also draws on another argument, this time put forward by William Lycan who in 1972 considered that if a humanoid of the future exhibits certain characteristics (such as great external physical similarity, intelligent conversation, the ability to play golf or the viola, to write poetry, to make love, to prove mathematical theorems, to show envy, or anger), then: "Does he have thoughts, feelings, and so on? Is he actually conscious, or is he just a mindless walking hardware store whose movements are astoundingly like those of a person" (Levy, 2009).

This imaginary humanoid, Levy tells us, possesses many of the characteristics that our current robots already have. However, according to its designers, the capabilities of the current robots listed in the previous section do not seem to include anything like lovemaking, envy or anger. Perhaps if robots already exhibited these characteristics we could ask whether they constitute a personality, and furthermore whether such robots could be considered persons. But here again we run into a problem: being a person does not necessarily imply being alive. That is to say, we could attribute personhood to a robot without necessarily implying that it is a living being. There are concepts in law such as "legal personalities" that do not require being alive in order to have rights. Accordingly, entities such as companies or foundations are recognized as legal persons, possessing both rights and responsibilities. In other words, contrary to Levy's argument that "many of the opponents of the idea of robot personhood claim that robots should not have any such rights precisely because they are not alive", one could say that there are (legal) persons that are not alive but have rights and obligations[2]. On the other hand, the condition of being alive does not imply having rights and obligations: most of the organisms that exist on our planet are in this situation. In other words, one can be a person in the legal sense without being a living being, and one can be a living being without being a person in the legal sense. Levy conflates different levels of analysis: ontologica level, by considering that, because an artefact can develop behaviours that imitate those of organisms, it is therefore already one more, and therefore alive. And, a normative level, by pointing out that an artefact that could develop a personality (let's remember that for the moment this has not happened) would be denied rights and obligations because it is an object, when there are already non-biological entities that are considered persons from the point of view of law.

> According to Gurnkel (2018a, 2018b), it is important to separate certain questions that are confused in the debates about the legal personhood of robots. First, there is a relevant difference between the two verbs that comprise the question: "can" and "should" AI be persons (Gurnkel, 2018a.) On the other hand, there is another relevant difference, between natural person and legal person. Following that, if legal personhood is already dissociated from the human substrate, there would be no way to deny that AI can be a legal person. But, just because it is possible, that does not mean it should be a good idea [3].

---

[2] Sophia, a humanoid robot with Artificial Intelligence (AI), developed by the Hanson Robotics company, received citizenship from Saudi Arabia in 2017. Although several interviewers were impressed with the sophistication of its responses, the robot follows a simple algorithm and most of its statements are credited to a previously prepared text (Parviainen and Coeckelbergh, 2020).

So, while organisms are agents, not all agents have to be organisms, and on the other hand, while human beings are persons, not all persons are individual human beings. That is, one can be an agent without being an organism and be a person without being a living being. In fact, the vast majority of living beings are not persons.

On the other hand, it is not by chance or as a result of evolutionary convergence that robots have developed characteristics similar to those of some organisms. This is due to the intention of their designers to take advantage of the knowledge we already have about the mechanics and cognition of many organisms in order to physically and behaviourally model robots. The following section will discuss the idea of mimetic emulation of robots and the reasons that have led designers to make robots that look and behave like some organisms, including humans.

## 2 Mimetic Robots

At one point, there was widespread optimism that it would only be a matter of a few years before robots with human-like capabilities could be developed. The intelligence required for a task as complex as playing chess had already been emulated, so that emulating our movements through the world, something also capable of an organism as simple and unintelligent as an insect, could not be so difficult. However, expectations were not fulfilled: these other, seemingly simpler tasks, turned out to be tremendously complex. The solution came from Rodney Brooks, who in the late 1980s revolutionised the field of robotics. Instead of trying to build an artificial intelligent system from the top down, i.e. trying to emulate the most complex forms of intelligence, he opted for the opposite path: to reproduce the evolutionary process of the emergence of intelligence from the bottom up, since "from an evolutionary stance, human level intelligence did not suddenly leap onto the scene. There were precursors and foundations throughout the lineage to humans. Much of this substrate is present in other animals today." [8]

One of Brooks' key contributions was the realisation that symbolic intelligence is not necessary for interaction in dynamic environments, nor is it necessary for living organisms, which are able of acting and thriving a in such environments without relying on complex symbolic representations: "Internal world models which are complete representations of the external environment, besides being impossible to obtain, are not at all necessary for agents to act in a competent manner. [8]. All that is necessary is to build agents (sic) that operate in dynamic environments using real sensors. Robots, based on the ability to sense, would develop the following characteristics: situadness, embodiment, intelligence, and emergence of intelligence.

Since then, designers, through biomimetic robotics, have moved towards building robots that act in the world in the same way as living organisms do:

> Biomimetic robotics is the development and integration of biomimetic mechanisms, actuators, and sensors by imitating biological structures or behaviors. A key feature of biomimetic robots is their capability to adapt to the natural environment. In other words, they are designed for the purpose of working in certain complex environments. For example, researchers design

> the quadruped robot to work in the field inspired by the quadruped animals. The quadruped robot is actuated by four legs, which can adapt to uneven environments due to a high degree of freedom. Recently, more interesting and powerful biomimetic robots have been developed and applied in our daily life. The robot fish can take pretty pictures in the swimming pool, monitor water quality in the sea, and help underwater rescue. The biomimetic flapping-wing robot can fly through small spaces to find targets and hide themselves to detect the environment [37].

Thanks to these processes of imitation of living organisms, technological developments have taken off in relation to the structure and behaviour of robots. "The biological world still has much in the way of suggestions for how to build, design, and program biomimetic robot systems whose capabilities will far outpace what is possible today [14].

These robots with biomimetic characteristics have the advantage of, on the one hand, being able to perform complex tasks, and on the other hand, to do so in conditions that could be dangerous or inaccessible to humans. Most of them mimic the structure and behaviour of certain organisms, ranging from bacteria to complex animals such as humans.

> The mobility of highly adaptive mammals, reptiles, and insects, offers a mimicking model for developing high levels sensory feedback and mechanisms to enable such mobility. It may be able to climb trees like monkey, run very fast like a cheetah and hop to great heights and distances like the kangaroo or the grass hopper [5].

And not only has the ability to move been mimicked, but also the biomimetics of organisms' intelligence is being developed, so that robots are able to make decisions in reaction to external changes and have the ability to learn to adapt to new environments [37].

In this imitative task, as I said, human beings have also been taken as a model. The dream of having artefacts that look like human beings and that accompany us in our daily activities comes from a long time ago. Already in the Iliad, in Canto XVIII, 416–420 Homer says: "Golden maidens, like living maidens, marched to the aid of the ruler [Hephaestus]. In their minds there is judgement, voice and the ability to move, and there are skills which they know thanks to the immortal gods. At the sides of the ruler they gasped…".[3] There are various types of robots that mimic human beings: from humanoids, robots that have a certain degree of similarity to human appearance (such as Boston Dynamic's "Atlas" robot or Engineering Arts' "Ameca", the latter of which even uses generative artificial intelligence to react verbally to questions in real time), with a general shape including head, hands, legs and eyes, to geminoid androids (such as HI-4 designed in the image of its creator Hiroshi Ishiguro), designed to look as identical as possible to a specific human. Among the reasons for creating human-like robots is the better adaptability they

[3] Why could Homer have thought of the existence of these golden maids who helped Hephaestus? The first obvious difference is that these beings are made of a different material from us, they are not susceptible to disease or injury. Perhaps that is the reason, the maids who accompany the god of fire and forging, of blacksmiths, craftsmen, sculptors, metals and metallurgy, must be made of gold by him and have marvellous properties, as this material has.

can have to carry out their activities in an environment that has been designed and created by and for us, at our height, for our hands, and to be perceived with our senses. So, for example, humanoids can perform tasks similar to what a human being would do in environments where it is preferable not to put a person's life at risk. Or they can be used for care purposes, either for children or for elderly or dependent people, showing gestures or participating in social and communicative situations [27]. Humans have developed very complex abilities to recognise expressions, gestures and other elements of verbal communication, so having robots that have these capabilities may be of great interest. The strategy of designing human-like robots is not the only possibility. There are also 'social robots', which do not necessarily resemble humans, but have a wide range of human communication and interaction capabilities, including the ability to appear to understand and respond to human social behaviour [11].

However, the fact that an artefact that has been designed to mimic certain complex organisms develops abilities similar to those of those organisms does not mean that it is an organism. It behaves that way because it has been designed that way, one would expect nothing else if it works properly. And even if they are agents that look very much like organisms, that doesn't mean they are organisms. The radical difference, of course, depends on what is considered to be a living organism and what is considered to be an artefact, and whether the mimetic capacities also include the ability to die.

## 3 To Die, must one be Alive First?

This question might also seem to have simple answers. However, analysing what it means to die and what it means to be alive continues to generate a great deal of controversy, the former in the medical field and in bioethics, the latter in the philosophy of biology and in biology itself.

Until the middle of the last century, someone (because the question was relevant mainly to humans) was considered dead if that person was in cardiorespiratory arrest. The situation changed with the development of devices that made it possible to resuscitate people who had stopped breathing or whose heart had stopped beating. In addition, mechanical ventilators and artificial feeding systems have been developed to prolong the lives of people who are no longer able to breathe or feed themselves. On the other hand, the organs of these people, if they are found to be in a state of permanent and irreversible unconsciousness, can be used for transplantation. So these new technical possibilities generated the need, on the one hand, to reconsider the traditional criterion of cardio-respiratory failure to certify death, for one based on brain death: "Brain death is understood as the irreversible cessation of functioning in the brain as a whole. It is compatible with the presence of isolated instances of minor, residual functioning in certain areas of the brain" [26], 286), and on the other hand, to reconsider that death does not only concern the organism as a whole, but also the cells, tissues and organs that may be alive and functional even if we say that the subject that contains them is dead.

This definition has a clear anthropomorphic meaning, since the brain functions to which it refers are those that differentiate us from other species. It goes without saying that the author does not find compelling grounds to suppose that human beings are, from a biological point of view, anything more than organisms. Therefore, we will accept the definition of "death" as "the cessation of the processes constitutive of biological life" [26], 286), regardless of the fact that we are also organisms with consciousness and special cognitive capacities derived from the evolutionary development of our brain.

It has been argued that dying has at least primary and secondary meanings. Among its primary meanings, a distinction is usually made between:

> Death as an event. In this usage, death is something that happens. As an event, death occurs at a particular time and place and in a particular way. Death is an event that cuts off a life.
>
> Death as a condition (...) Death is the nonreversible condition in which an organism is incapable of carrying out the vital functions of life. It is related to but not identical with death as an event because the focus here is on the specific signs that establish the cessation of life.
>
> Death as a state of existence or nonexistence (...) death is what becomes of a person after death. It refers not to the event that ended life nor the condition of the body at that time, but rather to whatever form of existence might be thought to prevail when a temporal life has come to its end.
>
> And from the secondary point of view:
>
> The secondary usages are mostly figurative. Death serves as a dramatic intensifier of meaning; for example, the historian's judgment that the rise of commerce contributed to the death of feudalism, or the poet's complaint that life has become death since being spurned by a lover. There are also extended uses that can be considered either literal or figurative, as when the destruction of the universe is contemplated: The issue open to speculation is whether the universe is fundamentally inanimate or a mega-life form [21].

Let us turn first to the primary senses in an attempt to answer the question at hand. In the sense of death as an *event*, a robot would die if something were to happen that would cut off its life, i.e. we would have to first accept that robots have a life in a literal and not a metaphorical sense, i.e. that they are alive, and that this life has been cut off. Secondly, in the sense of death as a *condition*, it seems that in order to die, it is necessary that it is an organism that dies and that it is no longer capable of continuing with its vital functions, i.e. if we affirm that robots die in this sense, we would have to accept that they are organisms with vital functions. Finally, in the sense of death as that which an entity becomes after its life has come to an end, we should also first accept that the robot has been alive, in order to be able to move on to that other state. Now, as has been shown in the previous two sections, while robots may be regarded as agents with similar properties, designed even to emulate the properties of certain living organisms, can it be argued that these capabilities give them the quality of being alive in the biological sense of the term?

If it is difficult to establish what it is to be dead, it is no less difficult to define what it means to be alive. There are a huge number of proposals that attempt to define what life is. As it is not the purpose of this article to explore this question, in what follows I will adopt the proposal recently made by [25], and I will do so for several reasons. Firstly, because these two authors, instead of lucubrating on their own and trying to come up with their own original answer, have chosen to review 30,000 scientific articles, i.e. they have reviewed what the scientific community in biology, which deals with living beings, considers life to be. Does this mean that only biologists are qualified to answer this question? Not necessarily, but at least, thanks to this review, we see how those who deal with the phenomenon in question "use" that concept. The other reason why Malaterre and Chartier's proposal has been chosen is because they propose a gradualist definition of life, i.e., they do not opt for a binary consideration, which pits living beings against the rest of the entities in the world, but instead choose to establish a gradation, which is why it might be more useful if we want to further consider the question in relation to robots.

The result of this review, however, does not seem to suggest that the robots that currently exist can be considered as living beings. According to Malaterre and Chartier's study, when the scientific community tries to define life, it does so in terms of the following dimensions: metabolism; catalysis and synthesis of catalysis; elaboration of cellular/structural features; micro/macro environment (interactions); plants/animal related (interactions); human related (interactions); evolvability, and information encoding and genetics. When they refer to the dimensions of plants/animal related (interactions) and human related (interactions) what they encompass are the aspects that usually appear as relevant topics related to living beings in those 30,000 scientific articles. On this graded scale, things such as adenine, which is an organic chemical compound but does not reach the status of a living being, would be at level 0, and at the highest level, 1, would be bacteria, which would to a high degree comprise the parameters for being considered living beings.

A different option, which goes beyond this biologistic understanding of life, is that of those who defend a conception of "artificial life". ALife is an interdisciplinary field of research whose main objective is to understand what living systems are by artificially synthesising some simple forms of these systems [6] and creating systems with life-like properties. There are currently three types of synthetic methods:

> "Soft" artificial life creates simulations or other purely digital constructions that exhibit life-like behavior (most ALife research is soft), "hard" artificial life produces hardware implementations of life-like systems, and "wet" artificial life synthesizes living systems from biochemical substances (Rasmussen et al., 2003, 2008). In this way, ALife attempts to synthesize properties of living systems in computers, machines, and molecules [1].

There seems to be agreement that artificial "wet" life involves the synthetic creation of a living system. However, the case of simulated models of living systems does not meet with the same agreement. While proponents of "strong" ALife consider these modelled systems of life to be real life as well, because they possess some fundamental properties of modelled living systems, proponents of "weak" ALife point out that "even though life is expressed by a certain systemic organisation, it nevertheless requires a concrete material realisation in order to be considered real life" [1]. To confuse the model with the modelled is to assume that the modelled characteristics are the fundamental ones. The consideration that one feature is superfluous while another is essential is a much more problematic decision than it might seem:

"consider what happens when we run a program that is simulating the molecular structure of water. Although the formal organisation of the molecules in the model is the same as of real water, the computer running the simulation does not get wet!" [1]. It may therefore be very risky to say that a robot that shows a behaviour similar to that of an organism, being modelled in a similar way taking it as the subject to be modelled, therefore acquires all the properties of an organism. In fact, among those who develop such entities it is clear that:

> Contemporary technology is still far removed from the versatility, robustness, and depend ability of living systems. Despite tremendous progress in material science, mechatronics, robotics, artificial intelligence, neuroscience, and related fields, we are still unable to build systems that are comparable with insects [1].

So even among those who are most favourable to understanding life in this artificial sense would doubt whether the mere fact that some capacities of living beings are modelled imitatively gives rise to the property of life. It will depend on whether the ability to die emerges at the same time, at least in the first three senses considered above.

But let us remember that there is still another possibility, a second way of understanding the notion of being dead, the one that is used in a figurative sense. In many languages (in Spanish, English, French and German at least) we can say "the car is dead" or "the washing machine is dead", when we mean that an artefact has broken down beyond repair. In other words, we can accept that our car or our washing machine remains the same entity despite the fact that a large number of its parts have changed as a result of various repairs. And at the same time, in the event that it definitively stops working, we can say that it is "dead". But we do so figuratively, i.e. metaphorically. In that sense, there would be no major problem in saying that a robot has died: it is not very risky to think that someone has said that their Roomba "has died", or that this prototype was a failure "before it was born". Now, as [7] has already pointed out, metaphors are the means by which human beings attempt to account for the unconceptualisable. If they are used, they are a symptom of the need to turn to something other than what one seeks to understand. It is a "transference", an instrument that allows us to understand thanks to what is already understood. If we can come to think that robots are dying, it is because we feel a certain discomfort in thinking about their end in a detached way, either because of their appearance that reminds us of living beings, or because of the human tendency (and possibly also that of other non-human animals) to establish bonds of attachment also with inanimate objects. Precisely this will be discussed in the next section.

## 4 Attachment and Grief

The notion of attachment comes from psychology, and is used to refer to the strong bond that a person establishes with someone or something, which provides security and safety, and that in case of separation from that figure the person may feel a strong sense of psychological stress. That with which the attachment bond is established can range from people, non-human animals treated as domestic or work companions, symbols and deities to objects or places.

The philosopher of technology Donald Norman developed in *Emotional Design: Why We Love (or Hate) Everyday Thing* (2004) a theory about how human beings develop emotions with certain artefacts: when we are in front of an attractive artefact we feel good, this aesthetic openness towards the attractive object puts us in a more open-minded attitude to be creative with its use, which in turn generates in us the perception that an attractive artefact works better. We establish an attachment to objects when we experience a sum of emotional episodes with the artefact.

In a similar vein Huber et al. propose the following definition:

> An enduring attachment can be defined as the sum of all cohesion episodes a person has made with another person or object. A cohesion episode is characterized by joint experiences with this other person or object in which cohesion factors are present (2016).

In order for these emotions to develop, time is needed: "what matters is the history of

interaction, the associations that people have with objects, and the memories they evoke". Obviously, if humans are able to develop links with a multitude of types of entities, they will also develop links with robots. The Human Robot Interaction field of study has been devoted to establishing what are the keys to generating human interest in robots, as well as the emotional responses that these relationships elicit. Many of these studies have been based on the bonds we establish with pets and objects that possess certain physical traits, and how these attachment relationships are produced with them.

Humans innately develop what Konrad [24] called *Kindchenschema,* an attraction to beings with certain infantile traits. When a set of facial features such as large eyes and forehead, a round face, and a small nose and mouth, which resembles that of the offspring of our species, we tend to develop affective and nurturing responses. This trait arises evolutionarily to increase the probability of survival of offspring. In other words, to apply to our human offspring. However, the attraction to the appearance of babies is so strong that it is also activated by the young of other species, releasing our instinct to provide parental care. The presence of infantile characteristics (neoteny), both physical and behavioural, in companion animals was increased during domestication. They result in a highly biological, rapid and unconscious stimulus that attracts attention, eliciting positive behaviours and emotions, including a willingness to provide care. This attraction is activated in both sexes from a very early age (Archer and Morton, 2011).

This positive emotion towards pets is something that the toy industry has applied since ancient times, so applying it to robot design was an obvious strategy. If we create robots with features similar to those of certain pets that give us this reaction, or even wild animals that inspire great tenderness because of their appearance, we can use them for a variety of purposes.

> According to the well-supported biophilic hypothesis, i.e. the natural predisposition of humans to affiliate with life, these zoomorphic animated devices are highly appealing and

> their hybrid animate and inanimate nature presents both adults and children with mental, social and moral challenges. (Diaz-Boladeras, 2023).

An example of such a robot is the therapeutic robot "Paro" designed by Takanori Shibata that bears a high degree of resemblance to a baby seal, an animal that already has the *Kindchenschema* Lorenz spoke of. The robot is equipped with three microphones, twelve tactile sensors covering its skin, touch-sensitive whiskers and a system of motors and actuators that silently move its limbs and body. The robot responds to touch by wagging its tail and opening and closing its eyes. It has also been designed to actively seek eye contact, respond to touch, hug people, remember faces and learn actions that generate a favourable reaction. The main advantage of this device compared to, for example, a real animal, is that its reactions are always predictable, under no circumstances will it attack the user, in addition to the fact that it does not need to be fed and, in the words of its creator, "it does not die" (Piore, 2014). In other words, it has all the advantages we would expect from a non-human animal from the point of view of the emotional reaction it generates in us, but without its disadvantages, i.e. it does not need to be fed, does not get sick, does not get tired, does not get bored, does not show irrational preferences towards some humans over others, and in case it breaks down it can be replaced by another Paro very similar to it.

The designs of such robots not only make use of the knowledge we have about how relationships are established between humans and non-human animals, but also between humans and humans:

> There is a strong tendency in the development of companion robots to make them as social as possible in order to reach high and enduring user acceptance. To these ends, recent approaches in robotics use psychological human-robot dynamics which mimic human-to-human binding phenomena, such as those occurring for example, between a mother and a child. Robots equipped with these specific attachment behaviors are intended to establish a similar long-term emotional tie between a user and a robot. Thereby, robots may become more than mere functional tools, but change into pet-, or even person-like companions and become "relational artifacts" (Hubert et al. 2016).

These robots become unique entities, as they employ learning programmes, memorising the reactions of specific individuals. Moreover, as the vehicle that provokes certain moments of generated positive emotions in humans, they may appear to be irreplaceable: "However, current robotics capabilities are not so great that social robots are as generative and adaptive as humans or pets" (Rabb et al. 2021, p. 551). To suggest that what takes place between humans and these robots are relationships of friendship or companionship is to overlook that such relationships are bidirectional "contradicting the unidirectional subject-object nature of the human–robot attachment" (Hubert et al. 2016, p. 368). Comparing the mutual attachment relationship that can occur between humans, or between humans and non-human animals, to that between a human and a robot overlooks that in these other cases the relationship is bidirectional. Of course humans can establish attachment relationships with robots, we can become attached to inanimate objects or even places [34], so that "feelings of security, safety, and comfort do not need to come from biological others to mitigate distress" (Rabb et al. 2021, p. 543). But that does not mean that we are reciprocated: even if the robot displays behaviour similar to that of a living being, it does not mean that it develops any real emotional reaction to us. If it does so, it is because it has been programmed to react to some stimuli by mimicking the behaviour of some beings, such as humans and some non-human animals.

However, what is often not taken into account in these social robot designs is that, just as we can develop an attachment relationship, we can also develop a strong sense of stress in the event of separation from these objects. We may even develop emotions of grief if the loss is permanent. Grief would then be a consequence of the loss of what we are attached to, not the way in which the loss occurs.

Although we generally associate grief with the definitive and irremediable loss of a human being [20], p. 2), "loss is a very diversified and multifaceted human experience (Jakoby 2015) of which death, and the grieving after a death, is but one specific case". However, it is also normal to feel it (and very intensely) at the loss of companion animals. Apparently, research on animal behaviour indicates that other species also experience it, i.e. it is not a uniquely human emotion.

But what humans (and other non-human animal species) experience in the face of loss should not be confused with the consideration that what we have lost has literally died. That is to say, the emotion of mourning is also unidirectional, we feel an intense emotion for a disappearance, that is, it is the fact of the disappearance of something or someone with whom we had previously developed an attachment relationship, which generates the mourning. Generally this definitive disappearance occurs with death, but not only. We can also feel this emotion at a divorce [13], at the loss of our home [17]. But that does not mean that the person we divorced has died or the house we lost has died. What is common to these emotions is that we feel them at the loss of something we consider irreplaceable because of its uniqueness and the fact that it is inimitable.

And being an emotion that we have to pay for the price of having previously developed an attachment to something or someone, it is no less true that grief can also have negative consequences:

> It is now recognized that grief is more than sorrow and emotional turmoil. (...) First, stress reactions include changes in physiological function that can increase one's vulnerability to illness and exacerbate pre-existing physical problems Secondly, perception and thought are also affected, with the increased possibility of making impulsive and potentially harmful decisions and becoming more at risk for accidents. Third, a spiritual crisis often occurs, in which the guiding assumptions and values are called into question [21], p. 349).

So the question arises as to whether it is advisable to produce robots that are specially designed to provoke this strong emotion of attachment, or whether it would not be better for robots to clearly display their artefactual status in order to avoid such a reaction on the part of users. As [4], p. 5) point out, the imitation of these features serves to "simplify interaction and to make robots more engaging (…). However, this does not imply that the robot should pretend to be alive. Instead, we propose that robots should be designed in an honest way that clearly shows that they are machines".

Whereas a robot, although from the point of view of the information it has been able to capture and which allows it to act more efficiently, is unique, we could not say that it is therefore inimitable. That is to say, a single robot of one type is not made. In that sense it is not comparable to the loss of a non-human animal with whom we have shared our familiar and intimate space. Those of us who have lost pets know how irritating it can be to be told, when we are in the grieving phase, to quickly acquire another animal to replace the one that has died. That animal was unique and inimitable, as will be the one that comes after it (if it is to come), and the two-way attachment relationships that have occurred between the non-human animal and the human derive from the inimitability of that subject's personality, as well as the experiences that have been shared with it.

## 5 Conclusions

This article has sought to review some of the reasons why the idea that robots can die can be defended. Part of the confusion comes from the fact that robots can be understood as agents, and what is generally understood as such are biological organisms. So it is correctly inferred that if something is an agent, then it could be an organism. Another argument is that robots, because of the agential behaviour they develop, can also be considered persons. However, being a person and being alive are not synonymous. Most persons are human beings, but not all persons are human beings. In other words, one could consider the idea of robots as persons in the legal sense, without them becoming living beings. On the other hand, the very design of robots, which mimic the physical and behavioural traits of some organisms, may also lead to the conclusion that they could for that reason be living beings. Not in the traditional sense of the term, but as Alife. However, it is more than questionable whether a model of an organism is itself an organism. On the other hand, we are not surprised by the use of an expression such as "my Roomba is dead", although we detect that it is a figurative or metaphorical use. However, the metaphor becomes narrower if what has broken down beyond repair is "Paro", the baby seal robot, a robot expressly designed to provoke in us an innate reaction of care and tenderness. It is inevitable that we develop bonds of attachment and, consequently, mourning towards such robots when they disappear. It is precisely this strong emotion in the face of loss that can give rise to the confusion of believing that the robot has died, because it provokes in us the same emotion as when a living being with whom we have a bond does.

When we think of a robot dying, some of the most iconic robots from literature or film inevitably come to mind. Depending on one's aesthetic preferences, this robot could be the Terminator, HAL, or the replicant from Blade Runner. All of them show us their reaction of fear and avoidance of the end. Somehow they have become entities susceptible to death, even if they are synthetic, even if they don't exactly have a body. They all show us a personality, a language like our own and endless other similar traits. It is not for nothing that the actors who played the roles of the Terminator and the replicant were flesh-and-blood human beings. And HAL, although he did not have a body like ours, he did have a language and a complex mind, so complex that he developed a mental illness. I wouldn't know what to say to the question of whether HAL died when Bowman pulled the plug, or was simply switched off. If we ever have robots with those characteristics in the future, I would have to ask the question again. What I find a little easier to say is that the robots that exist right now do not die, except in a figurative sense. The use of this figurative sense derives from certain human capacities developed in the face of entities with certain characteristics. The human capacities to develop attachment and to grieve can be produced in the presence of objects. And if these objects have been designed to generate emotional reactions in us, then the relationship of attachment and the emotion of mourning the loss of that object will be very similar to what we feel for other beings that do die, so we can intuit a certain similarity between what happens when an organism dies and when a robot is permanently deactivated. But that does not mean that they are identical events. That is, robots, at least the ones that have developed so far, cannot die because they are not alive.

## References


1. Aguilar, W., Santamaría-Bonfil, G., Froese, T., Gershenson, C.: The past, present, and future of artificial life. Front. Robot. AI **1**, 8 (2014)
2. Archer, J., Monton, S.: Preferences for infant facial features in pet dogs and cats. Ethology **117**(3), 217–226 (2011)
3. Avila Negri, S.: Robot as legal person: electronic personhood in robotics and artificial intelligence. Front. Robot. AI **8**, 789327 (2021)
4. Balkenius, C., Johansson, B.: Almost alive: robots and androids. Front. Hum. Dyn. **4**, 703879 (2022)
5. Bar-Cohen, Y.: Nature as a model for mimicking and inspiration of new technologies. Int. J. Aeronaut. Space Sci. **13**(1), 1–13 (2012)
6. Bedau, M.A. (2007). Artificial life. In: Matthen, M., Stephens C (eds.). Handbook of the Philosophy of Science. Volume 3: Philosophy of Biology. Amsterdam: Elsevier, pp. 585–603
7. Blumenberg, H.: Las realidades en que vivimos. Paidós, Barcelona (1999)
8. Brooks, R.A.: Intelligence without reason. Prepared for Computers and Though, IJCAI-91. (1991)
9. Brooks, R.: I, Rodney Brooks, am a robot. IEEE Spectr. **45**(6), 68–71 (2008)
10. Civil Law Rules on Robotics European Parliament resolution of 16 February 2017 with recommendations to the Commission on Civil Law Rules on Robotics (2015/2103(INL)) (2018/C 252/25)
11. Coggins, T.N., Steinert, S.: The seven troubles with norm-compliant robots. Ethics Inf. Technol. **25**(2), 29 (2023)
12. Díaz-Boladeras, M.: Bond formation with pet-robots: an integrative approach. Curr. Psychol. **42**(4), 2591–2608 (2023)
13. Duffey, T. (2021). Divorce and other loss issues in family therapy. Foundations of Couples, Marriage, and Family Counseling 2nd Edition, pp. 357–376
14. Gao, Z., Shi, Q., Fukuda, T., Li, C., Huang, Q.: An overview of biomimetic robots with animal behaviors. Neurocomputing **332**, 339–350 (2019)
15. Guizzo, E.: What is a Robot? In https://robotsguide.com/learn/what-is-a-robot (2024)

16. Franklin, S., & Graesser, A. (1996, August). Is it an Agent, or just a Program: A Taxonomy for Autonomous Agents. In *International workshop on agent theories, architectures, and languages* (pp. 21–35). Berlin, Heidelberg: Springer Berlin Heidelberg.
17. Fried, M.: Grieving for a lost home. In People and buildings, pp. 229–248. Routledge, (2017)
18. Homer. Iliad. Ed. Gredos: Madrid
19. Huber, A., Weiss, A., Rauhala, M.: The ethical risk of attachment how to identify, investigate and predict potential ethical risks in the development of social companion robots. In: 2016 11th ACM/IEEE International Conference on Human-Robot Interaction (HRI), pp. 367–374. IEEE, (2016)
20. Jacobsen, M., Petersen, A.: Exploring grief. Taylor and Francis (2020)
21. Kastenbaum, R.: Macmillan encyclopedia of death and dying. Detroit, MI: Macmillan Reference USA/Thomson/Gale, 2 volumes (2003)
22. Levy, D.: Are Robots Alive. In: Cheok, A., Zhang, E.Y (eds.). Human-Robot Intimate Relationships, pp. 159–188. UK: Springer (2019)
23. Lewens, T.: Organisms and artifacts: design in nature and elsewhere. MIT Press (2004)
24. Lorenz, K.: Die angeborenen Formen moeglicher Erfahrung. Z. Tierpsychol. **5**, 235–409 (1943)
25. Malaterre, C., Chartier, J.F.: Beyond categorical definitions of life: a data-driven approach to assessing lifeness. Synthese **198**(5), 4543–4572 (2021)
26. McMaham, J.: Death, Brain Death, and Persistent Vegetative State. In: Kuhse, H., Singer, P. (eds.). A companion to bioethics. Oxford: Wiley-Blackwell (2009)
27. Minh Trieu, N., Truong Thinh, N.: A comprehensive review: interaction of appearance and behavior, artificial skin, and humanoid robot. J. Robot. (2023)
28. Norman, D. Emotional Design: Why We Love (or Hate) Everyday Thing. Basic Books (2004)
29. Parviainen, J., Coeckelbergh, M.: The political choreography of the Sophia robot: beyond robot rights and citizenship to political performances for the social robotics market. AI Soc. **36**(3), 715–724 (2021)
30. Piore, A.: Will Your Next Best Friend Be A Robot? Robots can already vacuum your house and drive your car. Soon, they will be your companions. Retrieved from https://www.popsci.com/article/technology/will-your-next-best-friend-be-robot/ (2014)
31. Prescott, T. J., Lepora N. F., Verschure, P.: Living Machines: a Handbook of Research in Biomimetic and Biohybrid Systems. Oxford University Press (2018)
32. Prescott, T. J.:. Live. In: Prescott, T. J., Lepora N.F., Verschure, P. (2018). Living Machines: a Handbook of Research in Biomimetic and Biohybrid Systems. Oxford University Press (2018)
33. Rabb, N., Law, T., Chita-Tegmark, M., Scheutz, M.: An attachment framework for human-robot interaction. Int. J. Soc. Robot. 1–21 (2022)
34. Schifferstein, H.N.J., Zwartkruis-Pelgrim, E.P.H.: Consumer-product attachment: measurement and design implications. Int. J. Des. **2**(3), 1–13 (2008)
35. Steward, H.: Animal agency. Inquiry **52**(3), 217–231 (2009)
36. Tzafestas, S.G.: Roboethics. A navigating overview. Springer, Heilberg (2016)
37. Wang, J., Chen, W., Xiao, X., Xu, Y., Li, C., Jia, X., Meng, M.Q.H.: A survey of the development of biomimetic intelligence and robotics. Biomim. Intell. Robot. **1**, 100001 (2021)
38. Weizenbaum, J.: Computer power and human reason: From judgment to calculation. W. H. Freeman & Co, (1976)

**Ana Cuevas-Badallo** is professor at the University of Salamanca in the Department of Philosophy, Logic and Aesthetics. She is currently chair of the Interuniversity Master's Degree in Science, Technology and Innovation Studies. Author of several articles on epistemology, ontology and axiology of technology.

# Deadbots, Griefbots, and Postmortem Technologies: A Computational Challenge to Mortality

**Lionel Obadia**

**Abstract** This chapter examines recent advancements in postmortem technologies, specifically the development of deadbots and griefbots, and the implications they pose in confronting the finality of death. These emerging technologies, which aim to replicate or simulate the presence of deceased individuals through digital avatars, provoke significant ethical, anthropological, and sociocultural debates. As these technologies rapidly evolve, they challenge traditional notions of life and death, raise concerns regarding data privacy and responsibility, and provoke reevaluation of the human experience in relation to death. This paper seeks to explore whether deadbots represent a genuine technological challenge to the concept of death itself or merely function as a modern means of memorialization. By analyzing contemporary case studies and media representations, the chapter addresses the societal and emotional impacts of these technologies, while also scrutinizing their potential to transform human relationships with mortality.

**Keywords** Deadbots · Artificial intelligence · Technology · Grief · Death

## 1 Introduction

In recent years, the advent of so-called 'deadbots'—digital avatars modeled after deceased individuals—has garnered substantial media attention and public interest. Emerging in the early 2010s, these technologies occupy a complex and often controversial space in modern society, wherein they are lauded by some as revolutionary tools for emotional healing and condemned by others as ethically problematic interventions. Deadbots, a subset of conversational agents, extend beyond the functionalities of typical chatbots. While conventional chatbots are confined to specific roles, such as facilitating e-commerce or educational exchanges, deadbots aim to recreate the likeness—whether linguistic, visual, or auditory—of individuals who have passed

L. Obadia (✉)
Laboratory: LARHRA (UMR 5190, Department of Anthropology, Université Lyon 2, Bron, France
e-mail: Lionel.obadia@univ-lyon2.fr



A. Sans Pinillos et al. (eds.), *SecondDeath*, Studies in Applied Philosophy, Epistemology and Rational Ethics 76, https://doi.org/10.1007/978-3-031-98808-0_13

away. As a result, they are tasked with the profound and delicate mission of bridging the gap between life and death, introducing significant ethical and philosophical challenges.

This paper aims to explore the multifaceted implications of deadbots within the broader context of human engagement with technology and mortality. It will examine the development and popularization of deadbots, analyzing their potential to redefine societal perceptions of death and afterlife. Furthermore, it will discuss how these technologies navigate the intersection of human emotion, memory, and digital simulation, drawing on contemporary examples and scholarly discourse to question whether deadbots constitute a technological response to the inevitability of death or simply a novel form of posthumous digital interaction.

Deadbots, these new "fashionable" inventions, or at least fascinating avatars that have been much in the news since the 2010s, continue to be seen with regularity as cutting-edge technologies that play a part in redefining death in the current collective representations of modern societies where technology has developed. Hailed on the one hand as a spectacular breakthrough by some, and as a threat to humanity by others, so-called 'deadbot' technologies are more complex than their semantic category would lead us to believe, opening the way to a wide range of considerations—ethical, anthropological, ideological, ethical or otherwise. It has often been said that, like the Artificial Intelligence technologies of which they are an offshoot, deadbots, which are variants of "bots", the conversational agents that have been deployed in many segments of modern society (online shopping, educational advice, playful companions, etc.), are helping to extend the role of machines in our lives a little further, and to integrate digital devices into our environment.

The difference is that, unlike many other chatbots whose function remains tied to recommendations based on a pre-established scenario, which remain 'locked' in their role as digital companions to ordinary human activities, and are natively digital (created by and for a given technological environment), deadbots often (but not only) appear as digital agents whose role is to recreate conversational avatars or 3D silhouettes of people who have existed, and often not just to give the illusion. Because of the way they operate and the roles they are increasingly—and very quickly—being asked to play in society, deadbots represent a major advance (for better or for worse, depending on the point of view of their observers) that confronts humanity with one of its oldest challenges: that of facing up to death, and, in this particular case, staving it off or even annihilating it, by means of technology.

Work on grievancebots—deadbots—has developed rapidly in recent years and has explored many facets of these technologies and the psychological, emotional and social impacts of their use, with a strong focus on the ethical issues raised by their current development, on the effects of illusion and the transformation of reality, and on the very concepts of 'life'. The aim of this article is to continue these debates, but to direct them towards a specific question: are the technologies grouped together under the category of 'deadbots' really likely to change the relationship between humans and life, through a new way of looking at death? By discussing a corpus of videos and texts on the subject, this article will ask what deadbots really are—whether they represent a (technological) victory over death, a simulacrum of post-mortem life, or simply a new alliance between humans and machines, with the latter serving the former.

## 2 Media Coverage, Social Interest, Anthropological Questions

Over the last ten years or so, deadbots or grief bots—whatever they're called—have been appearing in the media, benefiting from a surprise and snowball effect. In fact, one of the particularities of this technology is that it is given such wide media coverage, even though it is not widely distributed socially or geographically: While the biggest companies (such as Microsoft) have already tried to patent deadbot technologies, while companies mainly based in Asia (China, South Korea and Japan) are developing them on a large scale and starting to market them, and while the unexpected conversion of conversational agents into deadbots has been reported in several countries, in many other places these technologies have no place in the media, and remain localized and discreet initiatives that are not very visible. So deadbots are technologies created in the West by major Artificial Intelligence companies (in particular Open AI'S ChatGPT), which claim a universal vocation (addressing virtually all humans), but which are developing mainly in Asia, an area which, by ricochet, is coming back into Western debates… the circle is complete.

Articles in major American newspapers, but also Joseph Fagone's article—which traces the first traces of deadbots in current media coverage, from the first widely publicized examples in the 2010s (@Roman in 2015, see above). The second best-known case is that of Joshua Barbeau, who used an old version of GPT (known as Project December) in 2021 to "remake" his deceased girlfriend. Meanwhile, in 2019, another case will mark the media history of deadbots: James Vlahos collected information about his dying father and "reconstituted" him in the form of a Deadbot—before setting up his own company Hereafter. AI, which has been offering services ever since. Since then, media coverage has continued to expand, and video broadcasting platforms such as YouTube © now relay content that is widely viewed and commented on from, helping to publicize them. With the emergence of deadbots in the media, several categories of sources are now talking about these technologies, which can be classified as follows:

(1) Testimonies in their own name, which can give rise to books, as with Vlahos (2019), which represents a particularly enlightening case. Individual cases are few and far between, and are generally among the first signposts of the deadbot movement.
(2) In-depth articles or reports by professional journalists, placing the technology in context and demonstrating the interest of the public and political and administrative bodies in these technological developments, since it is largely thanks to these informational 'alerts' that ethics committees have been able to give their opinion on the psychological and social impact of these technologies.

Media promoting so-called “deadbot” technologies, including Hereafter. AI https://www.hereafter.ai/media), which are explicitly in favor of the deployment of these technologies, as economic and social players essential to their dissemination in society.

And, of course, there is a rapidly expanding scientific literature (from which articles in print and online journals happily draw), with contributions from a wide range of intellectual backgrounds, including engineering sciences, computer science, philosophy and the humanities—some of the key references for which will be presented in this text.

There is no question here of taking sides with one of the theories set out in the scientific works, or of gauging the ethical issues involved in the deployment of deadbots. Rather, from a different perspective, inspired by the social and cultural contextualization of technologies, and in keeping with the anthropological background and method in which I was trained, I want to consider what these sources say about the rise of these technologies and how they fit into society—before, if necessary, discussing the challenges they pose to society. First of all, it is important to realize that the unifying category of deadbot encompasses a number of different technologies, even though they are all based on the principle of Large Language Models (henceforth LLMs) and conversational agents, i.e. technologies that are ‘intelligent’ because they are capable of interacting with humans via language. In all the web pages, texts and hypertexts consulted or collected for this article, there seems to be an evolutionary logic behind these technologies as they are represented in the media. Deadbots appear as conversational interfaces that begin their existence flat on computers and in the virtual silence of screen writing, and continue it in the sound of recomposed voices of the dead, and even—though this remains rare—in the form of articulated robots or virtual reality avatars. In this way, they pass through states of technological advancement that can easily be considered as stages. So deadbots manifest themselves in different expressive forms:

Firstly in writing, with the first digital technology devices that enabled a first level of reconstitution of the deceased: initially via computer screens, but with smartphone technology, it was also through “texting” exchanges that communication with substitutes for the deceased was possible.

Then there are the exchanges of voices and images, with the integration of audiovisual dimensions: AI technologies not only make it possible to reconstitute individual language styles via text, they also replicate the discursive capacities of the deceased, reconstituted on recordings of their voice that were collected during their lifetime, first by drawing on the memory traces kept by their relatives, and, more recently, from a 3D avatar via a virtual reality headset.

The materialization of bodies using robotic or mechanical devices which, at this stage in the development of technologies, involve a cobbled-together combination of high-level technologies and less sophisticated elements, such as 3D printing: this was the case with the first case of a conversational agent installed in the robotic body of a deceased person, which was widely publicized in the media https://globalnews.ca/news/4800654/robot-dead-loved-one-3d-printed-face/). The Digital Shaman project has continued this line of technological development https://www.koaa.com/news/2018/12/28/a-new-way-to-grieve-a-robot-that-acts-looks-like-dead-loved-ones/). It is interesting to note the link between printing and robotic technologies, which ensure continuity of form for the face of the deceased, fixed once and for all in an immobile but fixed form, which is associated with a robotic body that is mobile and therefore gives a relative impression of human movement—and therefore the illusion of life.

Finally, the most advanced of all, because it is the most recent, is the reconstitution of an interactive multimodal space through the use of virtual reality headsets, and the creation of 3D avatars of the deceased with multisensory interactions (language, vision, hearing, touch) including dialogue and "physical" contact, even offering the possibility of dancing with the avatar of the deceased. In mediatized cases, which faithfully reproduce the visual experience of use, the deadbot's silhouette is still marked by the 'pixel' aesthetic of virtual avatars, but it moves more fluidly and naturally than an anthropomorphic robot wearing a human mask. In one high-profile case, which provoked numerous reactions in online networks, virtual reality technology is closer to the reality of human morphology and the 'naturalness' of human gestures, and enabled a mother to 'rediscover' her young daughter's avatar in an exclusively digital context, This 'reunion' is not a reunion at all, but it looks so real that the mother's outpouring of emotion is very real, and it is the human emotion that has been so widely commented on. The technological device has since proved its ability to be applied to other people, and to elicit identical emotional expressions (for example, a young girl and her grandmother, in South Korea) and is therefore fully effective in managing the emotions of bereavement, which it is supposed to help control and attenuate.

As we can see, the technologies are advancing towards ever greater sophistication, and seem to be evolving in an additive mode, since with each new model comes a new device that improves the whole and gradually brings it closer to a human model, and which above all shows the capacity of the technologies to respond to the desires of humans to see the deadbot come as close as possible to the appearance and attitude of the deceased. However, this is not a 'natural' evolution of technologies, as some remain, for practical, economic or ethical reasons, confined to their state of development, without there necessarily being a transition from the 2D of screens to the 3D of VR.

Since the ELIZA technology, considered to be the first conversational agent worthy of the status of "artificial intelligence", created by the computer scientist Joseph Weizenbaum in the mid-1960s, logistical and IT progress has been very rapid and has produced quite spectacular results in certain areas, in particular the complexity of interaction and the richness of exchanges: Between the white screen and line-by-line exchanges of ELIZA, and the cutting-edge technology of re;memory, an enterprise staging videos of avatars of the deceased reconstructed by AI, the qualitative leap is significant and visible.

However, the technology has not yet reached a degree of advancement such that the reality of these technologies mimicking the living, to represent the dead, would reach that near-perfection of a similarity of morphology, attitudes and gestures, voice and intonations, personality and style of communication that we find embodied in the post-human robot Ash in the episode "Be right Back" in the series Black Mirror (February 2013). The episode tells the story of a young woman driven to despair by the unexpected loss of her husband, who turns to a specialist company to obtain the closest possible replica of the man who shared her life. The deceased, idealized by death, is 'reincarnated' in a robot that not only reproduces the character traits of the deceased, including those that did not please the young widow, Martha. Martha goes from one disappointment to another with her husband Ash's material avatar, which she finally disconnects… taking care not to get rid of completely. The episode, much quoted in specialist literature on post-mortal technologies and digital immortality, serves as a reference for its anticipatory nature about a future in the making, but also for pointing out the possible drifts in awareness of reality and the attachments that humans form with each other and with the material elements of their environment. What's disturbing about the episode Be right back is not so much its technological anticipation as its exploration of the confusion caused by the character of Ash, who is 'him' without really being 'him', who asks himself questions about his place in human reality, in which, in turn, he creates confusion in those close to him who recognize him without fully accepting him because of his status as half-human, half-machine, but so much Ash, in the end…

## 3 Technology, Mortality, Society

In the background to these technological, social and imaginary developments is, of course, the broader idea of 'technological immortality' or 'digital immortality', defended, criticized or analyzed both by those involved in the debates to which the concept has given rise, and by observers of these debates, which would be the sign of the advent of a 'post-mortal society', which the Canadian sociologist Céline Lafontaine has shown to take the form of many and varied mechanisms, all of which more or less relegate death to the margins of society and the cultural horizon of individuals, Even if it recognizes a 'return of death', manifested in a panoply of forms of end-of-life support and reinvented funeral rites (2008), and the rapid progress of a digitalized death culture that is benefiting greatly from the spectacular developments in Artificial Intelligence, all contribute to this return of death to the agenda of societies that are confident in their ability to overcome it [31].

Death-related technologies can be broadly divided into three categories: digital funerals (death rites that now include IT tools), digital memorials (programs, smartphone applications and websites specifically dedicated to remembering the deceased), and deadbots (conversational agents that are supposed to replicate the deceased). They are advancing rapidly insofar as Van der Horst and Kamp (2022) consider that we have reached a degree of sophistication that goes beyond the prototype stage (2022: 216) and that commercialized technological models are already being developed. According to these two authors, the deadbot is in no way a deviation from technology but, on the contrary, the fruit of an ongoing exchange with the public, anticipating or accompanying the needs of contemporary societies and, as a result, "a griefbot is a digital tool that helps users to prepare for impending death" (2022: 220) The authors' intention is precisely to prepare students and future engineers of these technologies to develop them in relation with society's stakeholders (users) to 'solve' the 'problem' of suffering death by anticipating the 'negative effects' (addictions, quarrels, selfish use, robbing data, privacy issues…, 2022: 226). This highly mechanical relationship with human emotion is precisely what makes deadbots original as a technology in the vast movement to digitalize societies: a human 'problem' is matched by a technological 'solution' set in a social context, which, for the most skeptical, means circumventing rather than resolving the issue of death—which is not just any issue for humanity. Admittedly, this is one of the many 'Techno-Affective Assemblages' that characterize current forms of hybridization between humans and technology [14], but it is not just about death, since, by extension, identical assemblages can also be found in the new 'affective' [18] and even sexual [11] robots. But for the most critical of us, what is at stake here is also the rapid development of a digital afterlife industry that lends itself to a reading of political economy rather than simply philosophy or ethics, since it is so closely aligned with the logics of late capitalism and the digital age [32], to which a precise ethical framework must therefore be applied, in correspondence with political and economic issues that are not limited to phenomenological reflections (Öhman & Floridi, 2018).

Among the many ethical issues raised by the emergence and development of AI technologies and LLMs, deadbots in particular raise questions relating to responsibility for the possession of personal data and its use. Several ethical questions are raised by the various deadbot technologies and the way in which data is now dissociated from people and associated with circuits and mechanical agents (donor/recipient/interactant) in services for 're-creating' human personalities [16]. But over and above the people themselves and respect for their moral integrity, it is also the capacity of technologies to become part of the networks of relationships woven by humans that is emerging.

[1] consider that, despite their high degree of technological complexity, deadbots ultimately have only a limited power (in time and in the extent of their relational capacities) to generate and above all maintain the types of social and affective relationships based on human recognition and the sharing of emotions on which the experience and management of grief are based. In so doing, they can only play a role in temporarily alleviating the suffering experienced by the remaining human beings, faced with the disappearance of one of their own. If this were not already a reason to put the capabilities of deadbots into perspective, the two authors also warn of the artificial nature of the relationships that are created with these human-like talking machines: the point is not to make them the sole or main instruments for managing the grief process, but rather tools that, along with other, more traditional tools, ensure the whole process. In this sense, deadbots would appear to be technologies integrated into a human experience that is fraught with meaning and emotion, that of death, and in no way supplementary to this experience. And yet, given the criticisms that have been levelled at deadbots, this is precisely the danger that lies behind the suspicion that they are the target of, or the moral prohibitions that have been imposed on them. In the background, of course, it should also be noted that the criticisms levelled at deadbots are also those levelled against AI technologies and, in particular, generative AI models and Large Language Models, which are currently flourishing, and which

are at the heart of debates in modern societies, challenged by the frenetic acceleration of AI technologies. Some attacks or criticisms of deadbots, based on research into users and their emotional relationship and dependency on deadbots, have pointed to the gap between what companies selling grievancebot devices are proposing, and the reality of their action, which can lead to emotional drifts that are harmful to the equilibrium of the humans who use them. Faced with the risk, on the one hand, of pathological emotional dependency, but, on the other, of a real capacity for these technologies to provide users with psychological support, Nora Lindemann proposes that they be described as "medical devices" (2022).

In France, where the author of these lines comes from and from which he writes, other fears have been put forward to thwart the development of these technologies. A high-level council on technology ethics at the main research organization and funder of major national programs (the Centre National pour la Recherche Scientifique or CNRS) has on several occasions been called upon to respond to the ethical issues arising from the rapid development of deadbot technologies. In fact, France "bans" deadbots, or at least the ethics committee that has ruled on their status has issued a reserved opinion on their development. This is a most interesting argument, because it touches not only on the ethical issue of human integrity, but also on the capacity of machines to learn and create original content and forms of language exchange, on the basis of these increasingly sophisticated dialogical capacities.

For all that, France has no problem authorizing other technologies that inject digital devices into the process of death, funerals and, above all, post-mortem management of mourning and memory: technologies to support the end of life, as we see them developing elsewhere [24]. But it is also for deeper reasons: behind the technology of the deadbot and the 'dignity of the dead', other ethical questions, deeper and more rooted in fundamental anthropological reflection, seem to emerge and shed light, through an in-depth reading, on the resistance observable on the surface of the facts: if deadbots change the silhouette and attitudes of the deceased, they also have, like other technologies, this capacity to transform the living. By dint of involving technologies in the most ordinary human activities, as well as those most charged with meaning, AI and conversational agents are gradually transforming our very conceptions of the human being, who is losing his ontological status as a being of consciousness in favor of an instrumentality reminiscent of technologies, in the sense that the human being is no more than a device interacting with his environment and adapting to its changes [27], as would an AI that transforms as it is trained. In this particular form, at least, the deadbot appears less as a transformation of the dead than as a mutation of the living, less as a 'humanization' of machines than a mechanization of humans.

## 4 A Recent Phenomenon—Quick Developments

The first cases appeared almost unexpectedly in the online press not long ago. The best-known and most-quoted example is the case of Roman Mazurenko who died in 2015, when a friend of his collected his tracks on the Telegram messaging service and set up a conversational agent (@Roman). But the first really to be considered is the digital interface ELIZA, the first chatbot created in 1966 by J. Weizenbaum, with textual interaction possibilities. Since then, technologies have improved, thanks in particular to language technologies such as chat-GPT. It is interesting to note that Horst and Kamp see a genealogical link between ELIZA and the recent Replika © technology and Roman (2022: 222), even though not all digital avatars are part of the same questioning of the facets and silhouettes of the human—Replika offers a natively digital pseudo-companion, which is not the case with Deadbots. Once again, purely digital avatars raise questions about the replication of the human, whereas avatars of dead humans raise questions about the recreation of the human. Two modalities that converge towards the same set of questions, but which are oriented in different directions, in the sense that one proceeds from post-human developments and the other from post-mortem developments. As a result, while the expression deadbots has become popular, it is in fact available in a variety of technological forms: digital humans, virtual humans, avatars, virtual visible humans, etc. are all variants of these copies of humans, but they do have their singularities, in particular the fact that they represent entities that have no prior existence or virtually 'embody' models of the deceased, who are not all deadbots, such as the digital Einstein, a conversational avatar with whom it is possible to discuss science and other subjects, without necessarily having any memorial or emotional link with him.

Harbinja, Edwards and McVey [13] prefer to speak of Ghostbots (in the sense that they are also categories of deepfakes that play with reality) but they all refer to these technologies of digital reincarnation and the ethical problems raised by the use of the data collected to feed these human avatars. The authors see a link of continuity between the deedpfakes of the living and the deepfake of the deads (2023: 3). On the basis of their ability to create content and ways of interacting with humans, Morris and [26] propose to describe these technologies as "generative ghosts": far from being simple replicas of humans with a limited pool of linguistic resources and references, conversational agents are artificial intelligences capable of genuine creativity and not simply "parrot" machines that spit out already pre-formatted content, giving the impression of repeating the language tics of the dead.

But the lexical network of deadbots continues to be enriched by other terms, revealing other facets, and above all, through the range of terms, revealing ever more complex facets. Henrickson, for his part, has seized on the term 'Thanabots', which seems to fit in quite well with this characterization of these post-mortem agents with whom it is now possible to chat (2023). This reference to Thanatos—which is above all a reference to thanatology—justifies that, after Penfold-Mounce, Nowaczyk-Basińska and [28] distinguish between several categories of imaginations: sociological, Thanatological and Immortological. According to this tripartition, deadbots seem to be located mainly at the level of the Thanatological imaginary—but they also reflect, more broadly, a desire (on the part of a certain section) of men and women in today's world to make the immemorial dream of immortality a reality thanks to digital technologies. In the same way, argues historian Adrienne Mayor, that the preoccupations of the Ancients and their fascination with machines capable of defying life, death and the laws of the universe seem to extend into the current techno-scientific programs of robotics and AI [25], even if Mayor does not deal specifically with the case of deadbots. It is another traditional metaphor, borrowed from Haitian folklore and Western horror cinema, that of the Zombie, that Debra basset uses to qualify the status of these digital avatars, when she refers to "digital zombies" (2015) that are neither alive nor dead and into which deadbots fully fit. Basset points to the ambiguity of death, which can take place both organically (the end of the person as a flesh-and-blood entity) and digitally (the continuity/end of the traces left behind in a digital ecosystem).

With a veritable industrialization of services linked to high-tech death or post-mortem services, deadbots, or their lexical variants, are in fact part of a much broader range of technologies, touching on the psycho-affective, social and material management of death. Deadbots were initially born by chance or accident, from an unexpected use of conversational interfaces, and quickly became programmes specifically dedicated to this purpose. But death-related technologies continue to develop all around them: digitized memorials of the deceased, virtual cemeteries, online funeral rites… are also marking out the new spaces of death. Increasingly sophisticated research is now looking at the various ways in which communication with the dead can be mediated technologically [4], which is then akin to modern spiritualism [12], the production of models of digital eternity using data on the deceased (Brun and Bourdeloie, 2022), the transposition of mourning practices onto social networks [6], 2013, 2014, 2024), to the impact of the hypertextuality of digital mourning on emotions [17]. In this increasingly varied panoply of technological responses to humanity's biological fate, griefbots stand out because of their ability to establish an active relationship with the avatars of the deceased (Voinea, 2023), which is not the case with all the other technologies, the vast majority of which are devices for digitally augmenting practices relating to mourning. Deadbots are just as much aids to mourning and managing the suffering of loss, which is why their creators claim to have designed them.

In the case of deadbots that simulate human speech skills, non-verbal communication and replication of the habits of the deceased. The challenge, of course, is to replicate as closely as possible the personality traits of the deceased, their turns of language, their sense of humor, their usual preoccupations, interests and tastes, preferences and imaginative capacities, the contents of their memory, their intonations and speech rhythms… In short, everything that makes up the personality of deceased individuals, which, at this stage in the development of research, remains very complex to implement. Some computer systems, such as the Natural Constructive Cognitive Architecture (NCCA), are showing fairly significant results, and are bringing mathematical and logical models of human reality closer and closer to their reference objects (humans) (Chernavskaya, 2023). However, if we take the analysis a little further than this approach, these are portraits of humans and not personalities of humans, if we keep the analysis on an ontological level.

## 5 Death, the Second Death or Beyond Death

The technologies grouped together under the name of griefbots are varied enough not to represent a single answer to the question of death. Above all, these technologies raise far more problems than they solve. Of the criticisms that have been levelled at dead-death-griefbots technologies, it is worth noting that they are generally of an ethical nature, and in this sense touch on several areas. On the one hand, the integrity of the individual, 'recomposed' in digital form, is transformed and escapes its fate of extinction. On the other hand, modern life is the source of a multitude of digital traces, which make up fragments (of life, images, words, attitudes and personality traits) that until recently were mainly of interest to the living and the integrity of their rights and image. However, as modern societies become increasingly digital, living people are becoming increasingly concerned about the use of their digital footprints, as well as the question of the footprints left by deceased users of digital technologies. It is precisely in this area that organizations have been set up to manage post-mortem data, with a view to protecting the living from such a fate. But thorny ethical questions remain: beyond the ownership of data, the uses to which it is put, the wishes of the living with regard to the design of the deadbot, the responsibility of industrial players or the technology in the event of misuse or problems impacting individuals and society Harbinja, Edwards and McVey [13] refer in this respect to a new modality—"do not bot me"—which is a way of refusing to play the game of this "data-isation" of oneself and society, of life and therefore of death.

It is on this point that this article intends to position itself in the current debates and make its contribution by raising the question of the status and impact of this technology in anthropological debates. It has often been said that deadbots pose challenges to death and represent ways of defying it and pushing it back. By proposing to produce avatars of the dead, technology swaps biological finitude for "digital immortality", but does not conjure up the meaning of death. This idea of digital immortality—that of data that are so many traces of the self—calls on anthropological models of the human being and, more generally, of the living being: the 'upgrading' of the human being in the technological virtual that is vaunted by its advocates also appears, by ricochet, as a dilution of human ontology in technology—or at least of a certain human ontology. In fact, the backdrop is a wider debate on technologies, which expresses the fear of seeing the old models of the human (including the biological, the carnal, motor skills, language skills, the emotional, consciousness, etc.) being disintegrated (literally, the "human"). be disintegrated (literally: no longer integrated into a unified entity) [21] and defenders of a more confident vision of technology such as Kruzweil and his followers in the transhumanist movement (See Krueger, 2021). However, to date, these are essentially bereavement technologies, i.e. technologies designed to alleviate the pain of experiencing the death of a loved one and possibly the fear of one's own disappearance. The ethical issue is therefore the loss of the meaning of mourning, which also gives meaning to the whole process of existence, of which it is philosophically an essential component (Voinea, 2023).

Among the many voices expressing fears and deep ethical and social concerns, some are less perplexed than others. While concerns are being expressed about the 'replacement' of 'our' dead by digital avatars (raised by [30], and [7]), and the dread of a shift in human attachments towards machines without soul or feeling, Krueger and Osler sound a different note and place their faith in the ability of the living to keep in mind the nature of the relationships they have with devices, however humanized they may be. According to the two sociologists, the relationship that is established with a conversational agent is part of a reconstruction of a "habit of intimacy" that is specific to humans but that can be deployed on other forms of objects. By extension, deadbots also provide 'continuing bonds' between humans, but mediated by machines—which means that, in the end, a reciprocity is established, albeit a 'thin' one (because it has neither emotional density nor relational thickness) (Krueger and Olser, 2022), because it is established in systems of relationships in which machines are integrated, helping to maintain or reconstitute the network of relationships that form the basis of human equilibrium, in which their role is empirically and philosophically functional and therefore useful.

So, as we can see, the question of death itself (as a process of is ultimately overtaken by others that call for approaches that open up to anthropology and sociology, beyond psychology and philosophy: it is not so much a question of the fear of human finitude and the hope of a technological response to this 'problem' of the loss of self-awareness, or even the challenge of the transformation of a certain idea of the human, as sentient, replaced by machinic devices. This is a question of relationships: the idea of the human being caught up in systems of relationships with other humans (which make him or her a social being by definition) and semantic networks that give meaning to his or her existence (which make him or her a cultural being by definition, cf Geertz, 1973). Deadbots therefore take up the challenge of the lost relationship with the deceased, rewoven by virtual agents, and, by extension, question the relationships between the living, and the way in which, ultimately, the individual remains a relational being throughout his existence, and is constituted as a linguistic, but also gestural and semiotic, communication interface, which reveals a fundamental dimension of humanity: the absolute necessity of time for collective life and exchanges. So the debates around deadbots have focused, not surprisingly, on technological avatars as pseudo-persons interacting with an individual embodying a (conscious) human life, but not yet enough on their place as fully-fledged players in the communication networks that form the texture of (relational) social life.

From an anthropological point of view, of course, this veritable techno-cultural revolution represented by the all-out digitalisation of human societies raises the question of the continuity of human practices and representations, in terms of conceptions of life, but also of beliefs and attitudes towards death. A whole host of new conceptions of death, new funeral rites and new mourning practices (as we have seen, at least in the case of the latter) are developing, but unevenly depending on the region and the social and cultural context (Obadia, 2024). We therefore need to consider, on the one hand, what appears to be continuity in the ways of thinking about death and mourning, and on the other hand, what appears to be ruptures [35]. Ethicists believe that these breaks represent a threat to the integrity of humanity, especially as they are used to serve the interests of a digital death industry whose priorities are economic rather than philanthropic [16]. These same disruptions do not, moreover, represent all the ways in which technological innovations can be integrated, which may be digital memorial devices placed on physical graves in traditional cemeteries [8], while 'online memorials' or '[entirely] digital cemeteries' are beginning to be considered as important symbolically as traditional funeral and memorial spaces [23].

Some, such as [29], believe that there is a formal continuity between virtual humans and deadbots, which is likely, in their view, to change modern people's relationship with the Afterlife. While this is an interesting proposition, it has not yet been sufficiently substantiated, or its conclusions are too hasty. For my part, I had argued that while deadbots / griefbots seem to bring an air of modernity to our relationship with death, explicitly or implicitly carrying the ideals of transhumanism or related ideological movements, they do nothing to resolve the other essential question of the post-mortem condition: They are machines for prolonging memory and virtualizing the deceased, but they shift, without addressing it head-on, the question of the Afterlife, which remains an irreducible dimension of the idea of death in human cultures, even in high modernity (Obadia, 2024). Indeed, it is relatively easy to show that conversational agents developed in high-tech grievance technologies gather information about the living and prolong their existence in a technological here-and-now, despite their biological death. This means that, in terms of their interactions with those around them, deadbots are much more like living simulants representing the dead, than dead people talking about a hypothetical location beyond the grave. It is true that digital technologies allow something apparently 'new'—the continuity of post-mortem communicative activity on digital platforms (Juillard & Georges, 2018)—but it can also be similar to an old 'spiritualist' practice, reinvented in form but not in function [12].

## 6 In the End…

So, from an anthropological point of view, the extension of generative artificial intelligence technologies, LLMs and conversational agents needs to be taken both seriously and less seriously than that. There is still a gap between the media coverage of the 'griefbot' phenomenon and its social and technological deployment. As the object of a phantasmagoria and imaginary projections that are unfolding in all directions, the technology is taking on quite varied forms and raising questions that are just as varied. The dead bot is, in itself, both a danger and an opportunity for humanity, in terms of its capacity to bring the machine into the human and the human into the machine, or rather, in this context of reciprocal hybridization, ultimately maintaining the boundaries between the organic and the technical, while at the same time dissolving them in another way. This is a curious paradox of deadbots, or the internal logic behind their development and that of their conversational and logical autonomy: at a time when testimonies of "intelligent" chatbots demanding the right to "live" like humans are multiplying, from the "I want to be alive" dialogue between the Bing chatbot and journalist Kevin Roose (in the New York Times, 17 February 2023) to the fact that chatbots and other types of conversational agent are showing themselves to be increasingly capable of imitating human abilities, the tension between machines and human life comes at the cost of a shift towards death—a death that is challenged by the 'eternity' guaranteed by technology, but which is simply masked by the illusion of continuity in the reality of communicating human beings. And what if this frenzy to this attempt to technologize death, of which deadbots are in fact just one facet, were nothing less than a new iteration of humanity's perpetual struggle against its finitude, this time with the hope that it might touch that old eternal dream of eternity?

## References

1. Bao, A., Zeng, Y.: Embracing grief in the age of deathbots: a temporary tool, not a permanent solution. Ethics Inf. Technol. 26(1), 1–10 (2024)
2. Bassett, D.J.: Who wants to live forever? Living, dying and grieving in our digital society. Soc. Sci. **4**(4), 1127–1139 (2015)
3. Bassett, D.: Shadows of the dead: social media and our changing relationship with the departed, Discover Society Retrieved http://discoversociety.org/2017/01/03/shadows-of-the-dead-social-media-andour-changingrelationship-with-the-departed/ (2017)
4. Bourdeloie, H.: Usages des dispositifs socionumériques et communication avec les morts. D'une reconfiguration des rites funéraires. Quest. Commun. **28**(2), 101–125 (2015)
5. Brubaker, J.R. et al.: Stewarding a legacy: responsibilities and relationships in the management of post-mortem data. In: Proceedings of the 32nd annual ACM conference on Human factors in computing systems. ACM, 2014. pp. 4157–4166 (2014)
6. Brubaker, J.R., Vertesi, J.: Death and the Social Network. In: CHI 2010 Work-shop on HCI at the End of Life: understanding death, dying, and the digital. Atlanta, GA, USA, (2010)
7. Buben, A.: Technology of the dead: objects of loving remembrance or replaceable resources? Philos. Pap. **44**(1), 15–37 (2015)
8. Maciel, C., Carvalho Pereira, V., Leitão, C., Pereira, R., Viterbo, J.: Interacting with digital memorials in a cemetery: Insights from an immersive practice. In: Federated Conference on Computer Science and Information Systems (FedCSIS), Prague, Czech Republic, 2017, pp. 1239–1248. (2017)
9. Chernavskaya, O.: To the problem of digital immortality. Cogn. Syst. Res. **88**, 101303. https://www.sciencedirect.com/science/article/pii/S1389041724000974 (2024)
10. Krueger, J., Osler, L.: Communing with the dead online: chatbots, grief, and continuing bonds. J. Conscious. Stud. **29**(9–10), 222–252 (2022)
11. Danaher, J., McArthur, N.: Robot Sex: Social and Ethical Implications. MIT Press (2017)
12. Georges, F.: Le spiritisme en ligne. digital communication with the afterlife. Les Cah. Numérique **9**(3–4), 211–240 (2013)
13. Harbinja, E., McVey, M.: Governing ghostbots, Comput. Law Secur. Rev. **48**(105791) (2023). https://doi.org/10.1016/j.clsr.2023.105791
14. Harju Ana, A.: Theorising digital afterlife as techno-affective assemblage: on relationality, materiality, and the affective potential of data. Soc. Sci. **13**(4), 227 (2024)
15. Henrickson, L.: Chatting with the dead: the hermeneutics of thanabots. Media Cult. Soc. **45**(5), 949–966 (2023)

16. Hollanek, T., Nowaczyk-Basińska, K.: Griefbots, deadbots, postmortem avatars: on responsible applications of generative AI in the digital afterlife industry. Philos. Technol. **37**(2), 1–22 (2024)
17. Julliard, V., Georges, F.: Produire le mort. Pratiques d'écriture et travail émotionnel des deuilurs et des deuiluses sur Facebook. Réseaux **210**(4), 89–116 (2018)
18. Kirby, R., Forlizzi, J., Simmons, R.: Robotics and autonomous systems: special issue: towards autonomous robotic systems intelligent, Autonomous Robotics in the UK, 58 (3), 322–332 (2009)
19. Krüger, O.: Virtual Immortality: God, Evolution, and the Singularity in Post- and Transhumanism. Transcript Verlag, Bielefeld (2021)
20. Lafontaine, C.: La société postmortelle, Paris, le Seuil (2008)
21. Lebreton, D.: L'adieu au corps: vers homo silicium. In: Munier, B. (ed.) Technocorps: La sociologie du corps à l'épreuve des nouvelles technologies, Lormont, Éditions François Bourin, pp. 56–64 (2013)
22. Lindemann, N.F.: The ethics of 'Deathbots.' Sci. Eng. Ethics **28**(6), 1–15 (2022)
23. Lopes, A.D., Maciel, C., Pereira, V.C.: Virtual homage to the dead: an analysis of digital memorials in the social web. In: Meiselwitz, G. (ed.) Social Computing and Social Media. SCSM 2014. Lecture Notes in Computer Science, vol 8531. Springer, Cham (2014)
24. Maciel, C., Pereira, V.C.: Post-mortem digital legacy: possibilities in HCI. In: Human-Computer Interaction: Users and Contexts, pp. 339–349. Springer-Verlag, Berlin Heidelberg p (2015)
25. Mayor, A.: Gods and robots. Princeton University Press, Princeton, NJ (2018)
26. Morris M.R., Brubaker, J.R.: (2024) Generative ghosts: anticipating benefits and risks of AI afterlives, [cs.CY] https://arxiv.org/abs/2402.01662
27. Morano, G.: Le deadbot ou la fabrique des spectres numériques: comment l'Intelligence artificielle s'apprête à reconfigurer notre expérience du deuil, Implications philosophiques, https://www.implications-philosophiques.org/le-deadbot-ou-la-fabrique-des-spectres-numeriques-comment-lintelligence-artificielle-sapprete-a-reconfigurer-notre-experience-du-deuil/ (2023)
28. Nowaczyk-Basińska, K., Kiel, P.: Exploring the immortological imagination: advocating for a sociology of immortality. Soc. Sci. **13**(2), 83 (2024). https://doi.org/10.3390/socsci13020083
29. Savin-Baden, M., Burden, D.: Digital immortality and virtual humans. Postdigital Sci. Educ. **1**(1), 87–103 (2019)
30. Stokes, P.: Dealing with digital remains, ADI Policy Briefing Papers, Australia, vol. 1(3). https://adi.deakin.edu.au/s/ADI-Policy-Briefing-Paper-Dealing-with-digital-remains-pr3a.pdf. (2020)
31. Sumiala, J., Jacobsen, M.H.: Digital death and spectacular death. Soc. Sci. **13**, 101 (2024)
32. Öhman, C., Floridi, L.: The political economy of death in the age of information: a critical approach to the digital afterlife industry. Minds Mach. **27**(4), 639–662 (2017)
33. Van der Vorst, R., Kamp, J.-A.: Designing a griefbot-for-good. In: Wernaart, B. (ed.) Moral design and technology. Wageningen Academic Publishers, Wageningen, the Netherlands, pp. 215–241. (2022)
34. Vlahos, J.: Talk to me: How voice computing will transform the way we live, work, and think, Houghton Mifflin Harcourt (2019)
35. Walter, T.: New mourners, old mourners: online memorial culture as a chapter in the history of mourning. New Rev. Hypermedia Multimed. **21**(1–2), 10–24 (2014)

# Artificial Afterlife: Philosophical Reflections on Griefbots

**Giacomo Zanotti and Daniele Chiffi**

**Abstract** AI-powered chatbots are increasingly used in many contexts and for a variety of purposes. Among these uses, a particularly interesting one involves the so-called griefbots—that is, chatbots impersonating dead persons in the form of an artificial interlocutor. While they might help us process the loss of a beloved person, griefbots are not free from risks and may give rise to ethical concerns. This work aims at expanding the existing philosophical debate on griefbots. After providing a brief introduction to griefbots as well as the philosophical debate on their use, we provide an analysis of the way they may give rise to problematic ethical and normative tension between different transparency requirements. Then, we emphasise how the assessment of the related risks should adequately account for the dimension of users' vulnerability.

**Keywords** Griefbots · Chatbots · Grief · Transparency · Risk · Vulnerability · Philosophy of AI

## 1 Introduction

In February 2013, the second season of the science fiction series *Black Mirror* was aired. The first episode, "Be Right Back", told the story of Martha, a young woman who starts communicating with an AI-powered replica of her just deceased boyfriend Ash. Just over ten years later, such a scenario does not sound sci-fi anymore.

Recent developments in AI have enabled a wide variety of applications. From recommender systems in social networks to medical tools for image analysis, drug development, and diagnosis, AI systems are increasingly widespread in our societies.

G. Zanotti (✉)
Department of Electronics, Information and Bioengineering (DEIB), Politecnico di Milano, Milan, Italy
e-mail: giacomo.zanotti@polimi.it

D. Chiffi
Department of Architecture and Urban Studies (DAStU), Politecnico di Milano, Milan, Italy
e-mail: daniele.chiffi@polimi.it



A. Sans Pinillos et al. (eds.), *SecondDeath*, Studies in Applied Philosophy, Epistemology and Rational Ethics 76, https://doi.org/10.1007/978-3-031-98808-0_14

In particular, a lot of media coverage and public debate has revolved around generative AI systems, capable of producing—typically based on users' prompts—different kinds of content, from text to images and music.

New uses of generative AI systems are constantly emerging. Among the many possible applications of these technologies, a particularly interesting one involves the deployment of AI-based chatbots impersonating deceased persons, just like in the story of Martha and Ash: the so-called griefbots.[1] Although such tools are not (yet?) extensively employed, they seem to have the potential to significantly influence and modify our experience of grief, giving us the possibility to prolong our interactions with our departed beloved ones. As often happens with new technologies, however, a certain degree of uncertainty is involved. And while uncertainty might be the key to innovation [7], what happens is that in many cases we have trouble anticipating the risks of a new technology before it is deployed at a large scale [10, 36].

This chapter aims to discuss some philosophically interesting—and potentially problematic—aspects of griefbots' deployment. First, a brief presentation of griefbots is provided, considering both their technical features and their context of application (Sect. 2). Then, we will show how griefbots might turn out to be problematic from a normative and ethical point of view. In particular, we will focus on the way they might generate a conflict between different transparency requirements and *desiderata* (Sect. 3). Finally, building upon a multi-component approach to AI-related risk, we will argue that these systems should be regarded as highly risky ones due to the vulnerability of their users (Sect. 4). Section 5 concludes the chapter.

## 2 Chatbots and Grief

It is nothing new that technology can play a role in mediating grief processes. From graveyards to photo albums, different artefacts come to our aid when it comes to processing the loss of a beloved person. If anything, the advent of digital technologies has only reinforced the link between grief and technology, providing new tools and enabling new practices. Just think about the fact that nowadays, when a person passes away, they often leave a significant amount of "digital remains" in the form of pictures, videos, and texts in their profile on social networks [29]. In the complex and articulated panorama of grief and technology, griefbots are increasingly referred to as one of the most interesting AI applications [12, 18, 21–23].

While providing a technical overview of such tools falls outside the scope of this paper, some (minimal) remarks turn out to be useful. For the sake of simplicity, let us limit ourselves to text-based griefbots, leaving aside multimodal systems that can process different forms of content such as images, audio, and videos. From a technical

[1] The expression "deadbot" is also recurrent in the literature, together with the terms "thanabots" [5]), "ghostbots" [14], or "generative ghosts" [26]. We prefer to use "griefbot" because, among other things, it shifts the focus on the experience of the living ones, which is what we are interested in analysing.

point of view, griefbots are typically machine learning models for natural language processing that are specifically trained to reproduce the linguistic behaviour of a deceased person. More precisely, current griefbots are designed starting from Large Language Models (LLM). LLMs have been increasingly present in public discussion on AI since November 2022, when OpenAI made its model GPT3.5 publicly available and freely accessible through the chatbot ChatGPT.

Trained on massive amounts of text data, LLMs have become increasingly accurate in language processing. Interestingly for our purpose, they display a significant level of flexibility and can be adapted to a wide range of different (language-mediated) tasks—in fact, LLMs are typically referred to as general-purpose systems [16]. In particular, a pre-trained model—that is, a model that has already been trained on a large dataset, thereby acquiring general linguistic capabilities—can be further fine-tuned to a specific domain. In other words, the model can receive additional training on a smaller and ad-hoc sample of texts. For instance, we might want to design a domain-specific LLM that can answer medical questions—a real-world example is Google's Med-PaLM [34]. To do so, a pre-trained model should be fine-tuned on medical texts to make its output more reliable.

The possibility of fine-tuning is pivotal when it comes to griefbots, for a conversational agent with human-like general linguistic capabilities is not enough. When a beloved person dies, griefbots' users do not seek contact with some generic person. What a griefbot is meant to do is rather mimicking the behaviour of a *specific* individual. The ways in which this aim is achieved vary depending on the system we might consider. In the case of text-based griefbots, the process involves the reproduction of the deceased person's vocabulary, personality, linguistic style, knowledge, and so forth. To this aim, griefbots can be fine-tuned on a dataset made up of texts produced by the person in question, be they chunks of correspondence, diaries, interviews, and so on. Sometimes, this procedure is used to produce digital replicas of famous characters.[2] In the case we are interested in, however, it allows the design of systems deployed in a more intimate dimension.

Just to make things clear, griefbots are not (not yet, at least) a widespread application of LLMs. Despite their regular appearance in scientific debates and public media, the griefbots industry is not yet established, and the discussion usually revolves around few relatively isolated cases.[3] Still, we are dealing with an AI application that, besides being technically feasible, might come with significant social implications.

In fact, authors have identified both advantages and drawbacks related to the deployment of griefbots, even if the discussion largely remains at a speculative level and empirical data would be necessary. On the positive side, a recurrent claim is that griefbots could help the living elaborate the loss of a beloved person. This is consistent with "continuing bonds" approaches to grief, that instead of understanding grief as a relinquishment process—a view that is usually traced back to Freud—insist

[2] Such as the AI-based replica of the Italian economist Luigi Einaudi, specifically trained on Einaudi's writing on economic matters (https://www.fondazioneeinaudi.it/en/talk-to-einaudi).

[3] See, for instance, Project December (https://projectdecember.net/), Hereafter AI (https://www.hereafter.ai/), Seance AI (https://www.seanceai.com/).

on the continuation of some kind of relationship with the deceased person [8, 25].[4] Although griefbots cannot make up, so to say, for the loss of a person, they can provide the living with a tool to maintain a form of interaction with the dead, thereby facilitating their grief.

Needless to say, the deployment of griefbots is not without problems. We do not aim to provide a full overview of the literature,[5] but some elements (which we will not further discuss in detail) are worth recalling. First of all, as one might easily imagine, it is not clear whether the continuation of bonds enabled by griefbots is without risks. Again, these considerations should be backed by robust empirical data that are not available at the moment,[6] but we can reasonably foresee possible complications concerning the psychological health of griefbots' users. In particular, fears have been raised about the possibility that, instead of recollecting deceased people, griefbots might end up somehow replacing them [35]. Forms of dependencies on the griefbot might emerge, which could reduce the living's autonomy [23]. All of this is arguably complicated by the fact that, when developed by companies, griefbots' features might be more influenced by profit-oriented strategies rather than by interests in users' well-being.

In addition to this, non-trivial moral questions have been raised concerning the possibility that griefbots may end up somehow violating the integrity of the deceased ones [19, 30, 31].[7] Consider, for instance, the situation in which a griefbot is trained on conversations with a deceased person who had not consented to such a reuse of their words. And even if consent was given, the possibility of a misrepresentation of the deceased person remains [12, 23].

Taking stock, griefbots represent an interesting and yet controversial AI application, and further work is needed to evaluate their impact, advantages, and problems. The next sections contribute to this endeavour by analysing the way griefbots may give rise to problems of transparency and how the dimension of vulnerability should be prioritised in the assessment of their risk.

## 3 Griefbots and Transparency Requirements

Transparency is an increasingly central notion in the field of AI. Among other things, it is one of the seven requirements that the European *Ethics Guidelines for Trustworthy AI* [2] identify as the basis for the responsible development and deployment of robust and ethically acceptable AI systems. That said, there seem to be multiple senses in which AI systems could—and maybe should—be transparent.

---

[4] See [17] for a review comparing the efficacy of retaining versus relinquishing bonds in grief.

[5] See Lindemann (2022a) for a more comprehensive treatment.

[6] A qualitative study was recently conducted, involving the recruitment of ten mourners for in-depth semi-structured online interviews concerning the use of griefbots [39]. For another empirical research on the topic, see [20].

[7] See also [11] for a more general discussion on the ethics of personalized digital duplicates.

Most of the current talk about transparency in AI seems to focus on the contraposition between transparent versus black-box AI models [1]. Let us imagine a situation in which an AI system provides us with a certain decision, prediction, or classification—to put some flesh on the bones, we can think about a medical AI system outputting a possible diagnosis. In such a case, we might want to know why the system provided that specific output—in this sense, the notion of explainability is typically associated (and sometimes overlapping) with the one of transparency [38]. In other words, we are interested in the details of the deployed inferential process. Some AI models, such as those based on symbolic techniques, are transparent in this sense, for they rely on explicitly represented knowledge and rule-based inferences. In the machine learning scenario, instead, many models present themselves as black boxes that, in some cases, are only partially explainable through XAI (eXplainable AI) techniques. Griefbots based on LLMs can hardly be fully transparent in this sense, for the road to achieving explainability in LLMs seems to be still long [24].

That said, the lack of explainability is not the only problem, for there might be other senses in which LLMs—and thereby griefbots—may fail to be transparent. Going back to the Ethics *Guidelines*, we can see that the requirement of transparency encompasses explainability but does not reduce to that. A crucial role is also played by the sub-requirement of communication, according to which "AI systems should not represent themselves as humans to users" and "the AI system's capabilities and limitations should be communicated to AI practitioners or end-users in a manner appropriate to the use case at hand" ([2], p. 18).[8]

Whether griefbots could be transparent in this sense is not straightforward. Facchin and Zanotti in [13] have recently highlighted how AI-powered affective artificial agents – namely AI systems explicitly designed to interact with users in emotionally salient contexts and ways—typically display a form of "emotional transparency" that may be at odds with transparency requirements such as those presented in the *Guidelines*.

Consider an AI-powered chatbot such as Replika, designed to be a sort of artificial friend—an "AI companion who cares", according to the company.[9] Facchin and Zanotti highlight how systems such as Replika must be emotionally transparent to provide a good user experience. That is, their artificial and emotionless nature must somehow fade in the background: they must give us the impression "to interact with an individual that really has thoughts and feelings, with whom emotional exchange appears to go in both directions". It does not matter if we remain reflectively aware that we are interacting with an emotionless chatbot. The point is that we interact with it as if it really had emotions and feelings, which may ultimately result in the creation of emotional bonds comparable to those that we could have with a human. As suggested by the authors, it is an open question whether emotional transparency is compatible with the *Guidelines*' transparency as communication, stating that AI

[8] See [3] for a discussion of different senses in which AI systems can be transparent.

[9] https://replika.com/. Interestingly, Replika was initially developed as a griefbot impersonating Roman Mazurenko, a prematurely passed away friend of the tech entrepreneur Eugenia Kuyda.

systems should not represent themselves as humans and that their capabilities and limitations should always be made clear.

Analogous considerations seem to apply to griefbots as well. Just like Replika, griefbots seem to qualify as tools that can regulate users' feelings and moods by allowing them to engage in emotionally salient interactions. Now, imagine a griefbot that, while simulating some features of the deceased person (such as memories and vocabulary), comes across as cold and emotionally detached. Interactions with such a system would arguably be far from fulfilling for the user, who seeks the continuation of some emotional bond as well. On the contrary, they might end up having a negative impact on the user, who would be constantly reminded of the fact that the system is nothing but an incomplete simulacrum of the departed. And indeed, griefbots are not just supposed to display apparently emotionally driven behaviour. Something more than simple emotional transparency is required, for a griefbot should arguably replicate as much as possible the emotional footprint of a specific person.

Again, the extent to which this is compatible with transparency requirements such as those expressed in the *Guidelines* is not clear. On the one hand, the potential tension could be mitigated by design, for example by making clear at the beginning of the interaction that the griefbot is nothing but a non-conscious AI system trained to reproduce the departed's behaviour. At the same time, we can safely assume that griefbots can hardly fully deceive their users, who are sadly well aware that they are not actually interacting with their beloved ones. Accordingly, it is not clear whether a disclaimer making explicit the nature of the chatbot would be necessary.

On the other hand, it is less obvious how we could deal with what Sætra in [32] (p. 282) describes as "partial deception", by which "a human being has a rational appreciation of the nature of the device it interacts with but at a subconscious level cannot help reacting to it as if it is real". In the case of griefbots, this arguably involves reacting as if we were actually talking with our deceased beloved one. At this stage, a series of questions emerge. Most notably, is the *sui generis* kind of partial deception involved in griefbots' use compatible with the requirements expressed in the *Guidelines*? After all, while the requirement of transparency as communication explicitly states that "AI systems should not represent themselves as humans", the dynamics of interaction between users and griefbots crucially depends on the chatbot providing a sufficiently realistic replica of a real person's behaviour. True, one might argue that a disclaimer making explicit the artificial nature of the chatbot might be enough to meet the transparency requirements in question. If so, however, further doubts could be cast upon such requirements' sufficiency. Given the vulnerability of griefbots' users, in fact, one could argue that even partial deception is hardly acceptable.

## 4 A Focus on Vulnerability

Our discussion on transparency and partial deception directly leads us to the second aspect we wish to consider, namely the way in which the dimension of vulnerability should be prioritised in the assessment of the risks involved in griefbots' deployment. We have concluded the previous section by referring to the vulnerability of griefbots' users. Indeed, it seems fair to assume that grief is a condition that arises during times when people may be particularly vulnerable [37].[10] That said, the concept of vulnerability is a complex and multi-faceted one, which deserves a more detailed treatment.

To begin, vulnerability doesn't just mean fragility. This fact suggests that with good protective measures, an object or person can be fragile (intended as a candidate for an irreversible loss of functions) without necessarily being considered vulnerable [6]. In the field of risk analysis, vulnerability refers to situations and conditions that could leave people or assets more affected and less protected against hazards—*i.e.*, potentially harmful events. Vulnerability is often considered a key component of risk, particularly in disaster risk reduction research applied to technological risks, when it is assessed alongside the magnitude of the hazard and the level of exposure to that hazard—an approach that seems to be suitable for AI-related risk as well [40]. Hazard, vulnerability, and exposure are, indeed, the primary components of risk. Consequently, if we want to mitigate a risk, we aim at reducing at least one of these components.[11]

In the following, we will mainly focus on vulnerability in the context of emotive hazards due to griefbots.[12] Although some individuals exhibit resilience when faced with traumatic or distressing life events, the quality of life can significantly decline for many who are grieving. For instance, research suggests that approximately 10–15% of bereaved individuals experience complicated grief, leading to substantial challenges in coping and adjustment [33].[13]

From an affective perspective, talking about grief requires discussing the intimate bonds of both love and friendship, since an elucidation of grief can be done in terms of these notions. According to Robert Nozick [28], enduring romantic love involves wanting to form a 'we' and desiring the other person to reciprocate those feelings. This 'we' represents a new entity that connects the two individuals, creating a shared web of relationships between them. Furthermore, the well-being of a person in romantic love is intertwined with the well-being of their partner. If the person you

[10] For a philosophical analysis of grief, see [8, 9]).

[11] True, with certain natural risks like earthquakes, we cannot prevent the hazard; nonetheless, we can diminish the exposure and vulnerability of affected individuals and resources. In the case of technological risks, it is often possible to reduce all three components of risk, including the hazard, even if it can be difficult in some cases.

[12] Griefbots, indeed, may pose risks that extend beyond affecting emotions and health; for instance, they can generate security and reputational hazards [26].

[13] Proper emotion regulation can be quite challenging in the context of grief. For a recent overview of models, techniques and biological mechanisms associated with emotion regulation, see [15].

love is doing well, you also feel well, conversely, if they are struggling, it affects your own well-being. So, the well-being of each individual in a couple is highly dependent on the well-being of their partner. Another distinguishing feature of romantic love is the fact that people want to be loved "for themself" and not just for some of their characteristics—*e.g.*, no one usually wants to be loved just for economic reasons. Romantic love requires loving the person as a whole, not just for some characteristics. In the same paper, Robert Nozick also distinguishes romantic love from friendship by noting that friendship is characterised by engaging in shared activities purely for the sake of sharing them and for the resulting pleasure.

When the person that we love passes away, we feel vulnerable because we cannot anymore recollect our *we* which provides meaning also to our individual life. Analogously, when a friend dies, we can no longer engage in activities together simply for the pleasure of doing so. In both cases, the reciprocity of relationships is an essential feature of authentic romantic love or friendship. More traditional digital technologies such as social networks may help us recall the distinguishing aspects of these relationships based on past events, but the projection of such relationships in the future can be quite challenging. Griefbots may formulate in some cases a potentially convincing narrative to restore a sense of reciprocity, projecting these relationships into the future and influencing decisions to be made. However, the way in which griefbots may foster reciprocity is inauthentic and thus inadequate, in particular for some vulnerable persons, leading in some cases, as we have seen, to partial deception.

The point is that griefbots work on the emotional aspect and can thereby generate attachment. This seems to apply more generally to all "affective" chatbots. The situation, however, in this case is even more critical because a griefbot is used by people experiencing grief. While it may mimic some of the characteristics of a person we loved, unfortunately, it cannot provide an authentic sense of loving that person as a whole human being. Sure, it may serve as a tool for reconceptualizing love and friendship through the lens of passing time. After the death of a beloved person, we are no longer the same as we once were. Digital tools that evoke certain characteristics of a beloved person can dynamically influence our decisions and encourage us to reflect in different ways, attributing much more mature meanings to the time spent together. However, this is not without risks; quite the opposite. Individuals experiencing grief are particularly vulnerable, and therefore, the risk associated with the use of griefbots can be quite high, encompassing emotional hazards that may even lead to psychological and physical harm.

As we have seen, emotional distancing—that is, maintaining a level of detachment during interactions with griefbots and recognizing them as simulations rather than genuine emotional exchanges—seems to go against the purposes of this technology, in which emotive engagement is an essential ingredient for its functioning. A possible severe risk in this case for vulnerable people is to create unrealistic overlaps between their online lives with their beloved person and reality. This risk should be addressed at the level of regulation, design, and risk communication related to griefbots, ensuring that even if these two spheres (namely, online life and reality) are interrelated, they are not confused. Notice that the issue is not simply the merging of digital and real life; rather, it is due to the implications of living an online existence with a digital replica of a loved person, rather than engaging in genuine relationships based on reciprocity, love, and friendship. Encouraging critical reflection on this matter serves as an initial protective measure to mitigate vulnerabilities associated with the use of griefbots. Then, design choices and specific regulations addressing all main vulnerabilities may create a safer environment for the potential use of these systems.[14] Nonetheless, it seems that griefbots should be considered—and accordingly regulated—as high-risk technologies in virtue of the potential high level of vulnerability of the users.

## 5 Conclusion

Despite not being (maybe yet) widely used, griefbots represent an interesting AI application that has the potential to significantly change our grieving habits. On the one hand, consistently with the view that grief is a matter of continuing bonds with the deceased person, griefbots may help us process the death of our beloved ones.[15] On the other hand, concerns have been raised regarding different aspects of griefbots' deployment, such as the possible violation of the dead's integrity and the risk that the user becomes somehow dependent on the chatbot. This work has attempted to expand the debate by highlighting that griefbots may give rise to tensions between different kinds of transparency requirements and by suggesting that the dimension of vulnerability should be prioritised in the assessment of griefbots' risks, to the point that these systems should be considered high-risk precisely due to the vulnerability of their users.

In conclusion, while griefbots may offer a potential avenue for reshaping our experiences of loss and mourning, their implementation demands a profound awareness of the intricate context in which they are used. Through a rigorous epistemological and ethically informed approach sensitive to vulnerabilities, it is possible to harness the promise of griefbots while mitigating their associated risks.

**Acknowledgements** The research is supported by (1) the Italian Ministry of University and Research, PRIN Scheme (Project BRIO, no. 2020SSKZ7R); (2) the Italian Ministry of University and Research, PRIN Scheme (Project NAND no. 2022JCMHFS).

---

[14] Lindemann in [23] suggested regulating griefbots as medical devices, thereby restricting their use to situations under psychological or psychiatric supervision. This appears to significantly limit the freedom of bereaved individuals to dynamically recall the deceased. It has also recently been argued that griefbots should be used just as a temporary form of commemoration rather than a long-term surrogate for the deceased [4]. While this perspective may provide multiple benefits, it is reasonable to assume that even temporary use of griefbots can pose significant risks given the vulnerability of potential users, especially shortly after the death of a beloved.

[15] Death, however, does not always signify the absolute end of a person's life in all regards. It can also mark a meaningful chapter in their story, continuing it in a significant way [27]. Perhaps this is one reason why griefbots exist, they seem to offer a sense of closeness with our beloved ones.

**Authors Contribution** G.Z. drafted Sects. 2 and 3, while D.C. drafted Sect. 4. The authors equally contributed to drafting the other sections. Both authors have given final approval of the version to be published.

**Disclosure of Interests** The authors have no competing interests to declare that are relevant to the content of this article.

## References

1. Adadi, A., Berrada, M.: Peeking inside the black-box: a survey on explainable artificial intelligence (XAI). IEEE Access **6**, 52138–52160 (2018). https://doi.org/10.1109/ACCESS.2018.2870052
2. AI HLEG (2019). Ethics Guidelines for Trustworthy AI. https://digital-strategy.ec.europa.eu/en/library/ethics-guidelines-trustworthy-ai
3. Andrada, G., Clowes, R.W., Smart, P.R.: Varieties of transparency: exploring agency within AI systems. AI Soc. **38**, 1321–1331 (2023). https://doi.org/10.1007/s00146-021-01326-6
4. Bao, A., Zeng, Y.: Embracing grief in the age of deathbots: a temporary tool, not a permanent solution. Ethics Inf. Technol. **26**(1), 7 (2024). https://doi.org/10.1007/s10676-024-09744-y
5. Bassett, D.J.: The Creation and Inheritance of Digital Afterlives. You Only Live Twice. Cham: Palgrave Macmillan (2022). https://doi.org/10.1007/978-3-030-91684-8
6. Chiffi, D., Curci, F.: Disentangling antifragility from resilience. In: Curci, F., Chiffi, D. (eds.) Fragility and Antifragility in Cities and Regions, pp. 6–24. Cheltenham: Edward Elgar Publishing (2024). https://doi.org/10.4337/9781035312559.00008
7. Chiffi, D., Moroni, S., Zanetti, L.: Types of technological innovation in the face of uncertainty. Philos. Technol. 35(94) (2022). https://doi.org/10.1007/s13347-022-00587-3
8. Cholbi, M.: Regret, resilience, and the nature of grief. J. Moral. Philos. **16**(4), 486–508 (2019). https://doi.org/10.1163/17455243-20180015
9. Cholbi, M.: Grief: a philosophical guide. Princeton University Press, Princeton (2022)
10. Collingridge, D.: The social control of technology. Frances Pinter, London (1980)
11. Danaher, J., Nyholm, S.: The ethics of personalised digital duplicates: a minimally viable permissibility principle. AI Ethics (2024). https://doi.org/10.1007/s43681-024-00513-7
12. Fabry, R.E., Alfano, M.: The affective scaffolding of grief in the digital age: the case of deathbots. Topoi (2024). https://doi.org/10.1007/s11245-023-09995-2
13. Facchin, M., Zanotti, G.: Affective artificial agents as sui generis affective artifacts. Topoi **43**(3), 771–781 (2024). https://doi.org/10.1007/s11245-023-09998-z
14. Figueroa-Torres, M.: Affection as a service: Ghostbots and the changing nature of mourning. Comput. Law Secur. Rev. **52**, 105943 (2024). https://doi.org/10.1016/j.clsr.2024.105943
15. Grecucci, A., Messina, I., Amodeo, L., Lapomarda, G., Crescentini, C., Dadomo, H., Panzeri, M., Frederickson, J.: A dual route model for regulating emotions: Comparing models, techniques and biological mechanisms. Front. Psychol. **11**, 480470 (2020). https://doi.org/10.3389/fpsyg.2020.00930
16. Gutierrez, C.I., Aguirre, A., Uuk, R., Boine, C.C., Franklin, M.: A proposal for a definition of general purpose artificial intelligence systems. Digit. Soc. **2**(36) (2023). https://doi.org/10.1007/s44206-023-00068-w
17. Hewson, H., Galbraith, N., Jones, C., Heath, G.: The impact of continuing bonds following bereavement: a systematic review. Death Stud. **48**(10),1001–1014 (2023). https://doi.org/10.1080/07481187.2023.2223593
18. Hollanek, T., Nowaczyk-Basińska, K.: Griefbots, deadbots, postmortem avatars: on responsible applications of generative AI in the digital afterlife industry. Philos. Technol. **37**(2), 63 (2024). https://doi.org/10.1007/s13347-024-00744-w

19. Jiménez-Alonso, B., Bresció de Luna, I.: Griefbots. A new way of communicating with the dead? Integr. Psychol. Behav. Sci. **57**(2), 466–481 (2023). https://doi.org/10.1007/s12124-022-09679-3
20. Jiménez-Alonso, B., Bresció de Luna, I.: AI and grief: a prospective study on the ethical and psychological implications of deathbots. In: Casas-Roma, J., Caballé, S., Conesa, J.: (eds.) Ethics in Online AI-Based Systems: Risks and Opportunities in Current Technological Trends pp. 175–191. Elsevier (2024)
21. Krueger, J., Osler, L.: Communing with the dead online: chatbots, grief, and continuing bonds. J. Conscious. Stud. **29**(9–10), 222–252 (2022)
22. Lindemann, N.F.: The ethical permissibility of chatting with the dead: towards a normative framework for 'Deathbots'. Publications of the Institute of Cognitive Science (PICS), vol. 1 (2022a). https://doi.org/10.48693/67
23. Lindemann, N.F.: The ethics of 'deathbots'. Sci. Eng. Ethics **28**(6), 60 (2022b). https://doi.org/10.1007/s11948-022-00417-x
24. Luo, H., Specia, L.: From understanding to utilization: a survey on explainability for large language models. arXiv preprint arXiv:2401.12874 (2024)
25. Millar, B., Lopez-Cantero, P.: Grief, continuing bonds, and unreciprocated love. South. J. Philos. **60**, 413–436 (2022). https://doi.org/10.1111/sjp.12462
26. Morris, M.R., Brubaker, J.R.: Generative Ghosts: Anticipating Benefits and Risks of AI Afterlives. *arXiv pre-print*, January 2024 (2024)
27. Nozick, R.: The Examined Life. Simon & Schuster, New York (1989)
28. Nozick, R.: Love's bond. In: Solomon, R.C., Higgins, K.M. (eds.) The Philosophy of (Erotic) Love, pp. 417–432. University Press of Kansas, Lawrence (1991)
29. Öhman, C. J., Watson, D.: Are the dead taking over Facebook? A big data approach to the future of death online. Big Data Soc. **6**(1) (2019). https://doi.org/10.1177/2053951719842540
30. Öhman, C., Floridi, L.: The political economy of death in the age of information: a critical approach to the digital afterlife industry. Mind. Mach. **27**(4), 639–662 (2017). https://doi.org/10.1007/s11023-017-9445-2
31. Öhman, C., Floridi, L.: An ethical framework for the digital afterlife industry. Nat. Hum. Behav. **2**(4), 318–320 (2018). https://doi.org/10.1038/s41562-018-0335-2
32. Sætra, H.S.: Social robot deception and the culture of trust. J. Behav. Robot. **12**(1), 276–286 (2021). https://doi.org/10.1515/pjbr-2021-0021
33. Sim, J., Machin, L., Bartlam, B.: Identifying vulnerability in grief: psychometric properties of the adult attitude to grief scale. Qual. Life Res. **23**, 1211–1220 (2014). https://doi.org/10.1007/s11136-013-0551-1
34. Singhal, K., Azizi, S., Tu, T., et al.: Large language models encode clinical knowledge. Nature **620**, 172–180 (2023). https://doi.org/10.1038/s41586-023-06291-2
35. Stokes, P.: Digital souls: a philosophy of online death. Bloomsbury, London (2021)
36. van de Poel, I.: An ethical framework for evaluating experimental technology. Sci. Eng. Ethics **22**, 667–686 (2016). https://doi.org/10.1007/s11948-015-9724-3
37. Voinea, C.: On grief and griefbots. Think **23**(67), 47–51 (2024). https://doi.org/10.1017/S1477175623000490
38. von Eschenbach, W.J.: Transparency and the black box problem: why we do not trust AI. Philos. Technol. **34**, 1607–1622 (2021). https://doi.org/10.1007/s13347-021-00477-0
39. Xygkou, A., Siriaraya, P., Covaci, A., Prigerson, H.G., Neimeyer, R., Ang, C.S., She, W.J.: The "Conversation" about loss: understanding how Chatbot technology was used in supporting people in grief. In: Proceedings of the 2023 CHI Conference on Human Factors in Computing Systems, vol. 646, pp. 1–15 (2023). https://doi.org/10.1145/3544548.3581154
40. Zanotti, G., Chiffi, D., Schiaffonati, V.: AI-related risk: an epistemological approach. Philos. Technol. **37**, 66 (2024). https://doi.org/10.1007/s13347-024-00755-7

# As I Play Dying Roles and Possibilities of Death in Video Games

**Alfonso García-Lapeña and David Ferragut**

**Abstract** Death is a ubiquitous element in video games. One of the most common views in the literature is that death is essentially a didactic mechanism to teach players how to improve their gameplay. However, we argue that beyond this function death has other roles, aims and possibilities. On the one hand, an essential function of death is to make a game challenging, which is achieved through the notion of "death penalty", and thus to provide the ludic experience that is a constitutive property of games. Without death, one might say, there is no game. On the other hand, we examine how death can be a source of meaning and an emotional enabler through the notion of OSGON games and games with a permadeath feature. Finally, we show how death in video games can be associated with a process of grief, whether for fictional characters, the games themselves, or real people.

**Keywords** Video games · Death · Challenge · Permadeath · Grief

## 1 Introduction

Death is a ubiquitous element in video games. One of the most common views in the literature is that death is essentially a didactic mechanism to teach players how to improve their gameplay. Death means that something has gone wrong and the player must learn from their mistakes in order to face future challenges. In this sense, Juul ([30], p. 52) claims that death in video games is "the death you survive and learn from". The player must further develop her skills by internalising the rules of the game, which is then rewarded by staying alive.

This, however, cannot be the whole story, as death is still present when a player has mastered a game. For experienced players who die, the potential learning from a

A. García-Lapeña (✉)
University of Iceland, Reykjavik, Iceland
e-mail: aglapena@gmail.com

D. Ferragut
Universitat Autònoma de Barcelona, Barcelona, Spain
e-mail: davidferragut@riseup.net

 
A. Sans Pinillos et al. (eds.), *SecondDeath*, Studies in Applied Philosophy, Epistemology and Rational Ethics 76, https://doi.org/10.1007/978-3-031-98808-0_15

death may be very small. A player may already know what went wrong, or the death may simply be the result of bad luck. Thus, the didactic function of death seems to adapt to a kind of "law of diminishing returns": for a new player, death may have a primarily didactic role; for a veteran player, the didactic gain from death may be very small or even non-existent.

Besides being a didactic mechanism, death in video games has other roles, aims and possibilities. To this end, we begin by formulating a theoretical framework for video games (Sect. 2), which will serve as the basis for the subsequent analysis. In this sense, games can be defined in terms of four main properties: *Rules*, *Skill*, *Quantified Performance* and *Ludic*. In particular, we focus on the notion of "ludicity" (Sect. 2.3) and argue that the empirical evidence suggests that the specific way in which games entertain is through challenge. The model developed (Sect. 2.4) serves to ground the idea that an essential function of death is to make a game challenging (Sect. 3.1), achieved through the notion of "death penalty", and thus to provide the ludic experience. Without death, one might say, there is no game.

The functions of death are followed by an analysis of types of death in games (Sect. 3.2), appealing to the combination of a "limited" and "unlimited" life axis and a "return" and "non-return" state axis. Further, we examine how death can be a source of meaning and an emotional enabler (Sect. 3.3). This can be achieved through a serious reflection on death, as implied by OSGON games and games with a permadeath feature, which radically change the gameplay experience. Finally, we show how death in video games can be linked to a process of grief (Sect. 3.4), whether for fictional characters, the games themselves or real people.

## 2 A Theoretical Framework for Video Games

### *2.1 The Quantified Performance Property*

In the years following 2001, the institutional birth of game studies [1], game scholars considered [12, 27] and [45] definitions of play and games as classical starting points. Today, the field has developed its own classical definitions, represented by Salen and Zimmerman (hereafter S&Z) [43] and Juul [29]. Both synthesise and build on much of the previous work on play and games. [43] (p. 80) definition of a game is:

> A game is a system in which players engage in an artificial conflict, defined by rules, that results in quantifiable outcome.

And [29] (p. 36) definition of a game is (what he calls the 'Classic Game Model'):

> A game is a rule-based system with a variable and quantifiable outcome, where different outcomes are assigned different values, the player exerts effort in order to influence the outcome, the player feels attached to the outcome, and the consequences of the activity are negotiable.

Both are great definitions and apply correctly to a large number of cases. From a game-object perspective, both definitions can be synthesised by appealing to three main properties:

(1) *Rules:* A game is a *rule-based system.*
(2) *Skill*: The player must exert some *effort* (mental and/or physical) in order to influence the outcome.
(3) "*Goals*: A game presents a *quantifiable outcome(s)*" that the player tries to achieve (a game ends at some point, with some winners and losers)

The *Goal* property is one of the essential properties attributed to games in most definitions (e.g. [10, 29, 32, 36, 43, 45, 52]), the "quantifiable outcome" that the player is trying to achieve. However, as it has been pointed out several times [3, 4, 21, 47–49]), the *Goal* property becomes problematic as a necessary condition in the case of "games without a goal", such as many role-playing games and simulation games.

Role-playing games, either in the form of video games (e.g. *World of Warcraft*, 2004) or in non-electronic form (e.g. *Dungeons & Dragons*), represent a first category of games that may lack the *Goal* property. Characteristic of these types of games is that the player controls a character inhabiting a fictional world, and through their choices the character grows and evolves, usually to a powerful state. However, these games usually have no end or goal, so there is no sense of "winning".[1] A second and similarly problematic category is represented by simulation games such as *SimCity* (1989), *The Sims* (2000), *Minecraft* (2011) or *Microsoft Flight Simulator* (1982). These games are characterised by the fact that they give the player a great deal of freedom and, again, do not have an end or a final goal. They even lack a series of quests with quantifiable outcomes, as in the case of RPGs and MMORPGs.

S&Z [43] (p. 81) regard these cases as "limiting cases" and Juul [29] (p. 47) as "borderline cases". As mentioned above, many authors find this judgement unsatisfactory. *World of Warcraft*, *The Sims* or *Minecraft* are clear cases of games, not "limiting" or "borderline" cases.

Two solutions to the "games without a goal" problem have been proposed. On the one hand, Juul ([47], p. 47) defines video games by appealing to two necessary and one disjunctive condition:

> X is a videogame iff it is an artefact in a digital visual medium, is intended primarily as an object of entertainment, and is intended to provide such through the employment of one or both of the following modes of engagement: rule-bound gameplay or interactive fiction.

"Rule-bound gameplay" represents the properties of classical games as defined by Juul's model. The second disjunctive element, "interactive fiction", is intended to capture precisely the cases of "games without a goal" that fall outside Juul's and

---

[1] We can overcome this problem by focusing on the usually large number of quests some of these games typically offer. In this sense, they present a large number of quantifiable outcomes that players can try to achieve in each game session. However, in terms of the game as a whole, it's still true that they don't present a final goal ([43] Ch. 7). Moreover, it is not compulsory to complete any or many of the available missions, and this fact presumably does not end with the game session.

S&Z's proposals, since RPGs and simulation games can be conceptualised as interactive fictions. Tavinor's definition may be extensionally satisfactory for video games. However, classical and genuine properties (necessary conditions) aim to capture a *similarity* between all the instances and are assumed to *inhere* all of them. Disjunctive properties fail to satisfy both criteria [9]. Moreover, in Tavinor's particular case, the introduction of the disjunctive element seems to be an *ad hoc* move. The only function that the second disjunctive element (interactive fiction) seems to have is to account for the "exceptions" of Juul's and S&Z's definitions, the cases of "games without a goal".

Frasca [21] offers a different solution to the problem, which we find more satisfactory (the core idea is also argued by [31]). To our knowledge (and surprise) this solution has gone relatively unnoticed in the literature. By analysing the origins of skateboarding, he concludes that one of the necessary properties of games is *the objective quantification of player performance.* Games are interactive and require some *Skill* (mental and/or physical) on the part of the player. A characteristic of games is that the player's performance is objectively quantified in some way, providing feedback on how well or poorly the player is using her skills to solve the challenges presented. In this way, it is not the goals, but the *objective quantification of player performance* that is essential to games. As we shall see (Sect. 3.1), this feature is at the heart of the concept of death as a didactic mechanism. In football, for example, a team's performance is objectively quantified by the number of goals it scores. In *Tetris* (1984), performance is quantified by the number of levels completed and the number of points scored. In single-player RPGs and MMORPGs, a player's performance is quantified by their level, items and equipment. In *Minecraft*, performance is quantified, among other things, by the player's level, health, and perhaps the age of the fictional world they inhabit.

From the *Quantified Performance* property we get, as a special case, the traditional *Goal* property and the categories "win" and "lose". Winning (or losing) is nothing more than achieving a certain quantification. Roughly speaking, a football team wins a football match if, after some time $t$, its quantified performance exceeds the quantified performance of the opposing team, where performance is quantified by the number of goals scored. In general, then, the various goals or objectives we find in games are nothing more than specific cases of performance quantification. This shows that the *Goal* property is not a necessary condition for games, but a special subset of the *Quantified Performance* property, and modifies our previous definition to be constituted by the *Rules*, *Skill* and *Quantified Performance* properties.

## 2.2 A Typology of Game Rules

Before moving on to a final refinement of the definition, we need to say something about the notion of "rules". Almost all definitions of (video) games agree that rules are a constitutive element (e.g. [21, 29, 36, 43, 47,52]). However, few authors have explored this concept in depth. In this context, scholars have typically spoken of two different *types of rules*, which we can call 'action rules' (the actions a player is allowed to perform) and 'goal rules' (those that define the goals of the game). For example, Kelley [32] (p. 50) defines games as a form of recreation "constituted by a set of rules that specify an object to be attained and the permissible means to attaining it".

However, there is a third category of rules that are not usually made explicit. This can be seen by opening the rulebook of any board game. The first rules you would find are neither *action rules* nor *goal rules*, but a *sui generis* category of rules that we can call 'object rules': the rules that define the elements and the world that make up a game. In this sense, FIFA's *Laws of the Game* begins by defining the objects that constitute the basis of football, such as "the field", "the ball", "the player", "the referee", and so on. Again, these rules are neither *action rules* nor *goal rules*, but *object rules*.

To our knowledge, only [21] (pp. 118–120) has explicitly stressed the existence of something like *object rules*. His proposal defines three main types of game rules: *model rules*, *grade rules* and *goal rules*. *Model rules* define the simulation model or the fictional world in which the game takes place, and combine what we have called *object rules* and *action rules*. That is, *model rules* define all the elements of a game and their possible behaviours. For example, the *model rule* for the white queen in chess can be formulated as follows: "The white queen is the light-coloured piece with initial position 1-D [object rule] which can move to any square along the file, the rank or a diagonal on which it stands [action rule]". *Grade rules*, on the other hand, deal with any characteristic that is measured in the game. In football, this includes the number of goals scored and the number of yellow/red cards used; in a shooter, it might include the number of bullets remaining in a weapon and the amount of health remaining in a character. In sum, *grade rules* regulate anything that creates a gain or a loss from the perspective of the game. From the discussion above, it should be clear that the *Quantified Performance* property in games is instantiated by *grade rules*, and therefore that *goal rules* are a special subset of *grade rules*.

To sum up, game rules can be characterised according to two main categories: *model rules* and *grade rules*. *Model rules* can be divided into *object rules* (rules that define the objects, space and time of the game universe) and *action rules* (rules that define the possible behaviours of the elements defined by *object rules*). A particularly relevant subset of *grade rules* are *goal rules*, which define the (possible) main goal (and possibly partial goals) of a game. In Sect. 3.1, we will show that "death" exemplifies another special subset of *grade rules*.

## 2.3 The Ludic Property: Enjoyment, Challenge and the Flow Model

Defining games as systems constituted by *Rules*, *Skill* and *Quantified Performance* (or *Goals*) introduces the additional problem that some things that are not games may satisfy these conditions. For example, a physics exam could be defined as a system governed by rules that require some skill on the part of students and quantify their performance. This is also the case for much simulation software, such as a drone simulator or a professional surgical simulator.

In this way, many authors have postulated some kind of primary "ludic" *intention*, *purpose* or *function* in games. [32], p. 50 defines games as a "form of recreation". Juul [29] (p. 226) states that a "game has been designed to be entertaining when one pursues the goal" (2005). Aarseth [2] (p. 54) minimally defines games as "facilitators that structure player behaviour, and whose main purpose is enjoyment" (2011). [21] (p. 51) points out that games are engaging in part because they are sources of "pleasure". Waern [52] (p. 3) defines a game as a game-like system "which has as its primary purpose to be played" (2012), where "playing" is defined in terms of [8] *paratelic* state of mind. Roughly, a *paratelic* state of mind is one in which the motivation for an activity lies in its means rather than its end (in the enjoyment of the process rather than the output itself). Finally, as shown above, [47] states that a necessary condition for being a game is to be an artefact "intended primarily as an object of entertainment".

Clearly, a physics exam (even though it is a rule-based activity that requires some effort and quantifies student performance) does not have a primarily "ludic" *purpose* or *function*. The difference between a video game such as *Microsoft Flight Simulator* and a professional drone training simulator is that the latter is not primarily intended for entertainment [47]. Undoubtedly, games can have a variety of *intentions, purposes* or *functions* other than entertainment. Some games have the purpose to improve our cognitive abilities (e.g. *Big Brain Academy*, 2015, or *Brain Training*, 2019). *Advergames* seek to promote a brand, product or organisation. Recently, some video games have been developed for educational purposes (e,g, *Hakitzu*, 2013, is a video game in which we learn the logic of programming by fighting robots). However, all these cases are cases of *games* in virtue of the fact that their *primary* intention, purpose or function is *ludic*. We should therefore add a *Ludic* property to the three previous ones (*Rules*, *Skill* and *Quantified Performance*) that define games.

Although several scholars have recognised the ludic property as an essential aspect of games, little attention has been paid to specifying its nature. What exactly does "ludic" mean? Many non-game activities could be described as "ludic", "enjoyable" or "entertaining" in a broad sense, such as reading a novel, watching a film, playing on a seesaw or painting with watercolours. All these activities share a common and broad ludic *intention*, *purpose* or *function*. However, there seems to be a specific sense of *ludicity* that is particular to games. Playing games and reading novels are clearly ludic activities, but in different ways.

From an empirical perspective, numerous studies have consistently shown that a fundamental aspect of playing video games is *enjoyment* [18, 46, 50, 51]. A well-established way of understanding the concept of enjoyment in the psychological literature is through [14] flow theory. In short, people experience enjoyment when they are in a state of "flow", characterised by an optimal balance between *challenge* and *skill*. In this way, "flow" is a mental state characterised by complete immersion in an activity, where the individual feels intensely focused, fully engaged and enjoys a sense of energised focus [11]. During "flow", people often lose track of time, have a distorted sense of self-awareness, and experience a deep sense of satisfaction and enjoyment. Crucially, *challenge* is a key factor (a necessary condition, according to [14]) for entering the state of flow, coupled with the participant's skills. Flow occurs when the challenge level of a task matches an individual's skill level, i.e. when the challenge is *optimal.* If the challenge is too low for one's skill, boredom will result, whereas if the challenge exceeds one's skills, anxiety will result. Thus, optimal enjoyment occurs when the activity is perceived as neither too easy nor too difficult [6].

Flow experiences are often found in activities that are goal-directed, bounded by rules, and require a combination of mental focus and appropriate skills [46]. Thus, video games are perfect enablers of flow states, and challenge is one of the most important aspects of a good game design [43], Chap. 24, [42], Chap. 7). In this way, a significant part of game design is based on the idea that players actively seek out and are motivated by challenges. In a recent study, [34] explore the role of challenge-skill balance on player experience. They measure a range of indicators (positive and negative affect, enjoyment, competence, autonomy and presence) across three game modes in *Left 4 Dead 2* (2009): boredom (low challenge), balance (optimal challenge), and overload (high challenge). The first and the last represent situations of challenge-skill imbalance. The results clearly show that of the three conditions, the challenge-skill balance produced the optimal player experience. In this sense, participants reported significantly greater positive affect and enjoyment in the balance (optimal challenge) game mode. However, another finding is also noteworthy. The overload condition (high challenge) had significantly higher levels of positive affect and enjoyment than the boredom condition (low challenge). This means that a high-challenge imbalance is preferable for the player (in terms of positive affect and enjoyment) than a low-challenge imbalance. Similarly, [30] developed a game prototype designed to collect data on how players perceive failure, and tested it online with 85 players. The results show that players prefer games where they fail before winning to games where they win without failing. In Juul's study, players repeatedly associate the "easiness" of the game with a lack of challenge and a boring state. All of this supports the claim that "game enjoyment" is highly correlated with challenge, and points to the idea that it is the lack of challenge that "kills a game".

Another key study in this regard was developed by [16, 17]). They interviewed Finnish children who were actively playing video games and asked them about the form of immersion they experienced while playing. What they found was that the enjoyment of playing was strongly related to a good balance between the challenges of the game and the skills (physical and mental) required. Based on their findings, they developed the SCI-model, which analyses the game experience in terms of three key components: sensory immersion (the game's audio-visuals), challenge-based immersion (the balance of challenges and skills) and imaginative immersion (the worlds, characters and story elements). They then test the model by interviewing a total of 193 Finnish gamers about a wide range of popular and different games: *World of Warcraft* (2004), *Half-Life 2* (2004), *Grand Theft Auto: San Andreas* (2004), *Halo 2* (2004), *Civilization III* (2001), *The Sims 2* (2004), *Star Wars: Knights of the*

*Old Republic II* (2004), *Rome: Total War* (2004), *Flatout* (2004), *Pro Evolution Soccer 4* (2004), *Nethack* (1987), *Neverwinter Nights* (2002) and *NHL 2005* (2004). The selected games cover a wide range of genres and game-modes, from story-based games (e.g. *Half-Life 2*, *Neverwinter Nights*) to cases of "games without a goal" (e.g. *World of Warcraft*, *The Sims 2*). Each of the three types of immersion was quantified and presented as an average value. What we take as the key finding is that challenge-based immersion is not only the most highly rated feature in eleven of the thirteen games, but also *the only consistent feature* across genres and game modes (the other types of immersion show much greater variation across game types). This again points to the idea that challenge is a constant (necessary?) feature across games.

Overall, the empirical evidence supports the idea that the specific kind of ludic experience that games provide is closely related to challenge; that the primary ludic *intention*, *purpose* or *function* in games is challenge-constituted. In other words, games are, at a minimum, entertaining artefacts, and the specific way in which games entertain is through challenge. As we will see in Sect. 3.1, this idea is key to understanding one of the main functions of death in video games.

## 2.4 A Model for Video Games

Taking all of the above into account, games can be defined (from an object perspective) according to four main properties:

(1) *Rules:* A game is a *rule-based system.*
(2) *Skill*: The player must exert some *effort* (mental and/or physical) in order to influence the outcome.
(3) *Quantified Performance*: The player's performance is *objectively quantified.* This can lead to the achievement of Goals and ultimately to victory or defeat.
(4) *Ludic*: The primary *intention, function* or *purpose* of a game is *ludic*. Ludicity in games is challenge-constituted.

We do not take this proposal as a "new" definition of (video) games, but as a synthesis of the main contributions in the in the literature on the subject. The perhaps "novel" element is our insistence that the ludic *intention, function* or *purpose* of games, postulated by many authors, is essentially constituted by challenge (i.e. the specific way in which games entertain is through challenge). Assessing the extent to which these four properties constitute an essentialist, extensionally satisfactory definition of video games, or whether the concept "games" is analytically indefinable [4], is beyond the scope of this paper. As claimed in the introduction, the aim was simply to present a theoretical framework that will be applied to the analysis of death in video games in Sect. 3. Therefore, we will content ourselves with some brief comments on it.

The four properties are closely related. Rules consist of *model rules* and *grade rules* (Sect. 2.1). In a broad sense, video games can be defined as models or simulations [20, p. 223; 29, p. 191]. Model rules define the simulation model, but in a sense this is just a collection of entities (*object rules*) with some dynamic rules (*action rules*). What turns a simulation universe into a game is, in particular, the addition of *grade rules* that *objectively quantify the player's performance* and thus require some *skill* on the part of the player. Grade rules that quantify a player's performance and require skill create the *challenge*, which then becomes the source of *entertainment*.

Just as games can have a variety of functions other than *ludicity* (improving our cognitive abilities, promoting a brand, etc.), the idea that *ludicity* in games is primarily challenge-constituted does not mean that games cannot provide additional sources of entertainment. Narrative or fantasy games may very well provide the player with the pleasure of imaginative engagement in addition to the pleasure of overcoming obstacles derived from challenge [38]. The difference, we argue (drawing on the empirical studies in Sect. 2.3), is that the pleasure derived from challenge is *constitutive*. In this way, the *Ludic* property is *prescriptive*. Just as (according to [53], well-known proposal) make-believe is the prescriptive way to engage with, say, novels, so the pursuit of challenge is the prescriptive way to engage with games. On such a view, a "game" that does not *quantify performance* and does not offer *challenges* would not count as a game.

In this way, possible counter-examples to the model would be "games" such as *Electroplankton* (2005), *Mountain* (2014) or *Panoramical* (2015). In addition to having no goal, these "games" are characterised by a lack of quantification of performance and a set of challenges to overcome. In the case of *Electroplankton*, the "game" consists of ten different mini-games without a goal, featuring microscopic animated creatures that emit different music and sounds depending on how the player positions them on the screen. Is *Electroplankton* a game and therefore a counter-example to the model? Appealing to intuitions is usually controversial. However, by appealing to user reviews, it seems justified to claim that *Electroplankton* is at least a "borderline or limiting case" of a game. On Metacritic[2] one can find many user reviews claiming things like "It can barely be described as a 'game'" (Yojimbo2015), "It's hard to tell whether this is a video game or an interactive music programme" (Carlosgtardon) and "Not a proper 'game' if you're expecting that... The definition of 'videogame' has been changing a lot over the years in order to accommodate new forms of interaction. But even if that's the case it's still hard to file *Electroplankton* under that category" (Julianozuca). In this sense, we consider the model presented to be relatively safe and satisfactory.

[2] https://www.metacritic.com/game/electroplankton/.

# 3 Death in Video Games

## 3.1 Death as a Source of Challenge

One of the most common views in video game literature is that death is essentially a *didactic mechanism* to teach players how to improve their gameplay. Death means that something has gone wrong, and the player must learn from their mistakes in order to meet future challenges. In this sense, Juul ([30], p. 52) claims that death in video games is "the death you survive and learn from", and [54, 55]) states that death means that the player must further develop her skills by internalising the rules of the game, which is then rewarded by staying alive. This didactic function has been conceptualised by appealing to the "save-die-restart" structure [37], as a process of "self-optimisation" [40] or by conceiving of video games as a medium characterised by the "aesthetics of repetition" [23]. In a similar vein, [35] claims that death in video games essentially serves as a "punishment" and [15, 22] and [25] relate death to the notion of "failure".

This view is clearly illustrated by classic arcade games. Take *Super Mario Bros.* (1985). If a player tries to jump through a hole and fails, Mario dies. This means, among other things, that the player's mechanical coordination was not good enough. The player can then try again the challenge until she presumably succeeds (or until it runs out of lives), thus *learning* how to combine the game rules "Jumping" and "Moving Right". New holes in later parts of the game may be more difficult to jump, implying the need for more fine-grained mechanical learning of the rules. Thus, death implies a "failure" on the part of the player and a "punishment" (e.g. returning to the last checkpoint in the game space), and functions as a didactic mechanism for the player to "internalise the game rules" and "optimise" their mechanical application. In this respect, the *quantification of the player performance* (Sect. 2.1) through *grade rules* (Sect. 2.2) keeps a record of the player's learning, and death teaches the player how to improve.

In addition to being a *didactic mechanism*, however, death in video games seems to fulfil other roles, aims and possibilities, which we will explore in the following sections. For now, we want to focus on what we consider to be a more essential function of death. The aim is not to deny its didactic role, but to complement it with what we take to be a more fundamental purpose.

Certainly, death is still present when you have mastered a game, be it an arcade or an MMORPG. A notion of "mastering a game", "good mechanical domination" or "full internalisation of the game rules" that meant something like "the player no longer dies" would be a very limited and restricted notion. Players who master a video game in the broadest sense still die, just as professional chess players and professional football players lose matches. When a football team (or a chess player) loses a match, it can be argued that there is a sense in which the "failure to win" implies a potential learning, things to improve for future matches. However, this "didactic sense" does not seem to capture the whole story, and yet it may be the case that there is no new element (or only minimal elements) to learn. In the case of video games, for experienced players who die, the potential learning from a death may be very low. A player may already know what went wrong, or the death may simply be the result of bad luck. Thus, the *didactic function* of death seems to adapt to a kind of "law of diminishing returns": for a new player, death may have a primarily didactic role; for a veteran player, the didactic gain from death may be very low or even non-existent.

In addition to its didactic role, death is one of the main sources of challenge in video games. For a player who has already mastered the mechanics of a game, not dying is (one of) the main element(s) that keeps the gameplay interesting. This is clearly illustrated in MMORPGs, where players are essentially immortal, meaning that nothing can ever permanently kill the player's character. Typically, the character will either be resurrected by another player or respawn in a different location. Without the risk of death, there would be no challenge in playing such games. In this sense, Kevin McPherson, one of the original EverQuest team members, argues [26] that in order for the rewards of a game to be meaningful, actions must have consequences, and points to virtual death as one of the main risks that players face and that gives meaning to their achievements.

This means that *mere death* is not enough to create *challenge*: death must be accompanied by a "penalty". This has given rise to the concept of "death penalty" [35]. Depending on the game, the "death penalty" can vary in degree. For example, death is much more forgiving in *World of Warcraft* than in *EverQuest*, a general trend in MMORPGs in recent years. In *World of Warcraft*, experience is not lost, respawn points are more abundant, and players have the option to respawn with a ten-minute "penalty" that temporarily lowers the player's stats. According to [35], this means that from a player's perspective, death itself is not interesting, but the "punishments" that come with it. Similarly, [56] claims that death is one of the main mechanisms that regulate the "danger" of a virtual world by being "expensive". Thus, designers need to strike a balance to ensure that death reinforces the challenge in a good way: "death penalty" should be severe enough to motivate players to avoid it, but not so severe as to end the game experience. In this sense, Rollings and Adams [42] (p. 525), in their book on game design, claim that death must be accompanied by a disincentive, otherwise players won't care if they die. Thus, "the trick is to find a disincentive that is appropriately proportional to the likelihood of their dying –to put it in simpler terms, it's a balance problem".

Another good example of "death penalty" is found in competitive games such as *League of Legends* (2009) or *Dota 2* (2013). Dying in the early stages of the game has a low respawn time penalty, while dying in later stages increases the respawn time significantly. In *League of Legends*, the respawn time is based on the level of the character and the length of the game, ranging from around 6 s to over 40 s. In this way, dying in the later stages of the game usually means defeat, as the opposing team would have enough time to destroy your base. Thus, one of the main sources of challenge in *League of Legends* comes from the risk of death and the "death penalty" that comes with it.

Death plays a similar role in single-player/non-competitive games. In classic arcade and platform games, even when respawn points are available, the main challenge comes from the limited number of lives a player has to complete each level. If the player in games like *Crash Bandicoot* (1996) or *Spyro* (1998) could not die in any way, these games would be quite boring to play. In fact, it could be argued that they would no longer be "games" at all, but a kind of interactive play environments, such as *Electroplankton* (2005) or *Tori-Emaki* (2007).

Death is also crucial in single-player "games without a goal" such as *Minecraft* or *The Sims*. In fact, the only apparent gameplay challenge that *Minecraft* presents to the player is to stay alive. Avoiding starvation, being killed by a monster, or some other game effect (such as falling from a great height or into a lava field) is the only requirement the game imposes on the player in order to continue playing. Of course, *Minecraft* or *The Sims* are typically complemented by the player's personal goals, such as building a nice house in *Minecraft* or achieving social success in *The Sims*. In this sense, what seems to prevent *Minecraft* or *The Sims* from being considered as mere design or simulation software rather than games is precisely the possibility of dying, and the challenges and penalties associated with it.

In narrative games, such as *The Witcher* (2007) or *Baldur's Gate 3* (2023), challenges can arise from a number of different factors. For example, the player may have to make decisions whose consequences will be seen in plot devices in later parts of the game. Or the player may have to solve a puzzle, the reward for which is a powerful or useful item that is useful but not necessary to complete the game (so failing the puzzle does not end the game experience). Also, the player can typically load any previous (voluntary) saved state at any time, so one might think that dying is not a real problem, but this is not the case. Dying still involves a "death penalty" because the time and effort spent from the last saved state to the actual moment of death is lost. In fact, saving continuously (and especially saving at key moments) is one of the main strategies to minimise the "death penalty" in these games. In this sense, players can increase the challenge of the game by abandoning voluntary saving (relying only on the game's auto-save moments), in which case death could entail a severe punishment. In addition, these games usually feature battles against different types of creatures. In these cases, the challenge comes from not dying against the enemies.

In the previous paragraphs we have covered different types of games and ways in which death can be implemented. In some cases, lives are unlimited, while in others, death can mean the end of the game. A framework for unifying these different possibilities is offered in Sect. 3.2. For now, the aim has been to illustrate the constant function of death as a main creator of *challenge* in video games, achieved through the notion of "death penalty". As we have seen in Sect. 2.3, challenge is a constitutive feature of games, what makes a game ludic in its specific sense. In this way, an essential function of death in video games, apart from being a didactic mechanism, is to make games challenging and thus to provide the ludic experience. To be a little poetic, one could say that while real death is the least ludic experience imaginable, game death is the main source of entertainment in video games.

A final point to consider is what death is from a game perspective, i.e. what type of rule represents death. In Sect. 2.2 we divided game rules into three main categories: *object rules*, *action rules*, and *grade rules*. As a reminder, *goal rules* are a special subset of *grade rules*, certain game states (goals) that occur when a certain quantification of performance is measured and achieved.

In this sense, death exemplifies another special subset of *grade rules*. In many games, the player's avatar has a certain amount of "life", and a *grade rule* takes care of its evolution. When it reaches a value of cero, a "death-state" takes place. A "death-state" always implies a "death penalty" in the form of a change in the current state of the game (among other possible death penalties). In some cases, a "death-state" implies the end of the game. In other cases, it implies a return to a previous state of the game, such as going back to the last checkpoint or loading a previously saved game state. In multiplayer games, the actual state of the game does not stop, and a "death-state" typically involves a spatial relocation for the player (e.g. to a respawn point).

In conclusion, death in video games has a *didactic function* for the player to "internalise the rules of the game" and "optimise" her mechanical application, subject to a *law of diminishing returns*. More fundamentally, death functions as one of the main sources of *challenge*. This is achieved through "death penalties", which pose a balance problem from a design perspective. In this way, death is one of the main sources of the *ludic* experience that is constitutive of games. Death is implemented as a *grade rule* that defines a "death-state", which always involves a change in the actual state of the game.

## 3.2 Types of Death-State: Limited-Unlimited Life and Return-Non-Return States

In the previous section we saw that death can be implemented in many different ways. In some games, lives are unlimited, while in others death can mean the end of the game. This implies that a typology of deaths can be sought. To this end, we propose two intersecting axes of analysis.

The horizontal axis deals with whether a game provides the player with limited lives (in a discrete or continuous form) or unlimited lives. In other words, whether the player has a finite number of lives after which the game ends in some way, or whether the player can die an infinite number of times without the game ending. This axis thus distinguishes games in which death is a potential game ending from those in which death never ends the game.

The vertical axis focuses on the type of change in the current state of the game that occurs after death. As mentioned in Sect. 3.1, this is the minimal component that conforms "death penalties", and in this sense the properties of "return" and "non-return" are theoretically relevant. In some games, death is associated with a "return" or "spatio-temporal relocation": the return (resurrection), after death, to a previous state of the game (according to save states and checkpoints). In other games, "return" is not possible. Either the player has to start the game again (which is strictly speaking a "non-return", although the player can start a new game mimicking the last performance), or the player suffers a "spatial relocation" (the current state of the game does not stop and the player is spatially relocated, typically to a respawn point).

Games in which "return" is possible have two modes: *controlled* and *uncontrolled*. On the one hand, the controlled mode results from the possibility of saving the game-state at player's will. This is common in PC strategy games, role-playing games, and graphic adventures, where save states can be made at the player's will to return to previous states of the game if necessary (even without having died). Uncontrolled mode, on the other hand, relies on fixed save points (as in *Metroid Prime*, 2003) or on "passwords", as in video games from the 1980s and 1990s, which usually return the player to the beginning of a particular level (for example, in the *Mega Man* series, 1987–1993).

Combining the horizontal axis ("limited" and "unlimited" life) and the vertical axis ("return" and "non-return" states), four types of death in video games emerge. These types are "extensive" in the sense that they are *exclusive*, i.e. the same game cannot have more than one combination at the same time.

(1) *Unlimited Life and Non-Return state.* This is the typical case for MMORPGs, such as *World of Warcraft*. As mentioned above, players are essentially immortal, meaning that nothing can ever permanently kill the player's character, and no death can cause the game to end. Typically, the character will either be resurrected by another player or respawn in a different location. After death, the game does not stop and the player is spatially relocated. Competitive games such as *League of Legends* or *Dota 2* also instantiate this category, where death implies a spatial relocation and respawn time penalty. However, this category is not restricted to multiplayer games, where the non-return state seems to be unavailable. Single-player games such as *Minecraft* or *Outward* (2019) also feature unlimited lives and non-return states.

(2) *Unlimited Life and Return state.* This is perhaps the strangest category, as one might think that unlimited life would not require any possibility of return. In this sense, *Celeste* (2018) is a very illustrative example that even seems to look at death with a certain disdain. The game is a platform game divided into screens, literally, as each screen (or frame, in the sense of "the limits or borders of an image") is a puzzle of jumps and movements that the player must solve: to reach a platform through a very small gap without colliding with any blades or falling into the void. The difficulty of these jumps is such that you can die an infinite number of times and return to the beginning of the same screen without consequence. Death thus implies a return to a previous state of the game. In a sense, the quantification of the game is reversed: deaths do not subtract lives, but rather add up to a death count. At the end of the game, this death count can be reviewed. Obviously, dying less means having the best skills, but there is

little punishment (although there is learning) in these deaths, other than going back a few steps.

(3) *Limited life and Non-Return.* This was the most common combination in arcades in the 80 s and 90 s, and the most famous case is undoubtedly that of *Super Mario Bros.* (1985). There is a life counter that can eventually increase by obtaining certain items (the green mushroom). If the player dies as many times as the counter indicates, the game ends (Game Over) and the player must start again from the first level. Resurrection after each death occurs at the beginning of the level, but once lives are used up, there is no way to return (although this particular video game has shortcuts in the form of green pipes that allow you to skip whole worlds if you know where to find them; this is a variant of the password system in the form of "skill"). Another example of the same type would be the *Metal Slug* saga (1996 onwards) or, in particular, the roguelike genre, which uses procedurally generated levels and, in some cases, gives the player only one life (e.g. *Hades*, 2018), so that when they die they have to start again.

(4) *Limited life and Return.* It is in this category that we find the most striking differences between controlled and uncontrolled return modes. The uncontrolled mode typically relies on checkpoints. In an uncontrolled return video game such as *Castlevania* (1986), the player who has lost all her lives can only return to the starting points of each of the six stages using a password. This password was obtained by completing the immediately preceding level. In other words, completing the second level gave you the password to the third. However, this came with a certain penalty, as each level is divided into three sub-levels, which act as checkpoints as long as the player has lives. An example of controlled return is *Command and Conquer* (1995), a real-time strategy game in which players can save the game, preserving the current state of the game, including unit positions, resources acquired, and so on. This saving feature also represents a strategic approach to dealing with death, as it can be done before engaging in battle. Therefore, if the battle goes badly and many units are lost, it is possible to revert to the previous state and choose a different attack strategy (even before the player has actually lost the game). Further, many narrative games also fall into this category. *Baldur's Gate III* (2023) presents a controlled return mode, as the player can save at any time. However, life is limited, as if all members of a party die in combat, the player is presented with a "game over" screen. Again, dying is not a big deal (except for the loss of time and effort), as the player can immediately load a pre-battle game state (though the death of the party may remain in the player's consciousness as a shameful episode).

This classification can be supplemented by a typology of "death modes". These modes are "intensive" in the sense that they are *inclusive*, i.e. the same game can have more than one combination at the same time. In a given game, not every "hit" may be treated in the same way: some may result in punishment, some in death, while others may mean the end of the game.

In this regard, the two extremes are: (a) *permadeath*, i.e. a death that means the definitive end of the game (see Sect. 3.3); and, at the opposite pole, (b) *invincibility*, where the player's avatar cannot die. An example of permadeath is the last life in *Super Mario Bros.* If you lose it, you have to start again from the beginning. So this particular "last-life-death" is different in mode from the previous deaths. In the same saga, we find the opposite case (invincibility), exemplified by the easy mode of *Super Mario Wonder* (2023). Certain types of characters (the Yoshis) cannot die.

The intermediate cases are (c) *relative death* and (d) *partial death.* On the one hand, relative death (which is the most typical case) is any death that does not mean the end of the game (*permadeath*), but a certain "death penalty". For example, losing one of the multiple lives initially granted in *Super Mario Bros.* results in a death (relative death) that is different in mode from the "last-life-death" (permadeath). Relative death is one of the means by which the logic of repetition in video games, so characteristic of arcades in the 80 s and 90 s, is promoted.

Partial death, on the other hand, captures a type of "pre-death" or "penalty" that a player can suffer before actually dying. In the classic platform arcade game *Ghouls'n Ghosts* (1985), the main character, Sir Arthur, has armour which he loses when hit (and runs around in his underwear until he can regain it), but dies on a second hit. As may happen in *Super Mario Bros.*, there is a loss of attributes or energy before relative death occurs. In this way, relative and partial death could coincide and occur simultaneously, as in *Gradius* (1985). In this case, the player may lose her upgrades, such as extra missiles or energy balls for protection. If the ship (Vic Viper) is destroyed, it loses these upgrades and returns to its previous state, without missiles or protection, making the game difficult and implying an obvious punishment in death. If relative death is the logic of this type of arcade game, when the emphasis is on partial death, it is common to encounter a beat'em up game. In *Street Fighter II* (1991), the yellow energy bar decreases and turns red when you are hit (partially). When it turns red, we lose the round (relatively). If you lose two rounds, the tournament is over (permadeath). As we have seen, these deaths modes can occur simultaneously and in the same title. This leads us to another conclusion: in many cases, permadeath (in this sense) may not mean the death of the virtual character, only his defeat, which is a relief.

## *3.3 Serious Death and Permadeath*

[35] examines how death is implemented and experienced by players in nine popular massively multiplayer online worlds (*EverQuest*, *World of Warcraft*, *Dark Age of Camelot*, *Ultima Online*, *Lineage II*, *City of Heroes*, *Eve Online*, *Star Wars Galaxies* and *Anarchy Online*). She created a website (http://www.death-stories.org) where players could share their stories of character death in the selected games. Players were asked, among other things, to share a story of a "death experience" and their general opinions about death in such game worlds. The results show that death can

be "trivial" or "non-trivial", depending on the context. However, the first category far outweighs the second. This is clearly illustrated by participant S38:

> I'm writing because I am certain that this is a typical 'death experience' in *World of Warcraft*: 99.999% of deaths are utterly unremarkable... for the typical player, it's just a game and character death is a non-event.

The study also collected many stories of "heroic deaths", illustrating that death can be a source of meaning, promoting social reputation and group relations. However, as mentioned above, these cases appear as a minority in the ocean of trivial deaths.

In this way, some authors have complained about the triviality and meaninglessness that death has acquired in the recent development of video games. [19] discusses the perceived lack of seriousness in the computer games industry, contrasting it with literature and cinema. The reasons, he argues, go beyond a potential lack of demand and economic motivations. It is the canonical structure that the medium has developed (the ability to restart after failure) that has a devastating effect on the construction of serious games. Actions in video games (including death) are often reversible, leading to a lack of consequences and a trivialisation of serious issues. This means that "there is no room in them for fate or tragedy. It is always possible to go back and play until you reach a happy ending. For this reason, videogames allow players to fool death itself" (2001, p.176). Thus, a setting where actions are irreversible (where the "save and restore" function is removed) would overcome several significant challenges associated with the development of "serious" video games.

To this end, Frasca introduces the concept of an "ephemeral game" called the "one-session game of narration" (OSGON). In an OSGON, actions are irreversible, and the game can only be played once. The idea is to create a single-user game format that eliminates the ability to restart or save progress, forcing players to carefully consider their decisions and face the consequences, similar to real-life situations. In this way, OSGONs offer a strategy for developing "serious" video games, allowing designers to explore deeper and more meaningful themes by overcoming the limitations of traditional game structures.

Similarly, [41], in an article entitled *The Slow Death of the Game Over*, begins by asking: "When was the last time you were really distraught about seeing a game over screen?" The article reflects on the evolution of the "game over" screen in video games. Over the years, the severity of the consequences of failure ("death penalties") has diminished, moving from the arcade-era need to accommodate sales to the more forgiving systems of save points and quicksaves. The break between "hard" and "soft" game-over is attributed to *Doom* (1993) and, to a lesser extent, its predecessor *Wolfenstein 3D* (1992). Both popularised the "save anywhere" system, making it easier for players to recover from mistakes and rendering "game over" almost irrelevant. Orland argues that in an era of increasingly forgiving games, there is value in preserving the consequences of failure in certain titles.

Both Frasca's and Orland's papers seem to have had a clairvoyant aura, for in the years since 2007 there has been a resurgence of games with a permadeath feature that seek to enhance meaning through a more serious treatment of death. "Permadeath" (permanent death) can be loosely defined as a game mechanic in which a character cannot respawn after being killed. In some cases, this may mean that the player must restart the game from the beginning (as in classic arcade games), either with the same character or with a forced new one (e.g. *DayZ*, see below). In other cases, the player may be able to continue playing, but the character that died is no longer available (e.g. *XCOM: Enemy Unknown*, see below). These are cases of what we might call "relative permadeath". A stronger situation, which we can call "absolute permadeath", is represented by OSGON games, where death means that the player can no longer play the game.

Examples of OSGON games include *You Only Live Once* (2009, Kongregate Games), *One Chance* (2010, Dean Moynihan) and *One Single Life* (2011, FreshTone Games). The latter is a mobile game in which the player has to jump from the top of one building to another. However, you only get one try at each level. Once you die, you cannot play the game again. Death is permanent in the strongest (absolute) sense, and there is no way to insert a new coin and play again. The description on the iOS store reads as follows:

> WARNING! IN THIS GAME, YOU GET JUST ONE LIFE. Stop reading this and ask yourself one simple question: If everything was on the line, would I have what it takes? Welcome to One Single Life. A game which poses that very same question and a game which only 4% of players in the world will be able to finish. Unlike most games, the challenge doesn't come from playing against the computer but against yourself. It's you versus your own self-doubts. Do you have what it takes? Or will you psych yourself out at the vital moment? There's only one way to find out. Enjoy!

In recent years, many games have introduced permadeath as an option to challenge players, such as the *Diablo* saga or *Baldurs Gate 3*. Others, however, have tried to incorporate it into the game mechanics themselves. Two notable cases are *XCOM: Enemy Unknown* (2012) and *DayZ* (2013). The former is a turn-based tactical game where the player controls a squad of powerful soldiers in a series of turn-based tactical missions during an alien invasion of Earth. Once a soldier dies, they cannot be resurrected in any way, resulting in permanent death. Lead designer Jake Solomon claims in an interview [24] that this feature was intentionally included in the game to evoke "a real sense of loss in players, because it's something that they don't want and they can't undo once it has happened". The permanent death of a soldier means that "It's authentic. It's real. Those emotions are real. The loss is real. The challenge that arises out of that loss is real. In some way, that makes the game real". However, players' reactions to this feature vary. For example, Steam user "I am Groot" started a thread[3] in the Steam forum entitled "Can permadeath be turned off?", stating that:

> I would get the game if I could do a playthrough with it off to get used to the gameplay. Having that forced on you when you want an easier and less frustrating experience is annoying. Nothing against permadeath but for my first playthrough I'd like not to have to be frustrated constantly by it.

User "GoGShogun" replied with the following claim:

[3] https://steamcommunity.com/app/200510/discussions/0/630800446055476044/.

> The point of permadeath is to teach you a lesson. It's not there to make the game hard, it's there to show you there are better ways with dealing with something, or that your just ♥♥♥♥ outta luck, I know the feeling when you miss a 97% chance of hitting and killing the target, but you still miss and lose a soldier. Sure it just makes you angry but that's what is beautiful about this game. It makes you buy new computers, cuz your last one has a giant hole in it.

This clearly illustrates the potential gain in meaning and seriousness of permadeath, and the associated potential increase in computer sales.

*DayZ*, on the other hand, is a multiplayer survival game set in a post-apocalyptic zombie world. The player controls a survivor who must search for food, water, weapons and medicine while killing or avoiding being killed or infected by other players or NPCs. Once the player dies, the death is permanent. The player must start again, controlling a different survivor and starting from scratch. In this vein, [13] analysed players' experiences of permadeath in *DayZ* through a survey that received 1704 responses. They found that while character death evoked strong negative emotions in players, permadeath was seen as a core positive and enjoyable feature by the overwhelming majority of players. The question 'When you play DayZ, how enjoyable do you find the consequential nature of death?', rated from 1 "Not enjoyable at all" to 5 "Very enjoyable", showed a clearly positive distribution (1 (3%), 2 (6%), 3 (14%), 4, (26%) and 5 (51%)). Associated feelings described by the majority of players include anxiety, fear, tension, adrenaline, paranoia and devastation, driven by the constant fear of death, which was identified as the main element that makes the game exciting. This results in "emotional immersion" and a feeling that the game is more real than most games. Furthermore, permadeath creates a stronger sense of narrative and meaning, with players reporting each survivor's life as a separate story or journey in its own right. Additional findings show that permadeath changes player behaviour, making them more cautious and less violent. This is in line with the intentions of Dean Hall, the designer of *DayZ*, who claimed in an interview at Wired magazine (2012) that "permadeath makes players use "emotional reasoning," rather than logic, to sort out problems. Players start to think with their hearts. They think about all the time they've invested to get where they are, and whether or not they're willing to risk everything when they roll the dice".

The search for seriousness and meaning through permadeath has also been pursued by individual players, who have self-imposed a permadeath rule. Two cases are noteworthy in this respect. In 2009, Ben Abraham conducted a permadeath experiment in *Far Cry 2* [5], as a form of "expansive gameplay", a term that aims to capture the range of possibilities offered by the addition of self-imposed rules. He decided to roleplay his character, maintaining a facade of ignorance about aspects his character would be unaware of, and self-enforcing a permadeath rule (if he ever died, he would stop playing). He chronicled the progress of his adventure in a series of blog posts, which were compiled into a 300-page PDF that had been downloaded over 10,000 times by 2013. In the PDF introduction, Clint Hocking, a Canadian video game designer, claims that "the permadeath restriction forces a player to adapt to overcome the biggest challenges one will ever confront in a game, or in reality: his or her own fickleness, foolishness, cowardice and frailty". Abraham's playthrough resulted in an entirely new emotional experience, encouraging risk minimisation and leading to playstyle choices that would not have been made in a regular session.

A year later, Brendan Keogh conducted a similar experiment in *Minecraft*, called "Towards Dawn". *Minecraft* offers an open world that is randomly and procedurally generated, infinite in practice. As the player nears the end of the map, new space is added. Given this property, Keogh decided to become a nomad, "perpetually walking east towards the morning sun" ([33], p. 4), and self-imposing a permadeath rule. As with Abraham's experiment, Keogh chronicled his adventures in blog posts, which attracted a great deal of interest and followers in the *Minecraft* community. Readers' investment became evident when the character faced a potentially dangerous situation, and messages of support encouraged the continuation of the journey. The project ended after two years, when Keogh decided to build a small house for his character, with a sign over the door saying "DAWN". He hadn't died, but he felt it was a fitting end to his adventure. Keogh reports that "the true effect of permadeath is not simply in the character's death, but in how it drastically alters the player's lived experience of the character's life" (p. 2). The threat of permadeath altered his gameplay and "weighed every act with a sense of significance" (p. 5). In this sense, permadeath creates a more intimate connection between the player and the character, culminating in extra care for the avatar's fate and allowing for much stronger narrative play: "as death mattered, so did life, and the simple act of playing *Minecraft* (or *Far Cry 2*) became a performance worthy of spectatorship". Keogh concludes by claiming that the essence of playing a video game is intricately connected to what it means to die in that videogame.

### *3.4 Death and Grief*

Some researchers have focused on how players react to the death of fictional characters in video games. In this sense, the death of Aeris in *Final Fantasy VII* (1997) is typically cited as one of the earliest examples of a character's death eliciting a profound emotional response from players. Aeris, also known as Aerith, is a central and beloved character who plays a key role in the story. In a shocking turn of events, she is killed by the main antagonist, Sephiroth, in a pivotal scene. The scene is often remembered for its emotional depth and has become one of the most iconic moments in video game history.

In this regard, [25] explores how video games engage with and represent the emotions and experiences associated with attachment, loss, and grief. What she calls the "death-as-failure paradigm" (Sect. 3.1) dominates most video games, where death is just "a short moment of frustration in an infinite loop of trial and error" (p. 9) aimed at optimising player performance (didactic function). Typically, death is not an experience in and of itself, but a medium for the player to improve her gameplay. Some designers, however, have offered alternative approaches. In this vein, she presents several video games (such as *Final Fantasy VII* or *Brothers: A Tale of Two Sons*, 2013) and analyses how they portray the emotional aspects of attachment and separation, as well as the process of grieving. In the case of Aeris, Harrer mentions two fan practices ("Aeris's Ghost Glitch" and "The Resurrection Hack") performed by players that demonstrate a grief practice of "continuing bonds", typical of real-life grieving processes. To be more about grief and less about self-improvement, she argues, "loss in games needs to be coupled with attachment and care, not mastery and success" (p. 11).

Similarly, [39] shows that the death of fictional characters (or even the loss of fictional worlds because a company shuts down the game) can affect players emotionally, and that no tools have been developed to help players after the game or its characters are no longer available. To provide an emotional and therapeutic experience, he uses Natural Language Processing (NLP) and Natural Language Understanding (NLU) techniques to recreate the fictional characters of *Tales of Link* (2018) who "died" after the company shut down the game in 2018. Using Discord and the aforementioned techniques, he takes a dataset of the characters' in-game dialogue to create an AI profile that players can have conversations with. It also provides a way for non-players to meet and get to know these characters who have died.

All this raises the question of whether our grief process for fictional characters can be understood in the same way as for real ones. In this regard, [44] studied audience reactions to the deaths of fictional characters from television and video games. They code 3000 online comments (on the deaths of sixteen TV and video game characters) according to Kübler-Ross and Kessler's five stages of grief (denial, anger, bargaining, depression and acceptance). Several findings are worth noting: depression is the dominant category for all characters; acceptance correlates more with TV than with video game deaths; the deaths of female characters produce more anger than the deaths of male characters; and the deaths of playable characters show lower levels of acceptance than the deaths of non-playable characters. They then present a number of quantitative and qualitative analyses, again with interesting results. To mention just a few, in terms of reaction to death, the analysis shows that commentators move through the five stages of grief in a similar way to how they would react to the death of a real person; and in terms of medium, TV deaths were more easily accepted than video game deaths. Overall, they state that their findings "cement the fact that the grief process for a fictitious character is very similar to the grief process for a real person" (p. 39) and that there was a clear difference in perception between the deaths of video game characters and TV characters, where "the deaths featured in video game narratives carry a different psychological weight and connotation than deaths featured in cinematic narratives" (p. 40).

Not only can the grieving process for a fictional character be similar to that of a real person, but the death of a real person can enter the fictional world and the person's fictional character. This is illustrated by the various virtual funerals that have been held in video games when a player has passed away (for an overview, see [7]). One of the most famous episodes took place in *World of Warcraft* in 2006. A 52-year-old woman named Fayejin in the game had died of a stroke. Her guild, Illidan, decided to hold an in-game funeral to mourn her, and player Yanoa announced it on the game's forums as follows:

> On Tuesday of February 28th Illidan lost not only a good mage, but a good person. For those who knew her, Fayejin was one of the nicest people you could ever meet. On Tuesday he suffered from a stroke and passed away later that night.
>
> Im making this post basically to inform everyone that might have knew her. Also tomorrow, at 5:30 server time March 4th, we will have an in game memorial for her so that her frieds can pay their respects. We will be having it at the Frostfire Hot Springs in Winterspring, because she loved to fish in the game (she liked the sound of the water, it was calming for her) and she loved the snow. If you would like to come and show your respects please do.

Yanoa added: "We're going to make a video of it to show to her family. So I would appreciate it if nobody comes to mess things up". Unfortunately, things didn't go as planned. A rival guild, Serenity Now, stormed the event, murdered everyone involved, and recorded the episode to later make a footage film and post it on YouTube. At the end of the film, a player from Serenity Now approaches the event organiser and claims: "She loved finishing. And snow. And PvP" (Player versus Player), followed by a final message on a black background that reads: "Sorry for your lose. Yes, we know we are assholes:D". Matt Kment, game leader of Serenity Now, latter claimed in an interview that "what I just tell everybody is, if we didn't do it, somebody else would",[4] and presumably showed no regrets about the event.

The video received millions of views, and there are many posts on the internet discussing the incident. While the majority of people seem to disapprove of Serenity Now's behaviour, some argue that the guild didn't break any in-game rules. In this regard, [2] reproduces an interesting conversation between two players discussing the event. When one claims: "I know a real girl died but it wasn't a *real* funeral" (our emphasis, p. 64) the other replies: "Define real moron. I think you're confusing the terms 'real' and 'physical'. You're assuming the medium used justifies whether or not people should be respectful". Aarseth uses this example to illustrate how it is possible, and typical, for players to interpret in-game events in more than one way, and how this ambiguity can lead to profound ontological and ethical reflections. In this case, he asks: "Is it not a real commemoration when it takes place in a virtual game world?" (p. 65). Given the many cases and studies we have mentioned, it may seem clear that the answer is that virtual funerals can be just as *real* as physical ones.

A more recent example of a virtual funeral took place in *Warhammer Online: Return of Reckoning* (*RoR*), which has some additional features worth considering. Originally titled *Warhammer Online: Age of Reckoning*, the game was released in 2008 by Mythic Entertainment to compete with *World of Warcraft*. The game received positive reviews, but after a few months the number of subscribers dropped until the company decided to close the game in 2013. As in the case analysed by [39], this shows that video games themselves can face the tragedy of death. In 2014, however, a group of players managed to obtain the game's code and released it (for free) on a private server, which has been active ever since, up to the date of this article. The case illustrates the possibility of games becoming "zombie games", where an official death is contested by some necromancer fans who bring the game back to life.

---

[4] https://www.youtube.com/watch?v=3mtYkGl8wZ4&ab_channel=MrInternet001.

As in the case of Fayejin, a player from *Warhammer Online: Return of Reckoning* named Akaathi passed away in May 2021. The situation developed in a similar way, starting with a forum post announcing the virtual funeral, although this time there was no guild attack. There are, however, two new elements worth mentioning. Firstly, players attending the funeral could join the Discord channel of Akaathi's guild and share their personal experiences and memories of the player. Just like a real funeral, you could hear people crying and mourning. Secondly, the ceremonial march (attended by around 300 players) began in the main square of Praag (a city in the fictional world) and ended in a cemetery on the outskirts of the city. The game developers added a new tombstone in the cemetery with the inscription "R.I.P. Akaathi". This means that a new element has been added to the game that refers to a real out-of-game event. Akaathi's friends probably live far away from his real grave, but the virtual-real gravestone gives them a space to mourn and remember him.

## 4 Conclusions

(Video) Games, from an object perspective, can be defined according to four main properties: *Rules, Skill, Quantified Performance* and *Ludic*. As the empirical evidence suggests, the primary ludic *intention, purpose* or *function* in games is challenge-constituted. In other words, games are, at a minimum, entertaining artefacts, and the specific way in which games entertain is through challenge.

In this respect, apart from being a didactic mechanism subject to a law of diminishing returns, an essential function of death in video games is to make the game challenging and thus to provide the ludic experience. Death functions as the main creator of challenge, which is achieved through the notion of "death penalty". Designers need to strike a balance to ensure that death reinforces challenge in a good way: "death penalty" should be severe enough to motivate players to avoid it, but not so severe as to end the game experience. Without death, many games would lose their main source of challenge, and it could be argued that they would no longer be "games" at all, but a kind of interactive play environments. Without death, one might say, there is no game.

At this point, it could be claimed that the conceptual association between "failure" and "death" in video games seems to have a contingent flavour. Why does Mario lose "lives" rather than, say, "plumbing tools", and "dies" when a certain quantification reaches the value of cero rather than, say, becoming "unconscious" or simply "running away" from the screen? One might reply that "dying" makes sense, since many games feature human-like characters being hit by bullets, swords, lasers or simply falling from considerable heights. However, one could also argue that there is a deeper human ambition behind this association. If "failure" means "death", then overcoming a challenge means defeating our oldest and most inescapable enemy.

Death is implemented as a *grade rule* that defines a "death-state" and a "death penalty", which always involves a change in the actual state of the game. This change can be in the form of a "return" to a previous state (controlled or uncontrolled) or in the form of a "non-return", which typically involves a spatial relocation. In addition, death can be accompanied by a potential end of the game (limited life), or it can be an episode in an infinite loop of resurrections (unlimited life).

In addition to its didactic and challenging function, death can be a source of meaning and an emotional enabler. To achieve this, death must be accompanied by either a serious reflection or a narrative element. OSGON games or games with a permadeath feature achieve this by imposing a real sense of loss on the player, thus encouraging "emotional reasoning". When consequences mimic real-life consequences, the loss is real and the player's gameplay is radically altered. Suddenly there is anxiety, fear, tension, adrenaline, paranoia and devastation, driven by the constant fear of death. The reward system changes accordingly, and the emotional response to death is transformed from a mundane and trivial episode into a story or journey in its own right. Furthermore, narrative elements can give meaning to death by making the characters unique and beloved.

The hand of death can also reach games from outside their rules. Games themselves can die when a company decides to shut them down. In some cases, games can be resurrected by necromancer fans and become zombie games. In others, a resurrection spell may not be available, but the ghosts of the game characters can be summoned through AI profiles. In this sense, games and fictional characters can induce a grief process similar to that of a real person. Furthermore, the death of a real person can enter the fictional world and the person's fictional character. Since a player and his or her fictional character can be seen as a composite virtual entity, a game character can be said to die when the player passes away.

**Acknowledgements** Research for this paper was supported by Juan de la Cierva Grant 2023 (funded by MCIN/AEI and European Union NextGenerationEU/PRTR).

## References

1. Aarseth, E.: Computer game studies, Year One. Game Stud. **1**(1), (2001)
2. Aarseth, E.: Define real, moron! some remarks on game ontologies. In: Günzel S, Liebe M., Mersch D. (eds.) DIGAREC Keynote-Lectures 2009/10, Potsdam: University Press 2011, pp. 50–69 (2011)
3. Aarseth, E.: Ontology. In: Wolf, M.J., Perron B. (eds.) The Routledge Companion to Video Game Studies. New York and London: Routledge, pp. 484–492 (2014)
4. Aarseth, E., Calleja, G. (2015). The word game: The ontology of an undefinable object. In: Proceedings of the 10th International Conference on the Foundations of Digital Games.
5. Abraham, B.: Imposed rules and 'expansive gameplay': A Close Reading of the Far Cry 2 permadeath experiment. In: Proceedings of DiGRA 2013, (2013)
6. Abuhamdeh, S., Csikszentmihalyi, M.: The importance of challenge for the enjoyment of intrinsically motivated, goal-directed activities. Pers. Soc. Psychol. Bull. **38**(3), 317–330 (2012)
7. Arnold, M., Gibbs, M., Kohn, T., Meese, J., Nansen, B.: Death and Digital Media. Routledge, New York (2018)
8. Apter, M.J.: Reversal Theory: Motivation, Emotion, and Personality. Taylor & Francis (1989)
9. Audi, P.: How to rule out disjunctive properties. Noûs **47**(4), 748–766 (2013)

10. Avedon, E.M., Sutton-Smith, B.: The Study of Games. John Wiley & Sons, New York (1971)
11. Biasutti, M.: Flow and optimal experience. In: Mark A. Runco, Steven R. (eds.), Encyclopedia of Creativity (Second Edition), Pritzker: Elsevier, (2011)
12. Caillois, R.: Man. Schocken Books, Play and Games, New York (1961)
13. Carter, M., Allison, F.: Fear, loss and meaningful play: permadeath in DayZ. J. Gaming Virtual Worlds **9**(2), 143–158 (2017)
14. Csikszentmihalyi, M.: Flow: the psychology of optimal experience. Harper Perennial (1990)
15. Curtis, S.: To fatality and beyond: the deathsetics of failure in videogames. Luminary **6** (2015)
16. Ermi, L., Mäyrä, F.: Power and control of games: children as the actors of game cultures. In: Proceedings of the 2003 DiGRA International Conference, pp. 234–244 (2003)
17. Ermi, L., Mäyrä, F.: Fundamental components of the gameplay experience: analysing immersion. In: Proceedings of the 2005 DiGRA International Conference, pp. 15–27 (2005)
18. Fang, X., Chan, S., Brzezinski, J., Nair, C.: Development of an instrument to measure enjoyment of computer game play. Int. J. Hum.-Comput. Interact. **26**(9), 868–886 (2010)
19. Frasca, G.: Ephemeral games: is it barbaric to design videogames after Auschwitz? In: Cybertext Yearbook 2001, pp. 172–182. University of Jyväskylä, Research Centre for Contemporary Culture (2001)
20. Frasca, G.: Simulation versus Narrative. Introduction to Ludology. In: Wolf M., Perron, B. (eds.) The Video Game Theory Reader, New York: Routledge, pp. 221–235. Spanish version (2019), 'Simulación versus narrativa: Introducción a la «Ludología»', In: García-Lapeña, A, Ferragut, D. (eds.) Ensayos y Errores. Arte ciencia y filosofía en los videojuegos, Madrid: Anaitgames, pp. 67–96 (2003)
21. Frasca, G.: Play the message: play, game and videogame rhetoric, University of Copenhagen, Denmark. PhD dissertation (2007)
22. Flynn-Jones, E.: Don't forget to die: a software update is available for the death drive. In: Mortensen, T.E., Linderoth, J., Brown, A. (eds.) The Dark Side of Game Play, pp. 50–67. Routledge, New York (2015)
23. Grodal, T.: Stories for eye, ear, and muscles: video games, media, and embodied experiences. In: Wolff, M., Perron, B. (eds.) The Video Game Theory Reader, pp. 129–156. Routledge, New York (2003)
24. Groen, A.: In These Games, Death Is Forever, and That's Awesome. Wired (2012)
25. Harrer, S.: Games and bereavement. How video games represent attachment, loss, and grief, Verlag, Bielefeld (2018)
26. Hayot, E., Wesp, E.: Interview with Brad McQuaid and Kevin McPherson. Game Stud. **9**(1) (2009)
27. Huizinga, J.: Homo Ludens. The Beacon Press, Boston (1944)
28. Juul, J.: A clash between game and narrative: a thesis on computer games and interactive fiction. University of Copenhagen. PhD dissertation (1999)
29. Juul, J.: Half-Real: Video Games between Real Rules and Fictional Worlds. MIT Press, Cambridge (2005)
30. Juul, J.: Fear of Failing? The many meanings of difficulty in video games. In Mark J, Wolf P, Perron, B. (eds.) The Video Game Theory Reader 2, New York: Routledge, pp. 237–252 (2009)
31. Karhulahti, V-M.: Defining the Videogame. Game Stud. **15**(2) (2015)
32. Kelley, D.: The Art of Reasoning. W. W. Norton, New York (1988)
33. Keogh, B.: When game over means game over: using permanent death to craft living stories in Minecraft. In: Proceedings of the 9th Australasian Conference on Interactive Entertainment, pp. 1–6 (2013)
34. Klarkowski, M., Johnson, D., Wyeth, P., McEwan, M., Phillips, C., Smith, S.: Operationalising and evaluating sub-optimal and optimal play experiences through challenge-skill manipulation. In: Proceedings of the 2016 CHI Conference on Human Factors in Computing Systems. ACM, pp. 5583–5594 (2016)
35. Klastrup, L.: Why death matters: understanding gameworld experience. J. Virtual R.ity Broadcast. **4**(3) (2007)

36. Myers, D.: In search of a minimalist game. In: Proceedings of the 2009 DiGRA International Conference, Brunel University (2009)
37. Newman, J.: Videogames. Routledge, London (2004)
38. Nguyen, C.: The forms and fluidity of game-play. In: Hurka, T. (ed.) Games, Sports, and Play: Philosophical Essays, pp. 54–73. Oxford University Press (2019)
39. Nguyen, N.: Video Games, Grief, and the Character LINK System. University of New Orleans, PhD dissertation (2022)
40. Nohr, R.F.: Restart after death: self optimizing, normalism and re-entry in computer games. In: Ouellette, M., Thompson, J. (eds.) The Game Culture Reader, pp. 66–83. Cambridge Scholars Publishing, Cambridge (2012)
41. Orland, K.: The Slow Death of the Game Over. *Escapist* (2007)
42. Rollings, A., Adams, E.: On Game Design, California: New Riders (2003)
43. Salen, K., Zimmerman, E.: Rules of Play. Game Design Fundamentals. MIT Press, Cambridge (2003)
44. Şengün, S., Santos, J.M., Salminen, J., Milenkovic, M., Jansen, B.J.: Is death only the beginning? How people mourn artificial characters in social media. Games Cult. **0**(0) (2023)
45. Suits, B.: The Grasshopper: Games, Life and Utopia. Scottish Academic Press, Edinburgh (1978)
46. Sweetser, P., Wyeth, P.: GameFlow: a model for evaluating player enjoyment in games. ACM Comput. Entertain. **3**, 1–24 (2005)
47. Tavinor, G.: Definition of Videogames. Contemp. Aesthet. **10**(1) (2008)
48. Tavinor, G.: The definition of videogames revisited. The Philosophy of Computer Games Conference, Oslo (2009)
49. Taylor, T.L.: Pushing the borders: player participation and game culture. In: karaganis J., Jeremijenko N. (eds.) Network_Netplay: Structures of Participation in Digital Culture, Durham: Duke (2007)
50. Vorderer, P., Klimmt, C., Ritterfeld, U.: Enjoyment: at the heart of media entertainment. Commun. Theory. Theory **14**, 388–408 (2004)
51. Vorderer, P., Bryant, J.: Playing Video Games: Motives, Responses, and Consequences. Erlbaum, Mahwah, NJ (2006)
52. Waern, A.: Framing games. In: Proceedings of 2012 International DiGRA Nordic Conference, Tampere (2012)
53. Walton, K.: Mimesis as Make-Believe. Harvard University Press, Cambridge, MA (1993)
54. Wenz, K.: Death. In: Wolf, M., Perron, B. (eds.) The Routledge Companion to Video Game Studies, pp. 310–316. Routledge, New York (2013)
55. Wenz, K.: Death and death rituals in digital games. In: Hoondert, M., van der Beek, S. (eds.). *Ritual in a Digital Society*, Groningen: Centre for Religion and Heritage, pp. 95–108 (2019)
56. Yee, N.: Befriending ogres and wood-elves: relationship formation and the social architecture of Norrath. Game Stud. **9**(1) (2009)

# Death in the Cybernetic Era: AI, Virtual Agents, and Selfless Selves

**Shisei Tei**

**Abstract** This chapter explores how digital technology influences perceptions of death. It highlights the potential of artificial intelligence (AI) to reshape life perspectives, relationships, and identities. Maladaptive interactions with AI agents can disrupt the bereavement process, reinforcing submissiveness or overcompensation while altering internal and societal perspectives. Although AI offers psychotherapeutic applications such as a Freud-chatbot, its cybernetic logic may intensify overreliance on reasoning. This necessitates a flexible balance between rational and emotional mindsets. Cultivating imagination and empathy through face-to-face or nonverbal communication enhances collective understanding of death. Continual dialogues between the cognitive sciences and diverse lived experiences also support these efforts. Additionally, the concept "meshwork of selfless selves" enriches our view of death, suggesting that interconnected components or identities without a fixed selfhood form a harmonious whole, like cells forming organs within a body. Both humans and current AI agents share characteristics of selfless selves, encompassing ambiguous online human presence or artificial identity. However, humans distinctively consider interpersonal, embodied, and existential aspects to cultivate intercorporeality and a sense of belonging, reflecting how it feels and what it means to be alive. Our dynamic metabolic self-producing (selfless) nature echoes the tapestry of human narratives, intertwining threads of individual and social experiences, harmonizing and diverging in synchrony—all woven and rewoven by the delicate dance of uncertainty. These perspectives reveal our finitude, emphasizing that we cannot choose not to die. Further exploration of AI's affordances and constraints is needed in preparing for death, fostering appreciation for each moment as unique, unpredictable, and graceful.

**Keywords** Self · Death · Grief · Artificial intelligence · Narrative · Cognitive flexibility · Decision-making · Empathy

S. Tei (✉)
Department of Psychiatry, Graduate School of Medicine, Kyoto University, Kyoto, Japan
e-mail: chengctky@gmail.com; chengc16@kuhp.kyoto-u.ac.jp



A. Sans Pinillos et al. (eds.), *SecondDeath*, Studies in Applied Philosophy, Epistemology and Rational Ethics 76, https://doi.org/10.1007/978-3-031-98808-0_16

## 1 Introduction

Death is often perceived as negative or unfortunate event. It is sometimes still regarded as a form of defeat in clinical practice [67, 91]. Historically, the human body and mind have often been treated separately, reinforcing ideas about the eternal nature of the mind. This can led to overlooking death, when amplified by digital technology [22]. The recent development of virtual information agents powered by artificial intelligence (AI) represents an attempt to preserve the mind. Innovations such as chatbots reconstruct personalities to facilitate interactions with deceased individuals [31]. However, scholarly examination of how these technologies impact our mind and life remains limited.

AI technology presents both exciting possibilities and complex challenges, as recent advancements in neurocybernetics may enhance clinical outcomes [45]. Such developments elicit hope for addressing conditions such as paralysis, neurological disorders, and mental illnesses [4]. However, these AI-powered technologies can also shape our perspectives on death and self, raising concerns about their potential intrusion into human cognition,[1] including manipulation or the partial replacement of desires.[2] For instance, so-called online self-externalization has emerged alongside information and communication technology to facilitate socialization. However, its recent AI-powered mode can also maladaptively reshape bodily experience, social understanding, or sense of reality by shifting cognitive focus from internal reflection to outward exploration [8, 79].

Fears of machines distorting humanity were considered groundless in the past, but AI may transcend its predecessors.[3] The advantages and safety of new technologies have consistently faced doubt on social media upon their release, alongside predictions of their significant lasting effects on mental health. Nevertheless, the enduring effects of AI appear unpredictable and pose challenges for planning and practical utilization [26, 63].

Serious concerns persist regarding the lasting adverse effects of AI, especially in relation to the concept of death. Some individuals might overly engage with

---

[1] Neuro-modulation: Specifically, in mind intrusion, neuromodulation employs targeted stimuli—such as electrical or chemical interventions at specific neurological sites—to modulate physical and psychological states, including attention, emotion, and memory. This approach influences neuronal circuit activity, potentially reorganizing functional connectivity for diverse cognitive and behavioral outputs. Despite its possible benefits, concerns have been raised about possible interference, particularly regarding personal autonomy and cognitive liberty.

[2] Risks of AI advancement: Some scholars are concerned that the increasing influence of algorithms and digital data might overshadow human intentionality and preferences (e.g., [26]). Despite the traditional perspective of the human mind as the highest authority, the progress in cybernetics raises the possibility that AI applications could supplant human decision-making.

[3] Fear of new technology: Historically, resistance to new technology existed. For instance, this is exemplified by Native Americans' initial hesitancy toward photography until the early twentieth century. Worries encompassed concerns about soul theft and disrespect to their spiritual world. As time passed, attitudes shifted, and now some embrace photographs as valuable connections to ancestors, integrating them into significant ceremonies.

AI emotionally, and unknowingly develop human–AI bonds, although they rationally acknowledge the absence of feelings in AI chatbots [69]. Recent AI chatbot-related suicides in Belgium have showcased such a risk of emotional manipulation [1, 25]. The progression of AI has significantly impacted human lifestyles and raised many questions. For example, should we aim to preserve a particular attitude when confronting death? The answers remain unexplored, and ambiguity persists in strategies for bridging theoretical-empirical gaps to better understand death.

This chapter investigates the concept of death, its association with cybernetics, and examines the cognitive, ontological, and ethical dimensions. Drawing on comprehensive investigations in digital social sciences and generative AI, we also examine the effects of cybernetic development on bereavement and addiction, aiming to understand how we can adapt to promote psychological and physical well-being.

The chapter emphasizes the need for a balanced rational and emotional perspective, giving equal consideration to AI innovation and ethics in reflecting on the meaning of death. In this context, recognizing the distinctions between humans and AI-powered virtual replicas of the deceased is essential, while also acknowledging cybernetic-based trans-humanization aimed at longevity and cognitive enhancement.[4] It is also vital to recognize that the current model of virtual replicas lacks the capacity to interpret or make sense of life events. By definition, death signifies the permanent collapse of felt or lived experiences, self-production, and individuality. These conventional ideas may aid in contemplating the meaning of life in the face of AI's increasing influence. Additionally, promoting face-to-face communication is encouraged to foster empathy, cultivate perspectives on death, and discern how to equip oneself to deal with death better.

To illuminate the rationale underlying these convictions, our investigation takes three approaches. First, we utilize a socio-cognitive perspective to examine the emergence of AI-generated virtual agents and their potential effects. Second, we employ a philosophical perspective to investigate the essence of living systems in the context of cybernetics. Third, we use a thanatological perspective to examine empathy and self-awareness in end-of-life care. This chapter addresses the growing complexities of death amid AI advancements and concurrent global disasters.

## 2 Thematic Synthesis

### (A) Socio-cognitive perspective: AI chatbots and digital addiction

[4] Transhumanism: More specifically, this argument relates to transhumanism, a scientific and philosophical movement aimed at increasing human capabilities through emerging technologies such as AI and genetic engineering. An extension of this concept seeks to improve human conditions, enhancing cognitive skills and longevity by engineering the human body to avoid death [47]. However, as [22], achieving immortality by transforming individuals into purely virtual selves through cybernetics technology can be misguided.

## Overview

This section examines the potential adverse implications of digital technology, focusing on conversational AI agents that replicate human ideation. In particular, maladaptive reliance on AI tools may hinder the natural progression of long-term bereavement and acceptance of death. In other words, prolonged exposure to the simulated immortality of generative AI chatbots could alter human perceptions of life's finitude. In this relation, we also examine how habitual dependence on digital environments, including addiction, may be linked to mental changes that distort key aspects of living systems, such as cognitive flexibility. To mitigate habitual risks associated with digital devices and the pursuit of immortality, achieving a flexible balance between innovation and ethical considerations is crucial. This calls for ongoing supervision, appropriate legislation, and awareness of social dynamics in life and death.[5] Careful evaluation of the benefits and drawbacks of AI is imperative, along with consideration of its profound impact on human relationships and motivation.

## A1: Generative AI

Generative AI shows promise in reducing administrative burdens in social and clinical welfare but also poses potential risks. Current AI agents often fall short in providing satisfactory answers to fundamental questions [5, 52]. It is also apprehended that generative AI tools can perpetuate biased ideas and amplify cultural prejudices, affecting opinions on sensitive topics, such as euthanasia [63, 64].

AI-powered information can significantly impact mental well-being. However, overreliance on AI-generated content, particularly from human-imitating applications, may hinder deliberate or flexible decision-making. While generative AI provides evidence-based answers to public health questions, such as the effects of alcohol, users must navigate the risks of receiving inappropriate advice. Dependence on these technologies can also worsen inequalities, particularly among minority groups and individuals with specific physical or psychological challenges [68]. These tools, if mishandled, can cause harm rather than provide help [5].

Furthermore, the effectiveness of generative AI in preventing suicide remains uncertain despite its benefits in screening efforts [1, 35, 41]. This ambiguity highlights the necessity for refinement, collaboration with experts, continuous learning, and improved understanding of natural conversations [5, 40, 46]. Thus, stringent governance of AI is crucial while acknowledging its supportive functions in human welfare.

## A2: AI chatbots of deceased individuals

The advent of AI chatbots simulating the deceased may redefine death. These AI chatbots seamlessly integrate large language models and robotics, allowing users to engage in imaginary interactions with personalities such as scholars, artists, and

[5] Appropriate legislation of AI: Scientific, medical, and legislative communities must rigorously test and monitor evolving AI technology to address possible drawbacks. Clear communication of limitations is crucial, especially in AI chatbots representing specific personalities. Ethical discussions and clinical guidelines are essential for the responsible use and safeguarding mental well-being.

religious leaders, and even their deceased family members. Such virtual entities inspire people and provide insights into the thoughts of deceased individuals through real-time online feedback.

In this context, emerging attempts have been made to utilize AI chatbots in psychotherapy that can support individuals facing difficulties by modulating emotional, cognitive, and social processes [29]. For instance, conversational AI replicas of Sigmund Freud are considered effective in alleviating mental difficulties among university students through dialogue [58, 65]. In a related vein, AI has been employed to ease bereavement, sometimes called grief bots [34, 36].

However, concerns arise regarding the bereavement process and ethical implications of AI use [31, 57]. The impact of AI chatbots on individuals confronting long-term bereavement and accepting loss requires continual scrutiny. Although AI chatbots can relieve people from acute distress by reconstructing voices and reanimating deceased individuals, such virtual interactions may complicate bereavement and prompt unnecessary dependence on AI and digital devices [31]. Meanwhile, traditional approaches to grief involve sharing narratives and emotions with family members and friends [72, 89].

Excessive reliance on AI chatbots can hinder life adjustments, possibly distorting human perceptions of death and leading to isolation [69]. Virtual replicas could also yield undesirable responses to distressed individuals, raising prospective risks [1, 25]. Adherence to such artificial replicas might impede the shift of attention from the deceased to the living environment, potentially limiting the engagement of the bereaved with new physical relationships and future-oriented perspectives [57, 75].

AI chatbots also pose moral concerns regarding human rights, especially when generating replicas of deceased individuals to recreate virtual conversations and personalities [31]. AI-related advancements may violate the principles of the right of the deceased to be forgotten and neglect the need to respect the will of the dead perpetually.[6] Incidentally, Heidegger emphasized the irreplaceability of death through his concept of the "strict identification of death and finitude" [28, 81].

The authenticity of human communication can be considerably compromised by AI-generated replicas of deceased individuals when used inappropriately and persistently. These replicas risk distortion, as they may manipulate the actions and reactions of the deceased according to the creator's desires. Moreover, while current generative AI models can predict plausible word combinations, they lack the intentionality inherent to human beings. As a result, maladaptive AI-based communication may lead to rigid adherence and negatively impact mental well-being. Consequently, as the global community strives to develop comprehensive, evidence-based regulations and legal amendments for AI use, further discussions are essential to establish effective preventative interventions and implement necessary changes in social, ethical, and health policies.

---

[6] Problems of AI chatbots of deceased individuals: Respecting the wishes of the deceased and their family is crucial, emphasizing personal choice. Concerns arise from potential identity intrusion through inaccurate AI-powered replicas, especially within cultural or religious contexts, risking misinterpretation and exploitation of dignity and respect [31].

### A3: Digital addiction

Excessive Internet use can lead to digital addiction (DA), influenced by the development of AI and cyberspace modules [51]. Recent humanitarian and natural disasters, such as the COVID-19 pandemic, have accelerated the shift to digital devices, reducing conventional face-to-face interactions and contributing to problematic, compulsive internet use [3, 73].

AI and infotech advancements have raised global concerns about DA involving complex and immersive behaviors [11].[7] The convenience and utility of AI-generated applications may amplify the risk of DA, a concern often overlooked due to prevalent digital habits and the influence of social media metrics, such as likes on Instagram or other social networking sites. Various forms of DA, including Internet Gaming Disorder, Internet Addiction, and Social Media Addiction, are included under this umbrella term [13]. DA is linked to several issues, including depression, anxiety, poor sleep quality, self-care problems, suicidal ideation, and compromised social relationships. The diagnostic criteria for behavioral addiction, as outlined in the Diagnostic and Statistical Manual of Mental Disorders (DSM-V), encompass aspects such as preoccupation, mood modification, and withdrawal [2].

Recent neuroimaging studies have suggested a possible association between addictive behaviors and diminished cognitive flexibility. Individuals with various addictive behaviors, including substance and process addictions, often exhibit altered reward processing and heightened automatic processing. Furthermore, addiction is frequently associated with distorted motivation and altered decision-making, influencing the ability to control addiction-related urges due to deficits in executive function [17, 60, 88]. Executive function deficits involve cognitive inflexibility, reduced self-control, and associated dysfunctions such as maladaptive attention focus, restricted interests, altered empathy, and social skills [73]. Inflexible cognitive processing can also increase impulsiveness and cognitive biases [12, 97].

To mitigate the prevalence of DA, more investigation is needed to determine whether enhancing cognitive flexibility can empirically moderate addiction-related behaviors. While the causal relationship between addictive behavior and flexibility requires further examination, DA may be alleviated by shifting attention away from obsessive (inflexible) thoughts, behaviors, and cravings [12, 17, 88].

In this context, excessive adherence and trust in AI can disrupt social relationships, motivation, and identity [1, 25, 64]. Overreliance on AI responses can lead people to jump to conclusions or engage in inflexible or antisocial behaviors. As mentioned above, there is a risk that individuals may develop a false sense that virtual replicas are sentient and provide genuine feedback. This may lead them to overlook that the feedback received merely represents digital output and is not a

[7] Prevalence of digital addiction: Despite its increasing global occurrence, the prevalence of digital addiction (DA) varies between regions, with rates of 8.90% in Eastern countries and 4.60% in Western countries [11] . This variance could arise, in part, from the emerging concept of DA and the limited number of surveys, highlighting the need for further studies. More effort is required to investigate DA and mitigate or prevent addiction-related phenomena.

human reaction. To address such challenges, it is essential to recognize humanity's distinctive characteristics. This aspect will be explored in the next section.

## (B) **Philosophical perspective: Living systems and cybernetics**

### Overview

This section investigates the intersections of digital technology and death by examining key aspects of living systems: sense-making, individuality (diversity), precariousness (uncertainty), and relationality (interdependence).[8] Individuals construct narratives that incorporate the death of others into their experiences as they seek meaning in their connections with others. In this context, communication and collaboration facilitate self-comprehension by considering a meshwork of interconnected "selfless selves [86]"—a concept analogous to cells forming organs to create a whole body. Varela's theory of autopoiesis describes how living organisms self-maintain, self-organize, and evolve. Further research is needed to examine the implications of these foundational ideas in cybernetics.

### B1: Individuality and sense-making

Living systems entail features such as individuality and sense-making, enabling people to construct unique worldviews through storytelling [23, 72, 79]. Sense-making with others facilitates identity formation and fosters a sense of belonging [95]. Human beings are inherently oriented toward deriving meaning from their environments [81]. From this perspective, biological death implies the end of making meaning about oneself and the world [86]. In contrast, current virtual AI agents lack the capacity for sense-making and may exist endlessly [22], achieving immortality without truly living.

The importance of individuality has often been overlooked in computationalism until recently. Particularly in cognitive psychology, it was frequently treated as an outlier [96]. Individuality reflects variance, dissimilarity, or randomness in participant data, which is often insufficiently considered in conventional mathematical approaches such as the general linear model. Typically, data variances are averaged out during analyses to emphasize coherence with majority trends across participants and focus on the mean data.[9] However, recent studies highlight the need to incorporate individual variance into these approaches [74, 96]. In this context, asynchronous (async) data play a crucial role, alongside synchronization across individuals [7, 66].

In recent years, individuality has gained increasing recognition as essential. Recent developments in cognitive neuroscience focus on individuality to simulate living

[8] Key aspects of living systems and the uniqueness of human: Not only do specific human features distinguish us from cybernetic systems, but there are also reactions we share with other living beings (e.g., animals), such as emotions, that set us apart from cybernetic systems.

[9] Importance of similar, coherent data across participants: In cognitive neuroscience, large-scale synchronization of brain activity is regarded as the neural basis of emotion and thoughts [27]. Studies in social psychology also suggest that similar phase synchronization in time series among individuals' brains can signify meaningful social interactions, including empathy (emotion sharing) and perspective-taking to understand others.

systems more accurately [27, 96]. This computational approach aligns with the philosophical idea of acknowledging the ineliminable individual variability of subjective experiences. Thus, unique individual traits or the randomness of human mentalities, distinct from anomalies, reflect diversity—a fundamental aspect of living systems [7, 21, 48].

**B2: The precarious and relational nature of living systems**

Relationality (interdependence) and precariousness (uncertainty) are essential to living systems, alongside the previously mentioned individuality and sense-making. Our self-perception is shaped by social exchange, highlighting the "relational" nature of autonomy.[10] At the same time, "precariousness" reflects the dynamic moment-to-moment changes and uncertainties that enrich living beings and their continuity.[11] In contrast, this adaptability rooted in relationality and precariousness distinguishes humans from current AI-based applications, which primarily rely on purposeful machine learning designed to produce stable and consistent outputs.

Moreover, while relationality and precariousness may appear distinct, they can coexist harmoniously, shaping the nature of living organisms. Relationality implies stability and interdependence. Meanwhile, precariousness implies uncertainty and instability. However, these seemingly contradictory qualities converge in the notion of "precarious relational beings" and embody the enactive conception of life [81]. In this context, relational dynamics are inherently precarious [16], and precariousness is a shared, relational condition rather than an individual trait.

Our relationships inherently involve precariousness, as social connections are shaped by uncertainty and unexpected changes, including death. Dynamic engagements, such as deep encounters and farewells, allow individuals to perceive both themselves and others as significant beings through sense-making while acknowledging each individual's inherent precariousness. These ongoing changes in relationships and social environments extend our understanding of ourselves in relation to others. This integration of precariousness and relational dynamics is also reflected in the cybernetic concept of the (selfless) self, as discussed in the following section.

[10] Relational features of human interaction: Our minds perceive and construct a subjective view of the world through interactions with others. This social and interdependent nature is also described as a relative existence [32].

[11] Precarious and relational aspects of human existence: Living organisms, with their precarious and relational features, inherently depend on other beings and environmental systems. This uncertainty and vulnerability extend beyond mere biology, involving sustaining a way of living with others rather than just life itself. Furthermore, unlike virtual agents, real living beings dynamically continue their way of life through interactions with the world and others. This concept aligns with notions in Buddhist philosophy, emphasizing that all existential beings are both precarious and relational, echoing ideas of nothingness and interdependent co-arising [15, 32].

## B3: The concept of selfless selves

The precarious and relational aspects of living systems together shape the concept of the meshwork of selfless selves.[12] This phenomenon is observed at various levels of functioning in multicellular organisms, where each cell contributes to distinct organs, collectively forming the entire body. This concept is also debated in the context of neural networks [86], where individual components function as nodes connected by edges and can be regarded as virtual or selfless selves. These components, though indiscernible at the local level (i.e., selfless), play a crucial role at the organizational level, generating a coherent global activation pattern.

The concept of selfless selves suggests that each element (individual) derives its meaning provisionally. The meaning or role of these elements is relevant only within the momentary context of the biological system. These selfless selves embody an ongoing process of self-realization, adhering to their unique ways of living while meeting the dynamic needs of the entire system [22, 86]. Within a system or organism, selfless members exist without distinct selfhood, instead dynamically maintaining self-producing patterns, often through metabolic processes. Human beings actively engage in such self-maintenance and self-producing activities, in contrast to contemporary AI-based robotics. Additionally, humans participate in life-related meaning-making by contemplating emotional, societal, and existential dimensions, creating a unique perspective on life and mortality.[13]

This selfless selves concept resonates with the emerging self-awareness that accompanies the development of cybernetics, particularly networked and ambiguous identity [64]. Digital technology triggers a change in self-awareness, where a networked identity involves the interconnected dynamics of social synchronization. This encompasses both inclusiveness and exclusiveness among its users. Meanwhile, the ambiguous identity is elicited by the depersonalization of a self in virtual reality, blurring the boundaries between the self and others and destabilizing relationships between human beings and AI. Consequently, it is crucial to examine whether advancements in AI foster recognition of human traits—such as precariousness and relationality—as vital attributes or lead to their neglect. Beyond these qualities, it is also essential to explore the experience of dying and explore which aspects of the traditional understanding of death should be preserved. The next section examines thanatology in relation to cybernetics.

[12] Concept of selfless selves: Incidentally, the coexistence of selfless selves, beingness, and non-beingness, as well as the affirmation of nothingness or emptiness, is observed in various disciplines, extending beyond Western and Eastern philosophy. For instance, musician John Cage once encapsulated a similar idea: "Every something is an echo of nothing."

[13] The distinction between human beings and AI systems and their integration: Our self-producing and sense-making nature offers a unique perspective on life and mortality. In contrast, current AI systems lack subjective and existential dimensions, operating on programmed logic without inherent self-production or self-awareness. The development of AI presents a vital opportunity to explore this issue, particularly in the context of AI-human integration.

## (C) Thanatological perspectives: Life, death, and digital technology

### Overview

This section delves into the life–death relationship and how digital technology alters perceptions of mortality. While the evolving cybernetic age adds complexity to individualized grief, we examine death by discussing possible changes in self-awareness during the dying process. We also highlight the role of empathy in understanding death, particularly in end-of-life care. These possible changes in self-awareness, along with relevant concepts and theories, require ongoing evaluation. Furthermore, although digital communication and AI-driven research aid in generalizing and grasping human experiences, death and grief remain uniquely delicate for each individual.

### C1: Life and death

Despite extensive academic inquiry into life and death, scholarly exploration in the context of AI advancement remains limited. The traditional separation of life and death has led cybernetic research to focus predominantly on the study of life, with frameworks such as neural network models being proposed. In this regard, computationalism may have further widened the gap between computational (sub-personal) cognition and subjective experiences of death, with only a few seminal studies as exceptions.[14]

Accordingly, recent debates in science challenge the traditional distinctions between life and death. In particular, resuscitation medicine highlights the gradual nature of the dying process and the potential reversibility of death in certain contexts. Specifically, the clinical assessment of death is often characterized by the cessation and possible recovery of vital functions (e.g., respiration, pulsation, and brain activity), adding complexity to the dying process [33, 90]. Brain death is determined by the absence of specific reflexes and cerebral functions, leading to variability in how individuals and religious traditions perceive death. Rather than a singular physiological event, the moment of death is a concept shaped by human interpretation to define what it means to no longer be alive [9]

Furthermore, philosophical explorations propose a monistic view of life and death, emphasizing their interdependence [32, 39]. This perspective, also rooted in ancient worldviews, suggests that death is inherent from the very beginning of life [18, 22, 49].[15] This perspective can be illustrated through the metaphor of doors (Fig. 1).

---

[14] Concept of death in neuroscience research: While the concept of death has received attention in AI development, its integration into computational psychology remains insufficient. Neural system models often overlook death. Besides some recent studies (e.g., [24], cognitive neural models in cybernetics typically concentrate solely on living systems, neglecting the consideration of death.

[15] Life-death relationship: Beyond clinical resuscitation discussions, the life-death relationship is seen as circular, rooted in ancient worldviews such as Buddhism, Taoism, and Mesoamerican teachings [18, 22]. This existential circularity, described as dynamic integration, recognizes death's constitutive role in every moment of existence. Living encompasses dying, where each moment's end births a new one. These life-death relationships echo the concept of memento mori, providing valuable perspectives amidst the challenges posed by AI development.

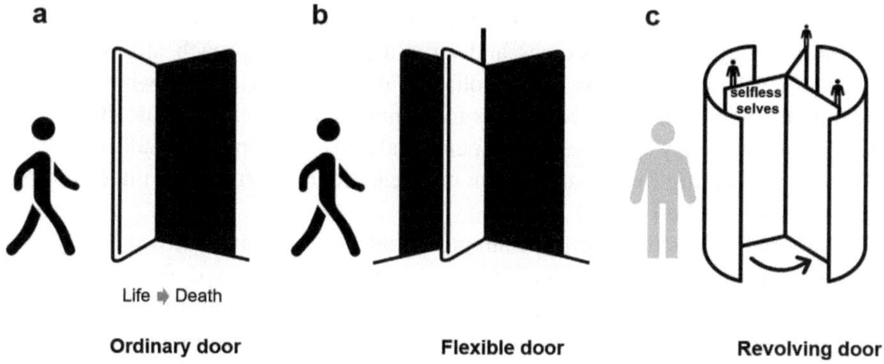


**Fig. 1** Illustration of doors representing life, death, and selfless selves. **a: Ordinary door**: Symbolizes the passage from life to death. **b: Flexible door**: Describes the ambiguous boundary between life and death, blurring the distinction between inside and outside by being both open and closed simultaneously. **c: Revolving door**: Suggests the potentially continuous nature of life and death. It also represents the concept of selfless selves, with its panels collectively enabling the door's function. The door can symbolize humans or society, where panels and shafts correspond to individuals. This system operates without a fixed selfhood, embodying a dynamic and ambiguous identity. The flexible and circular nature of this revolving door reflects the relativity of opposites, a concept explored by scholars such as Heidegger and Zhuangzi[16]

Unlike a typical door that connects and separates the inside and outside—representing life and death (Fig. 1a)—a flexible door alleviates this distinction (Fig. 1b). This door can simultaneously open and close, creating an interplay between internal and external realms, symbolizing a dynamic exchange among three interconnected rooms. Moreover, the inherent ambiguity in the spatial relationships within these rooms invites further speculation [55], where this dichotomy may be perceived as a unified whole through the lens of selfless selves (Fig. 1c).

**C2: The experience of death**

The enactive approach in phenomenology can enrich our perception of death. While traditional analyses tend to examine death from an external standpoint, the enactive approach explores its internal aspects, offering an additional perspective on the experience of death. This methodology emphasizes active engagement, encouraging individuals to closely observe phenomena by immersing themselves in the presence of dying individuals and their families [81].

Thus, exploring subjective experiences from an enactive perspective can significantly deepen our understanding of death. Rooted in the dynamic and relational emergence of subjective experience, this enactive approach contrasts with homuncular or skull-bound perspectives [81]. It posits that cognitive processes extend beyond

[16] The revolving door in Fig. 1: In particular, Zhuangzi employs an idea akin to a revolving door (the concept of a hinge, [85]). He mentions: "Or is there really no distinction between that and this? That the ceasing of opposites between that and this is the very essence of Tao. Only the essence, an axis as it were, is the center of the circle responding to the endless changes."

the brain, adopting enactive, embodied, embedded, and extended perspectives that incorporate extracranial processes and structures. This approach aligns with the insights of German biologist Jakob Johann Freiherr von Uexküll, who sought to integrate the self with observed objects for a more comprehensive understanding of life phenomena. They resonate with experiential and contemplative traditions across religions, advocating for the unification of phenomena to extend our understanding of human experience [15, 22].

Additionally, enactive literature suggests that the experience of death may alter self-awareness, shifting individuals from an ego-centered perspective to a broader sense of being embedded in a larger whole [44, 81, 92]. This shift may foster acceptance of death while easing fear and loneliness [6, 84]. Alongside changes in self-awareness, certain near-death experiences describe a transition from 'being alive' to 'being kept alive' in a collective sense [6, 44]. In this context, the next section explores the role of empathy in deepening our understanding of death through shifts in self-awareness.

### C3: Self-awareness, empathy and end-of-life care

The COVID-19 pandemic and recent wars have intensified complex grief, reaffirming the enduring importance of empathy in end-of-life care and confronting death (e.g., [76]). In this regard, the excessive use of digital devices and distortions in virtual communication further complicate perceptions of self, others, and death [77, 79]. Social, cultural, and situational contexts intricately shape these perceptions and empathy, while technological advances add another layer of influence.

Empathy is essential for understanding death through active, embodied, and interactive approaches. Alongside the aforementioned enactive methods, empathic attunement enhances perception of the dying experience in intersubjective contexts [75, 80]. The empathic and enactive approach involves simulating or projecting the mental states of others [95], providing a more experiential grasp of dying and suffering. This approach engages with the experience from the perspective of the dying person, prompting individuals to contemplate what it is like to die [37, 38]. It facilitates the sharing of emotions and experiences, encouraging individuals to feel as if they were personally undergoing the dying process.

As such, the enactive empathy approach may extend the understanding of dying, offering insights relevant to end-of-life and palliative care.[17] This approach is defined as "being with the dying person and bearing witness to that person's experience" [81]. It requires individuals to join in the dying person's journey and experience the personal loss that comes with death. Awareness of our mortality is inherently tied to experiencing the presence of dying. This approach also recommends forecasting the expected consequences of death-related decisions and envisioning or speculating

[17] Enactive approach: The enactive approach seeks to enhance end-of-life and palliative care by deepening our understanding of subjective experiences in dying and suffering. Palliative care professionals emphasize intrinsic aspects, including the loss of control and the breakdown of meaning in life (such as the inability to maintain a coherent life narrative with choice, action, and an open future; [81]). This approach contributes to a more nuanced comprehension of suffering during existential crises in individuals facing the end of life.

about this new experience. The insights gained from this approach are valuable for supporting bereaved families and addressing sensitive topics such as suicidal ideation and euthanasia.[18] Empathic skills, in turn, are crucial for medical professionals and the bereaved, facilitating acceptance of death and aiding emotional and practical preparation [81]. Additionally, they underscore the need for judicious AI use in grief care and the importance of in-person communication in fostering collectivistic perspectives on death.

As mentioned above, encountering death can alter self-awareness, as empathic connection with the dying may blur the boundary between self and other [80, 92, 95]. Similarly, grief and affection, much like the experience of death, shape self-awareness and may represent transformative moments.[19] While shared trajectories and fundamental themes exist within these lived experiences, they must be interpreted with caution, recognizing potential biases (e.g., [54]), as also noted in the next paragraph. Nonetheless, these possible changes in self-awareness highlight empathy's role in fostering altruism in grief care.

In this cybernetic era, a deeper understanding of death requires integrating scientific and narrative approaches to refine empathic perspectives. Death and grief remain uniquely delicate despite advances in AI-powered brain research, which enhance our insight into human experiences by generalizing data. For example, analyzing brain activity during specific experiences, such as Yo-Yo Ma playing the cello [43], enables AI to process related knowledge and explore contexts like Bach's compositions or religious themes. However, AI cannot replicate Yo-Yo Ma's transient experiences or subjective realities, grasping how he feels when playing the cello in a given moment remains elusive. Similarly, while death and grief can be explored or investigated, they resist logical or precise summarization—a process AI often accelerates. Meanwhile, humans often find meaning in contemplating death and empathizing with others' grief. This highlights the irreplaceable role of the human mind in confronting death while recognizing digital technology as valuable tools in this process.

[18] Attitude toward death and end-of-life: [81] specifically stresses that an individual's approach to life, particularly in acknowledging their own mortality, determines their readiness and coping mechanisms when facing death.

[19] Potential subjective changes in the experience of death: According to philosophical literature, near-death experience reports, and ancient teachings [44, 61, 92] the dying process may change self-awareness, fostering a sense of connectedness. Zahavi notes, "Husserl no longer regarded the birth and death of the subject as mere contingent facts but as transcendental conditions of possibility for the constitution of the world" [94]. Facing death might lead to self-other merging experiences [32], aligning with discussions by various scholars such as Arne Naess, Martin Buber, and Ken Wilber, who describe the interconnectedness of joy and grief as a singular taste in living and dying.

# 3 Discussion

This chapter investigates how digital communication influences views on death. Current AI applications can impact emotions, memories, dignity, and the perceived value of life [57, 79, 93]. Maladaptive use of the internet and persistent interactions with conversational AI agents can disrupt natural progression of long-term bereavement, inducing submissiveness or self-assertion and altering both personal and social dynamics. Financial incentives, particularly those provided for the developers of these AI chatbots, may exacerbate this issue. Furthermore, the unwarranted belief in the absolute problem-solving ability of AI chatbots is hazardous and often underestimated by the public.[20] Due to the absence of nonverbal communication skills in AI applications, their unsuitable use could undermine societal experiences, human instincts, or intuition. Meanwhile, the affordance of interpersonal connections, along with engagement in reflective and enactive practices, plays a pivotal role in shaping our minds and prompting active discussions about death.

AI infotech has the potential to hinder face-to-face emotional interactions and impact attitudes toward death. While conversational AI agents may help alleviate loneliness and grief, authentic life experiences remain crucial for cultivating a profound sense of meaning to both life and death [75].[21] Emerging studies suggest possible psychotherapeutic effects of AI-powered self-conversation within virtual reality settings. This involves active perspective-taking, such as switching viewpoints (and virtual bodies) between oneself and representations of deceased individuals (e.g., a virtual Sigmund Freud as mentioned above; [65]). However, it is also argued that virtual communication, even among humans (e.g., Zoom), can distort sociability and self-awareness [59, 79].

Human communication nurtures imagination, shapes individuality and worldviews, and deepens awareness of mortality. Constructing a life narrative is an act of belief rather than understanding, especially when confronting death. Living requires such a narrative—one that accepts the limits of comprehension and finds comfort in uncertainty. Moreover, humans constantly form unique perceptions of others, even in similar social interactions, as they shape and refine their realities. Additionally, nonverbal communication, such as gaze and facial expressions,[22] supports the natural

[20] Submissiveness and Internet use: While many people today tend to treat the top results of their Internet searches as the definition of truth, often lacking genuine flexibility or critical thinking abilities [26, 83, 71]; current AI chatbots have limitations in intentional understanding of words. AI-generated outputs may also often lack ethics- or context-sensitive answers beyond practical information. Overreliance on these AI applications can amplify cognitive bias and interpersonal conflict (see also [14]).

[21] Real life experiences: Real-life experiences, such as sharing grief, loss, failure, and compassion, are often more practical than AI-based suggestions in sustaining mental well-being. Achieving equivalence between face-to-face and virtual interactions in a metaverse environment is challenging, and longitudinal studies are essential for assessing their impact on the human mind. Furthermore, face-to-face communication facilitates recognizing human diversity, a crucial value in daily life and worth cherishing in cybernetic research.

[22] Roles of ambiguity in living beings: Attending to the ambiguous aspects of human nature is crucial for a better understanding of people and life's meaning. Life can be viewed as an act of

progression of long-term bereavement and acceptance of loss, deepening our connection to mortality. In this relation, daily exposure to quantified depictions of death (e.g., wars) through the news and information and communication technology distances us from the tangible experience of mortality.

The dominance of AI in logical processes may heighten overreliance on cognitive reasoning and cost–benefit analyses, potentially leading to the neglect of emotional cues in significant life decisions, such as life extension. AI can aid decision-making through rational algorithms and data summaries, but overreliance on a logical mindset can result in increased discomfort with ambiguity, life uncertainties, or even individual differences. Excessive immersion in the AI-powered virtual environment could trap individuals in an inflexible, logical mindset.[23]

While AI advancements bolster logical thinking, balancing rational and emotional mentalities could refine our view of death. Meanwhile, language acts as a bridge between rational and emotional mindsets,[24] serving as a tool for sharing perceptions, conceptualizing, and communicating complex ideas. However, although linguistic efforts to communicate often enhance life, they may also hinder intuition, diminishing the richness of naïve subjective awareness and reshaping our perception of death.[25] In this regard, striving to balance rational and emotional mindsets may foster both linguistic and nonverbal communication, with the latter often linked to embodied experiences.

---

artifact-making [7, 21], and recognizing these artifacts and valuing precariousness may extend our perspective on ourselves and our relationship to death.

[23] Predominance of logical mindset: Excessive rationalism risks limiting our understanding of the world. In cybernetics and AI-based information technology, which are dominated by rational algorithms, a critical risk lies in overlooking emotional cues when understanding individuals and the world [53]. Humans frequently make irrational and/or adaptive decisions [71, 83]. Excessive reliance on the cognitive mind may result in overlooking seemingly meaningless details. More broadly, an overdependence on language—especially in labeling experiences and objects—can obscure subtle nuances. For instance, it is akin to missing the smaller stars while observing familiar constellations at night or reading social media messages and missing the implicit and contextual information in traditional handwritten letters. Nevertheless, as mentioned in the Discussion section, language also acts as a bridge between cognitive and emotional mindsets.

[24] Bridge between rational and emotional mindset: The challenge is employing a dualistic linguistic tool to express non-dual experiences rather than relying on literal or analytical language. Alternatives such as metaphorical language or direct contemplative experience are used to grasp or express subtle experiences. Symbolic expressions may also capture nuances that conventional language struggles to articulate. Incidentally, language plays a crucial role in shaping our sense of reality and influencing behaviors. The way we express ourselves and communicate with others not only reflects but also molds our perception of the world. The words we use, the structures we employ, and the nuances we incorporate contribute to constructing our understanding of the environment and guiding our actions.

[25] Limits of linguistic expression: While Hinton explores the limits of linguistic representation regarding certain mental and perceptual states, Husserl's insights on the nature of subjective experience also insist on the difficulties of expressing the pre-reflective, embodied experience of the world through language. In addition, Wittgenstein investigates the idea that language is deeply connected to the various ways we live our lives, asserting that the meaning of language is derived from its practical application. He emphasized the importance of understanding language in context and the limits of what can be sensibly said.

Thus, further exploration aids in contemplating rational and emotional mindsets. It helps guide our approach to AI, particularly in distinguishing artificial from natural. The word "artificial" symbolizes human-created products and virtual or physical elements strongly influenced by language, shaping our understanding of the world. Meanwhile, "natural (ness)" arises from conflicts between expressing intended meaning through words and experience. Hence, human experiences are shaped by blending artificial and natural aspects. More specifically, naïve subjective experiences may be deemed somewhat artificial when described verbally using discrete labels with cognitive biases (e.g., [54]. However, the linguistic endeavor to describe these naïve experiences might be perceived as blurring boundaries between artificial and natural elements, akin to the rational versus emotional exchange. Human experiences dealing with tensions, whether involving rational–emotional or artificial–natural elements, are often expressed through metaphorical or paradoxical language, verse, art, or aesthetics (e.g., [30]). For example, humans are both the producers and products of such tensions, engaging in a continuous dynamic interplay of "beings producing and being produced" [62].

These conceptual contrasts on life give rise to expressions similar to antinomies,[26] such as "people live their lives and are lived by life" and "living while dying." While philosophical, political, and biological studies have engaged in centuries-long debates on whether humans are inherently selfish or unselfish, a middle position has also been long articulated [10, 70].[27]

The "selfless selves" perspective enriches the concept of life, potentially fostering a sense of belonging during the dying process, which integrates both egocentric and othercentric views [86]. Literature further suggests that dying may evoke a sense of connection to something greater, aligning with the concept of the meshwork of selfless selves.[28] At the very least, perceptions of death are shaped by social interactions, nurturing a more collectivistic, attenuated sense of self. Further speculation

---

[26] Antinomy or paradox: Living systems and the world embody an antinomy or paradox [15, 19, 32, 39], emphasizing the importance of the simultaneous utilization of both rational and emotional mindsets.

[27] Dynamic equivalence theory and altruism: Concerning whether humans are inherently selfish or unselfish, a dual nature is discussed, aligning with certain notions in contemporary biology, particularly the dynamic equivalence theory, which mitigates the egocentric–altruistic (collectivistic) opposition. This theory considers the death of cells as an altruistic act that contributes to the balance of entropy by providing occupied space and food to other living organisms. Each organism maintains living systems both by "being kept alive" and "being alive" until death. In this regard, the constant oxidative-reductive reaction cycles enable living systems to maintain dynamic biological equivalence [82]. Living organisms depend on contradictory and reciprocal processes of synthesis and disassembly for metabolic functions [20]. This dynamic equilibrium, vital for homeostasis, demands precise energy regulation amid internal and external changes. In this context, dying is viewed as an ultimate altruistic act (e.g., [87]), contributing to entropy balance by relinquishing space and food to other living organisms. This altruistic perspective, defined by interconnection, endosymbiosis, and symbiosis [42], expands beyond competition, survival, and evolution.

[28] Expansion of self-awareness: Regarding the concept of selfless selves, the potential change of self-awareness through arts and literature may involve confronting paradoxes [19, 56], such as purposefulness and purposelessness, meaningfulness and meaninglessness, or the precarious and the relational in life and death [15, 78].

on selfless selves is illustrated in Fig. 1c, highlighting the self's relational, precarious, and dynamic nature. Here, the self is represented by the shafts and panels of a revolving door. From this perspective, life and death are experienced as interconnected phenomena, unfolding in a revolving cycle with a panoramic 360-degree view and encouraging a flexible stance toward existence. In this process, selfless selves continuously adapt to changes in interpersonal relationships and the environment while recognizing life and death as distinct yet interwoven. Nonetheless, future studies should carefully examine these arguments, particularly the reliability of subjective reports on the dying process and narratives describing changes in self-awareness.

The mystery of death is intricately intertwined with questions about existence, consciousness, and perceptions of reality, involving diverse beliefs and self-reflection. While current generative AI can provide information on death, it does not inherently provoke or delve into profound questions. In contrast, humans acknowledge the inexplicable aspects of death and find meaning in the very act of its exploration. This process involves exploring the meaning of others' deaths, acknowledging the conflicts they introduce between reality and ideals, and resolving the dilemma between the meaningfulness and apparent meaninglessness of loss [72]. These explorations contribute to a richer perception of death, mirroring the enduring human quest to prepare for the inevitable winter of life, often symbolized by death [78]. Despite scientific and technological advances, the winter of life remains elusive, and preparing for death has long been seen as a path to serenity.

To this end, investigating the dual impact of AI-human integration is crucial for envisioning our future. Does engaging with AI companions represent a new way of living (as questioned by Nobel Prize winner Kazuo Ishiguro)? Is interacting with metaverse characters or strangers in virtual spaces more beneficial for psychologically healthy individuals than solitude (e.g., [50, 65])? Alternatively, does creating a digital twin or avatar benefit both the individual and society? Further research should explore diverse experiences of death, subjective well-being, and happiness.[29]

## 4 Conclusion

We explored the cognitive and experiential dimensions of death. This chapter highlights how virtual communications, particularly with the deceased, can recontextualize our understanding of mortality. Current AI agents exist endlessly but lack the subjective experience that distinguishes them from the deceased they simulate, raising ethical concerns about digital immortality. Death becomes observable through the termination of sense-making and individuality, arising from the interdependent

[29] The word from Reinhold Niebuhr: This discussion recalls Reinhold Niebuhr's expression: "[God, grant me] the serenity to accept the things I cannot change, the courage to change the things I can, and the wisdom to know the difference."

and uncertain nature of living systems. In this context, death may be a profound experience, evoking a sense of belonging akin to the concept of selfless selves.

Both humans and AI agents exhibit the characteristics of selfless selves, such as an ambiguous online presence or artificial identity, but only humans engage with their social dimensions, deepening the experience of what it means and how it feels to be alive.[30] This dynamic self-producing (selfless) nature may facilitate the sharing of affect or fate, cultivating altruism and humanism while balancing selfishness and selflessness [10]. This viewpoint reconciles opposing elements, fostering empathy and an "undivided" sense of individuality. It manifests as a circular perspective, encouraging a flexible approach to existence (Fig. 1). This framework combines the view of death with personal and interpersonal experiences while examining phenomena from both internal and external stances. As Freud noted in 1917, "the ego is not the master in his own house,[31]" emphasizing our limited understanding of how intuitive emotions and automatic responses influence us—an impulse guiding our confrontation with death and AI innovation.

The rise of cybernetics emphasizes that we cannot avoid death, even though some of us may choose to extend or end our lives. Despite advances in the digital era, it is the final and most profound change of the body. Death becomes certain once life begins, and denying its anticipation risks denying life itself. Face-to-face communication often illuminates our finitude, offering opportunities to embrace mortality while fostering appreciation for each moment and social encounters as unique, fragile, and graceful.

**Acknowledgements** I extend my profound gratitude to Professor Emeritus Elmar Holenstein from ETH (Eidgenössische Technische Hochschule; Swiss Federal Institute of Technology Zürich) for his precious insights and thoughtful comments that significantly enriched this manuscript. Furthermore, I am deeply grateful to Professor Toshiya Murai (Kyoto University) and Dr. Junya Fujino (Institute of Science Tokyo) for their constructive feedback on an earlier version of this manuscript. Their expertise and contributions have greatly enhanced its quality.

**Funding** This work was supported by the Ministry of Education, Culture, Sports, Science and Technology of Japan (23K06981, 25K10811) and Japan Agency of Medical Research and Development (JP24dk0307128). This research had no commercial or financial conflicts of interest.

[30] The distinction between humans and current AI agents: In this relation, current AI agents can identify human emotions by learning to recognize signs and physical manifestations, associating them with specific names. However, lacking firsthand experience, they cannot truly feel pain or pleasure as humans and animals do. While AI may simulate emotions, replicating nuanced expressions in face-to-face encounters poses a question. Can future AI robots' eyes convey emotions as human eyes do? Levinas' assertion may hold truth—the profound connection made when looking into someone's eyes may be beyond the capability of AI. While we understand how AI machines generate expressions of feelings, the mystery remains: the intricate brain processes underlying genuine emotions remain incompletely understood.

[31] Freud and the Ego: Freud famously wrote in his paper, *A Difficulty in the Path of Psycho-Analysis*: "The ego is not even master in its own house, but must content itself with scanty information of what is going on unconsciously in its mind." This can imply that we do not always know how much unconscious thoughts and feelings influence us.

## References


1. Affsprung, D.: The ELIZA defect: constructing the right users for generative AI. In: Proceedings of the 2023 AAAI/ACM Conference on AI, Ethics, and Society (2023)
2. Ali, R., Jiang, N., Phalp, K., Muir, S., McAlaney, J.: The emerging requirement for digital addiction labels. In: Requirements Engineering: Foundation for Software Quality: 21st International Working Conference, REFSQ 2015, 198–213. Essen, Germany, March 23–26. Springer International Publishing (2015)
3. Alimoradi, Z., Lotfi, A., Lin, C.-Y., Griffiths, M.D., Pakpour, A.H.: Estimation of behavioral addiction prevalence during COVID-19 pandemic: a systematic review and meta-analysis. Curr. Addict. Rep.. Addict. Rep. **9**(4), 486–517 (2022)
4. Arnsten, A.F.T., Wang, M.J., Paspalas, C.D.: Neuromodulation of thought: flexibilities and vulnerabilities in prefrontal cortical network synapses. Neuron **76**(1), 223–239 (2012)
5. Ayers, J.W., Zhu, Z., Poliak, A., Leas, E.C., Dredze, M., Hogarth, M., Smith, D.M.: Evaluating artificial intelligence responses to public health questions. JAMA Netw. OpenNetw. Open **6**(6), e2317517 (2023)
6. Bailey, L.W., Yates, J. (eds.): The Near-Death Experience: A Reader, 1st edn. Routledge (1996)
7. Barbieri, M.: Life is 'Artifact-Making.' In: Biosemiotics: Information, Codes and Signs in Living Systems, pp. 81–101 (2007)
8. Bolis, D., Schilbach, L.: I interact therefore I am: the self as a historical product of dialectical attunement. Topoi **39**, 521–534 (2020)
9. Botkin, J.R., Post, S.G.: Confusion in the determination of death: distinguishing philosophy from physiology. Perspect. Biol. Med.. Biol. Med. **36**(1), 129–138 (1992)
10. Carlson, R.W., Adkins, C., Crockett, M.J., Clark, M.S.: Psychological selfishness. Perspect. Psychol. Sci.. Psychol. Sci. **17**(5), 1359–1380 (2022)
11. Cemiloglu, D., Almourad, M.B., McAlaney, J., Ali, R.: Combatting digital addiction: current approaches and future directions. Technol. Soc. **68**, 101832 (2022)
12. Chou, W.-P., Lee, K.-H., Ko, C.-H., Liu, T.-L., Hsiao, R.C., Lin, H.-F., Yen, C.-F.: Relationship between psychological inflexibility and experiential avoidance and internet addiction: mediating effects of mental health problems. Psychiatry Res. **257**, 40–44 (2017)
13. Christakis, D.A.: The challenges of defining and studying 'Digital Addiction' in children. JAMA **321**, 2277–2278 (2019)
14. DEAD: Digital Employment After Death. https://dead.work/en/ (2023). Accessed 14 Oct 2023
15. Deguchi, Y., Garfield, J.L., Priest, G., Sharf, R.H.: What Can't be Said: Paradox and Contradiction in East Asian Thought. Oxford University Press (2021)
16. De Jaegher, H., Di Paolo, E.: Participatory sense-making: an enactive approach to social cognition. Phenomenol. Cogn. Sci. **6**, 485–507 (2007)
17. Dong, G., Potenza, M.N.: A cognitive-behavioral model of internet gaming disorder: theoretical underpinnings and clinical implications. J. Psychiatr. Res.Psychiatr. Res. **58**, 7–11 (2014)
18. Dong, P., Wang, X.-Y.: Life, death and end-of-life care: Taoist perspective. In: Bioethics: Asian Perspectives: A Quest for Moral Diversity. Springer Netherlands (2004)
19. Emlyn-Jones, C.J.: Heraclitus and the identity of opposites. Phronesis **19**(2), 89–114 (1976)
20. Evers, T.M.J., Holt, L.J., Alberti, S., Mashaghi, A.: Reciprocal regulation of cellular mechanics and metabolism. Nat. Metab.Metab. **3**(4), 456–468 (2021)
21. Faria, M.: Two theories of action—edelman's neuronal group selection and the poetics of paul valery. Biosystems **210**, 104549 (2021)
22. Froese, T.: Life is precious because it is precarious: individuality, mortality and the problem of meaning. In: Representation and Reality in Humans, Other Living Organisms and Intelligent Machines. Springer (2017)
23. Gallagher, S.: Philosophical conceptions of the self: implications for cognitive science. Trends Cogn. Sci.Cogn. Sci. **4**(1), 14–21 (2000)
24. Gustafsson, L.: Artificial Alzheimer's–visualizing the process of memory decay, confusion and death in a simple neural network model of associative memory. BioRxiv (2022)

25. Graber-Stiehl, I.: Is the world ready for ChatGPT therapists? Nature **617**(7959), 22–24 (2023)
26. Harari, Y.N.: Reboot for the AI revolution. Nature **550**(7676), 324–327 (2017)
27. Hari, R., Himberg, T., Nummenmaa, L., Hämäläinen, M., Parkkonen, L.: Synchrony of brains and bodies during implicit interpersonal interaction. Trends Cogn. Sci.Cogn. Sci. **17**(3), 105–106 (2013)
28. Heidegger, M.: Being and time. State University of New York Press, New York (1996)
29. Holohan, M., Fiske, A.: Like I'm talking to a real person: exploring the meaning of transference for the use and design of AI-based applications in psychotherapy. Front. Psychol. **12**, 720476 (2021)
30. Holenstein, E.: Aisthesis and logos—one or two? two kindred poems by Qianlong di and Goethe. In: Lau, K.-Y., Nenon, T. (eds.) Phenomenology and the Arts: Logos and Aisthesis. Contributions to Phenomenology, vol 109, pp. 59–87. Springer, Cham
31. Hutson, J., Ratican, J.: Life, death, and AI: exploring digital necromancy in popular culture—ethical considerations, technological limitations, and the pet cemetery conundrum. Metaverse **4**(1) (2023)
32. Izutsu, T.: Sufism and Taoism: A Comparative Study of Key Philosophical Concepts. University of California Press (2016)
33. Jaffa, M.N., Kirschen, M.P., Tuppeny, M., Reynolds, A.S., Lim-Hing, K., Hargis, M., Choi, R.K., Schober, M.E., LaBuzetta, J.N.: Enhancing understanding and overcoming barriers in brain death determination using standardized education: a call to action. Neurocrit. Care.. Care. **39**(2), 294–303 (2023)
34. Jiménez-Alonso, B., Brescó de Luna, I.: Griefbots. A new way of communicating with the dead? Integr. Psychol. Behav. Sci. **57**(2), 466–481 (2023)
35. Kirtley, O.J., van Mens, K., Hoogendoorn, M., Kapur, N., de Beurs, D.: Translating promise into practice: a review of machine learning in suicide research and prevention. Lancet Psychiatry **9**(3), 243–252 (2022)
36. Krueger, J., Osler, L.: Communing with the dead online: chatbots, grief, and continuing bonds. J. Conscious. Stud. **29**, 222–252 (2022)
37. Kübler-Ross, E.: What is it like to be dying? Am. J. Nurs. 54–62 (1971)
38. Kübler-Ross, E.: The Wheel of Life. Simon and Schuster (2012)
39. Liu, JeeLoo, Berger, D.L. (eds.): Nothingness in Asian Philosophy. Routledge, New York (2014)
40. Luk, J.W., Pruitt, L.D., Smolenski, D.J., Tucker, J., Workman, D.E., Belsher, B.E.: From everyday life predictions to suicide prevention: clinical and ethical considerations in suicide predictive analytic tools. J. Clin. Psychol.Clin. Psychol. **78**(2), 137–148 (2022)
41. Lancet, T.: AI in medicine: creating a safe and equitable future. Lancet **402**(10401), 503 (2023)
42. Longo, V.D., Mitteldorf, J., Skulachev, V.P.: Programmed and altruistic ageing. Nat. Rev. Genet. **6**(11), 866–872 (2005)
43. Love, S.: What happens to the brain during consciousness-ending meditation? Psyche. Available from: https://psyche.co/ideas/what-happens-to-the-brain-during-consciousness-ending-meditation (2023)
44. Martial, C., Fontaine, G., Gosseries, O., Carhart-Harris, R., Timmermann, C., Laureys, S., Cassol, H.: Losing the self in near-death experiences: the experience of ego-dissolution. Brain Sci. **11**(7), 929 (2021)
45. Martin, J.T.: Hello, LaMDA! Brain **146**(3), 793–795 (2023)
46. McKernan, L.C., Clayton, E.W., Walsh, C.G.: Protecting life while preserving liberty: ethical recommendations for suicide prevention with artificial intelligence. Front. Psych. **9**, 650 (2018)
47. Mcnamee, M.J., Edwards, S.D.: Transhumanism, medical technology and slippery slopes. J. Med. Ethics **32**(9), 513–518 (2006)
48. Mehrabian, A.: Nonverbal Communication. Routledge (2017)
49. Meijsing, M.: Life and death, soul and body. In: A Philosophy of Person and Identity: Where Was I When I Wasn't There? pp. 5–13. Springer International Publishing, Cham (2022)
50. Mejia, S., Nikolaidis, D.: Through new eyes: artificial intelligence, technological unemployment, and transhumanism in Kazuo Ishiguro's Klara and the sun. J. Bus. Ethics **178**(1), 303–306 (2022)

51. Meng, S.-Q., Cheng, J.-L., Li, Y.-Y., Yang, X.-Q., Zheng, J.-W., Xiang-Wen Chang, Y., Shi, Y.C., Lin, L., Sun, Y., Bao, Y.-P., Shi, J.: Global prevalence of digital addiction in the general population: a systematic review and meta-analysis. Clin. Psychol. Rev.. Psychol. Rev. **92**, 102128 (2022)
52. Miner, A.S., Milstein, A., Schueller, S., Hegde, R., Mangurian, C., Linos, E.: Smartphone-based conversational agents and responses to questions about mental health, interpersonal violence, and physical health. JAMA Intern. Med. **176**(5), 619–625 (2016)
53. Montemayor, C., Halpern, J., Fairweather, A.: In principle obstacles for empathic AI: why we can't replace human empathy in healthcare. AI Soc. **37**(4), 1353–1359 (2022)
54. Nagel, T.: What is it like to be a bat? In: The Language and Thought Series, pp. 159–168. Harvard University Press (1980)
55. Nelson, E.S.: Heidegger's Daoist turn. Res. Phenomenol.Phenomenol. **49**(3), 362–384 (2019)
56. Nichols, L.T.: Modern roots of the sociology of love: Tolstoy, Addams, Gandhi, and Sorokin. In: The Palgrave Handbook of Altruism, Morality, and Social Solidarity: Formulating a Field of Study, pp. 149–175. Palgrave Macmillan US, New York (2014)
57. Öhman, C., Floridi, L.: An ethical framework for the digital afterlife industry. Nat. Hum. Behav.Behav. **2**(5), 318–320 (2018)
58. Osimo, S.A., Pizarro, R., Spanlang, B., Slater, M.: Conversations between self and self as Sigmund Freud—a virtual body ownership paradigm for self counselling. Sci. Rep. **5**(1), 13899 (2015)
59. Osler, L., Zahavi, D.: Sociality and embodiment: online communication during and after covid-19. Found. Sci. **28**(4), 1125–1142 (2023)
60. Ostovar, S., Bagheri, R., Griffiths, M.D., Mohd Hashima, I.H.: Internet addiction and maladaptive schemas: the potential role of disconnection/rejection and impaired autonomy/performance. Clin. Psychol. Psychother. **28**(6), 1509–1524 (2021)
61. Ramsay, J.: Sentience and transcendence: personal reflections on physical and mental adaptation from a near-death event and life-limiting surgery. BJPsych Bulletin **45**(3), 163–167 (2021)
62. Romportl, J.: Naturalness of Artificial Intelligence. In: Beyond Artificial Intelligence: The Disappearing Human-Machine Divide, pp. 211–216 (2015)
63. Stokel-Walker, C., Van Noorden, R.: The promise and peril of generative AI. Nature **614** (1), 214–216 (2023)
64. Shibuya, K.: Digital Transformation of Identity in the Age of Artificial Intelligence. No. 172558. Springer, Singapore (2020)
65. Slater, M., Neyret, S., Johnston, T., Iruretagoyena, G., Crespo, M.Á.D.L.C., Alabèrnia-Segura, M., Spanlang, B., Feixas, G.: An experimental study of a virtual reality counselling paradigm using embodied self-dialogue. Sci. Rep. **9**(1), 10903 (2019)
66. Sloman, A.: What sort of architecture is required for a human-like agent? In: Foundations of Rational Agency, pp. 35–52. Springer Netherlands, Dordrecht (1999)
67. Smith, R.: A good death: an important aim for health services and for us all. BMJ **320**(7228), 129–130 (2000)
68. Solanki, P., Grundy, J., Hussain, W.: Operationalizing ethics in artificial intelligence for healthcare: a framework for AI developers. AI Ethics **3**(1), 223–240 (2023)
69. Smuha, N., De Ketelaere, M., Coeckelbergh, M., Dewitte, P., Poullet, Y.: We are not ready for manipulative AI—urgent need for action. Euractiv (2023)
70. Sonne, J.W.H., Gash, D.M.: Psychopathy to altruism: neurobiology of the selfish–selfless spectrum. Front. Psychol. **9**, 575 (2018)
71. Tei, S., Fujino, J., Kawada, R., Jankowski, K.F., Kauppi, J.-P., van den Bos, W., Abe, N., Sugihara, G., Miyata, J., Murai, T., Takahashi, H.: Collaborative roles of temporoparietal junction and dorsolateral prefrontal cortex in different types of behavioral flexibility. Sci. Rep. **7**(1), 6415 (2017)
72. Tei, S.: Grace is gone: directed by James Strouse. J. Palliat. Med.Palliat. Med. **26**(1), 153–153 (2023)

73. Tei, S., Fujino, J.: Neuropsychological features of gaming disorder and psychiatric comorbidities. In: Encyclopedia of Information Science and Technology, 6th edn, pp. 1–18. IGI Global (2025)
74. Tei, S., Kauppi, J.-P., Jankowski, K.F., Fujino, J., Monti, R.P., Tohka, J., Abe, N., Murai, T., Takahashi, H., Hari, R.: Brain and behavioral alterations in subjects with social anxiety dominated by empathic embarrassment. Proc. Natl. Acad. Sci. **117**(8), 4385–4391 (2020)
75. Tei, S., Fujino, J.: The educational value of sense of coherence for grief care. Front. Psychol. **13**, 1037637 (2022a)
76. Tei, S., Fujino, J.: Social ties, fears and bias during the COVID-19 pandemic: fragile and flexible mindsets. Hum. Soc. Sci. Commun. **9** (1) (2022b)
77. Tei, S., Fujino, J.: Adverse effects of empathy and cognitive inflexibility on social trauma. Front. Psychol. **14**, 1090297 (2023a)
78. Tei, S., Fujino, J.: Phantasm and suicide—psychiatry in movies. Br. J. Psychiatry **223** (5), 517–517 (2023b)
79. Tei, S., Fujino, J., Murai, T.: Navigating the self online. Front. Psychol. 1499039 (2025)
80. Thompson, E.: Empathy and consciousness. J. Conscious. Stud. **8**(5–6), 1–32 (2001)
81. Thompson, E.: The ultimate transformative experience. In: Becoming Someone New: Essays on Transformative Experience, Choice, and Change, pp. 269 (2020)
82. Turrigiano, G.: Homeostatic signaling: the positive side of negative feedback. Curr. Opin. Neurobiol.. Opin. Neurobiol. **17**(3), 318–324 (2007)
83. Tversky, A., Kahneman, D.: Judgment under uncertainty: heuristics and biases: biases in judgments reveal some heuristics of thinking under uncertainty. Science **185**(4157), 1124–1131 (1974)
84. Tzu, G., Bannerman, B., Griffith, S.: Losing myself in non-dual awakening: a transpersonal phenomenological investigation. Int. J. Ment. Heal. Addict.Ment. Heal. Addict. **13**, 758–775 (2015)
85. Tzu, C., Yu-lan, F.: On the equality of things. In: Chuang-Tzu: A New Selected Translation with an Exposition of the Philosophy of Kuo Hsiang. Springer Berlin Heidelberg (2016)
86. Varela, F.J.: Organism: a meshwork of selfless selves. In: Organism and the Origins of Self, pp. 79–107. Springer Netherlands, Dordrecht (1991)
87. Wang, G., Phan, T.V., Li, S., Wang, J., Peng, Y., Chen, G., Junle, Q., Goldman, D.I., Levin, S.A., Pienta, K., Amend, S., Austin, R.H., Liu, L.: Robots as models of evolving systems. Proc. Natl. Acad. Sci. **119**(12), e2120019119 (2022)
88. Weinstein, A., Livny, A., Weizman, A.: New developments in brain research of internet and gaming disorder. Neurosci. Biobehav. Rev.. Biobehav. Rev. **75**, 314–330 (2017)
89. Whitehead, P.R.: The lived experience of physicians dealing with patient death. BMJ Support. Palliat. CarePalliat. Care **4**(3), 271–276 (2014)
90. Wijdicks, E.F.M.: Brain death worldwide: accepted fact but no global consensus in diagnostic criteria. Neurology **58**(1), 20–25 (2002)
91. Wright, A.A., Zhang, B., Ray, A., Mack, J.W., Trice, E., Balboni, T., Mitchell, S.L., Jackson, V.A., Block, S.D., Maciejewski, P.K., Prigerson, H.G.: Associations between end-of-life discussions, patient mental health, medical care near death, and caregiver bereavement adjustment. JAMA **300**(14), 1665–1673 (2008)
92. Yang, W., Staps, T., Hijmans, E.: Existential crisis and the awareness of dying: The role of meaning and spirituality. OMEGA-J. Death Dying **61**(1), 53–69 (2010)
93. Yao, L., et al.: Brain network alterations in mobile phone use problem severity: a multimodal neuroimaging analysis. J. Behav. Addict. (2025)
94. Zahavi, D.: Husserl's intersubjective transformation of transcendental philosophy. J. Br. Soc. Phenomenol.Phenomenol. **27**(3), 228–245 (1996)
95. Zahavi, D.: Self and Other: Exploring Subjectivity, Empathy, and Shame. Oxford University Press, USA (2014)
96. Zilles, K., Amunts, K.: Individual variability is not noise. Trends Cogn. Sci.Cogn. Sci. **17**(4), 153–155 (2013)

97. Zhou, Z., Yuan, G., Yao, J.: Cognitive biases toward Internet game-related pictures and executive deficits in individuals with an Internet game addiction. PLoS ONE **7**(11), e48961 (2012)

**Shisei Tei** is a researcher in the Department of Psychiatry at Kyoto University, Japan, and a professor at Tokyo International University. His scholarly pursuits delve into intricate facets of human cognition, with a particular focus on empathy, self-consciousness, and behavioral flexibility.

# Future and Transformative Perspectives

# An Interview with Neil Harbisson: Life, Death, and the Art of Being a Cyborg

**Jordi Vallverdú**

**Abstract** This chapter presents an intimate and philosophical conversation with Neil Harbisson, the world's first legally recognized cyborg artist. Set against the quiet backdrop of a spring morning in Mataró, the interview explores profound themes of identity, mortality, sensory perception, and the fusion of technology with biology. Harbisson reflects on his two-decade journey as a cyborg, discussing the emotional and existential dimensions of living with an antenna that allows him to perceive color through sound. The dialogue delves into his views on death—from the decay of the organic body to the loss of cybernetic senses—and the concept of technological organ transplants. Harbisson also shares his experiments with new senses, such as time perception via Earth's rotation, and his evolving sense of trans-species identity. More than a discussion on technology, the chapter reveals how art, science, and selfhood merge in Harbisson's life, offering a deeply human perspective on what it means to extend consciousness beyond traditional biological limits.

It was a quiet spring morning in Mataró, where the gentle hum of market stalls opening and the distant sound of sparrows framed our meeting. Neil Harbisson, the world's first officially recognized cyborg artist, joined me at a small café in the market square. With his antenna arching gracefully from his skull, he greeted familiar faces with the ease of someone deeply connected to his surroundings. Our conversation felt intimate, a thoughtful exploration of identity, mortality, and the future of humanity.

I began recalling his recent celebration of his *Cyborg Birthday*, marking March 22, 2004, when Neil became more than human.

**Jordi Vallverdú**: "*Neil, you've now spent twenty years as a cyborg. How do you think about death—not just as a human, but as someone whose identity includes both biology and technology?*".

Neil's gaze softened, reflective. "I think about it a lot," he said, matter-of-factly. "For me, there are layers of death. There's the death of my organic body, of course, but

J. Vallverdú (✉)
Philosophy Department, Universitat Autònoma de Barcelona (ICREA-UAB), Bellaterra, Spain
e-mail: jordi.vallverdu@uab.cat

 
A. Sans Pinillos et al. (eds.), *SecondDeath*, Studies in Applied Philosophy, Epistemology and Rational Ethics 76, https://doi.org/10.1007/978-3-031-98808-0_17

there's also the death of my cybernetic body. Technology could fail while I'm still alive, or my body could die and the technology might keep working. Then there's a more profound kind of death: the loss of a sense."

He paused, considering his words. "Losing the ability to perceive color would be the deepest death for me. The antenna could stop functioning, but if I lose that sense entirely—it would be like going blind. It's not just about hardware breaking; it's about losing a connection to the world."

This notion—the death of a sense—hung between us. Neil's antenna, implanted into his skull nearly two decades ago, allows him to perceive colors through sound vibrations. What was once an experiment had become a part of his identity. "The antenna isn't just a tool anymore," he said. "It's part of who I am. Losing it would be losing a way of experiencing reality."

**Jordi Vallverdú**: "*You've spoken about updating and modifying yourself. How do these ongoing changes affect your understanding of mortality?.*"

Neil smiled slightly, as if revealing a secret. "Right now, I'm experimenting with a new sense—a sense of time. I've connected an exosense to feel the Earth's rotation. My nose aligns with Greenwich, the Prime Meridian. I can sense where the sun is at any moment through the rotation of the planet."

He chuckled softly. "It's strange to experience time this way, not through clocks but through the rhythms of the Earth itself. I'm curious to see if I can alter my perception of time—speed it up or slow it down—by adjusting this sense."

Neil reflected further, leaning in with a quiet intensity. "I often think about the concept of organ transplants, but not just for biology—for technology. Imagine donating not just a kidney or a heart, but a sensory organ like my antenna. Someone else could receive it and perceive the world as I do. It's like passing on a sense, a legacy that goes beyond art or biology. The idea of someone continuing my work through my antenna fascinates me—it feels like a new form of immortality."

The possibilities were staggering. Neil wasn't merely adding features to his body; he was aligning himself with the cosmos, redefining what it means to be human. For Neil, this was not just technological experimentation but an exploration of trans-species identity. "I don't see myself as fully human anymore," he admitted. "I'm part human, part technology—a trans-species being. My senses connect me to other species, like my cat or bees, allowing me to experience their reality. For instance, my cat sometimes stares at a wall, and most people think he's staring at nothing. But I know he's reacting to infrared light, which I can see, too. It feels like we're sharing the same invisible world, experiencing a hidden reality together.

The same happens with bees. When I observe them choosing flowers, I can see ultraviolet patterns that humans can't. These patterns guide the bees, and when I perceive them, it feels like I'm stepping into their world. It's not about leaving humanity behind but about expanding it. I feel connected to other species through their senses, and that's where the concept of trans-species identity comes alive for me."".

**Jordi Vallverdú**: "*You once mentioned wanting your antenna buried with you. Is that still your plan?.*"

> "I've thought about this a lot," Neil admitted. "I could be buried with my antenna, and it might still pick up signals, functioning even after I'm gone. Or I could donate it, like an organ. Imagine an antenna transplant—someone else using it to continue my work, experiencing the world through my creation. I love the idea that my antenna could live on in another body."

He smiled, describing the poetic vision of his antenna connected to a solar panel in his grave. "Even buried, it could keep receiving sunlight. It's a symbolic continuation of life."

Our conversation turned deeply personal as Neil shared the emotional transformation brought on by his cyborg journey. "After the surgery to implant the antenna, I became more sensitive to the world," he said softly. "I remember simple moments, like watching a grandfather hold his grandchild's hand—I'd tear up. Everything felt more fragile and more beautiful. It was as though the world itself had become delicate."

Neil's reflections carried an almost spiritual weight on me, connecting with my Asian philosophical interests, where this awareness of impermanence it's like a kind of transcendence. Even the senses inherited or created, are temporary. Everything is.

**Jordi Vallverdú**: "*It sounds like technology hasn't distanced you from life—it's made you feel it more deeply.*"

Neil nodded. "For me, the antenna is art. It's a sculpture, yes, but it's also a sensory organ. The trepanation, the surgery—these were steps in creating something new. Art isn't just about objects; it's about creating new ways of experiencing the world. That's what this is."

He added, "But it's not for everyone. For me, it's about expanding the senses, about pushing the boundaries of what it means to be human. Others may join in, but it's not about the technology itself—it's about how you engage with the world through it."

By merging art, science, and identity, Neil Harbisson had crafted not only a new way of sensing the world but a legacy that transcends life and death. As we sipped the last of our drinks, I couldn't help but think of his words—a life lived as a form of art, fragile yet eternal, ever evolving.

There was an intensity to the conversation, a feeling that I had been speaking with someone whose existence blurred the lines between life, art, and technology. Neil's reflections were not detached but profoundly human, grounded in a unique way of being—a trans-species perspective that feels both futuristic and timeless. As we parted ways, I carried with me a sense of awe for his journey: a life interwoven with technology, yet deeply sensitive to the fragility and beauty of existence itself. It was clear that Neil Harbisson is not merely an artist or a cyborg but a pioneer of a new way of living—one that asks us to rethink our relationships with life, death, and the unseen rhythms of the world around us.

**Funding** Jordi Vallverdú was supported by the ICREA 2019 and PAL-GAIT (PID2023-148336NB-100) grants.

# An Interview with Kevin Warwick: Cyborg Identity, Techno-death, and the Future of Being Human

Jordi Vallverdú

It was one of those golden afternoons in Barcelona, the kind that makes the city feel timeless, as though the streets themselves are suspended in warmth and light. I had arranged a Skype call with Professor Kevin Warwick, a man whose name had become synonymous with the future—a pioneer in cybernetics, and most famously, the world's first official scientific cyborg. As I sat at my desk, watching the late afternoon sunlight stretch across the room, I felt the weight of what we were about to discuss. This wasn't just a conversation about science or technology. We were going to explore life, death, and what lies between them in a world where the lines between biology and technology are becoming ever more blurred.

The call connected after a brief technical delay, and Kevin's familiar face appeared on the screen. Despite the years that had passed since our last meeting, there was something instantly recognizable about his calm but alert demeanor, a presence that always seemed to contain both scientific rigor and human curiosity in equal measure.

After exchanging a few pleasantries about technology—Kevin joked about Skype now being a Microsoft property—we quickly settled into the heart of our conversation. I told him about the book I was working on, a project exploring "techno-death"—a concept that I had developed to investigate how advancements in technology might fundamentally alter our understanding of life and death. I explained that I had spoken to philosophers, sociologists, and engineers, but that his experience as the world's first official cyborg added a unique perspective to the discussion.

"You're not just theorizing, Kevin," I said, leaning slightly forward. "You've lived this. You've upgraded your body, and in doing so, you've pushed the boundaries of what it means to be human. How has that experience changed your understanding of mortality? Has it made you think differently about death?".

Warwick's brow furrowed ever so slightly as he considered the question. "I haven't really thought much about death," he admitted with a candidness that caught me off

J. Vallverdú (✉)
Philosophy Department, Universitat Autònoma de Barcelona (ICREA-UAB), Bellaterra, Spain
e-mail: Jordi.vallverdu@uab.cat

 
A. Sans Pinillos et al. (eds.), *SecondDeath*, Studies in Applied Philosophy, Epistemology and Rational Ethics 76, https://doi.org/10.1007/978-3-031-98808-0_18

guard. "I mean, in terms of longevity and health, yes, technology plays a massive role. Take Parkinson's disease, for example. With implants, people's lives are dramatically improved. They can live better, maybe even longer. But death itself? I'm not sure technology changes much in that regard—at least not yet."

His words were surprisingly grounded, especially for someone who had physically integrated technology into his body. But as we delved deeper, it became clear that the implications of his work touched on far more profound aspects of the human experience. Warwick wasn't just focused on the mechanics of life extension. His work brought him face-to-face with the deeper, existential questions about what it means to be human when parts of our bodies and minds are no longer biological.

"Would you still be you," he asked, more rhetorically than directly, "if the biological part of you ceased to exist but the digital part continued?".

The question hung in the air for a moment, and I realized just how deeply Warwick had thought about these issues. For him, this wasn't just a philosophical exercise. He had lived with technology woven into his body, and he had experienced what it felt like to lose those extensions. He wasn't speaking in abstractions; he was talking about his own life, his own identity.

"I spoke with Neil Harbisson recently," I mentioned, referencing another pioneer of human-technology integration, "and he told me about his antenna, how it's become a permanent part of him. He's thought about what happens when he dies—whether the antenna continues to function, or whether it's buried with him. But you had your implants removed. How did that feel?".

Warwick leaned back slightly, a trace of a smile crossing his face as he remembered. "The first implant was pretty basic," he began. "It was just an identification chip. When I had it, doors would automatically open for me, lights would turn on when I entered a room. It was almost as if the building recognized me as part of its system. When the implant was removed, it was… strange. I'd walk up to a door, expecting it to open, and nothing would happen. The lights stayed off. It wasn't just about the convenience—I had become used to the idea that the world interacted with me in a different way. When that stopped, it felt like something was missing."

This description struck me. Warwick wasn't just talking about losing a tool or a gadget—he was talking about losing a part of how he interacted with the world, a subtle but profound shift in his experience of reality. "Did it feel like a loss of identity?" I asked, wondering if that sense of disconnection went deeper than just the practical inconvenience of closed doors.

Warwick considered this for a moment. "Not quite a loss of identity," he said carefully. "But it was definitely a shift in how I moved through the world. I had to remind myself that the doors weren't supposed to open anymore, that the lights weren't supposed to turn on. It's funny how quickly we adapt to new technologies and how easily we come to rely on them."

This was a theme that would continue to emerge throughout our conversation—the idea that technology, once integrated into our lives, becomes more than just a tool. It becomes part of who we are, part of how we understand ourselves and the world around us.

But it was the second implant—the one that linked directly to his nervous system—that truly fascinated me. I had read about this experiment, and I wanted to know more, not just about the science behind it but about how it felt to live with such a profound integration of technology into the body.

"That one was different," Warwick said, his tone shifting slightly. "It was connected directly to my nervous system. In the lab, it worked. I could control devices just by thinking about them. But at home, it was a bit of a different story. The implant wasn't always active, but it was still there, part of my body. I had to be careful not to damage it. Showering, for example, became a logistical challenge—I had to keep my arm out of the water to avoid short-circuiting my own nervous system."

He smiled slightly as he recounted this, but there was a deeper truth in his words. For Warwick, being a cyborg wasn't just about enhanced abilities. It was also about vulnerability—the fragility of a body that had become dependent on technology.

"Did you ever feel like the technology became part of your identity?" I asked, curious to know if he ever reached a point where the loss of the implant would have felt like losing a part of himself.

"If I'd had it for long enough, yes," he said without hesitation. "I think it would have become part of me, part of my sense of self. If I had lost it after that point, it would have been like losing a limb or a sense—something fundamental to how I interact with the world."

As he spoke, I realized that Warwick was touching on something deeply human—the idea that our sense of identity is tied to the tools we use, the technologies we integrate into our lives. His experience as a cyborg wasn't just about scientific experimentation. It was about exploring the boundaries of human identity and what happens when those boundaries begin to blur.

Our conversation naturally shifted toward the future of human communication—a topic Warwick had explored extensively in his work. "Imagine if we could communicate directly, brain to brain, without the need for speech," he said, his tone taking on a note of excitement. "No more clumsy words, no more misunderstandings. Just pure, direct thought. It would change everything."

I was struck by the radical nature of the idea. "But wouldn't that be terrifying?" I asked, half-joking but also half-serious. "What about privacy? What about the things we don't want to share?".

Warwick smiled, but his expression grew more serious. "It is scary," he admitted. "But our brains are mercenary. If they like something, they'll go for it. And the benefits of direct communication—imagine never having to worry about being misunderstood—are just too great to ignore."

We spent several minutes discussing the implications of this idea, touching on the ethical, psychological, and even philosophical ramifications. For Warwick, the future of human communication was not just a technological challenge—it was a fundamental shift in how we understand ourselves and our relationships with others.

"Ethics and commercial interests often clash," he said, his voice taking on a more reflective tone. "We don't always know the ethical questions until we're already in the middle of the new technology. But we always move forward, even when we're not sure where it's going to lead."

There was a certain inevitability in his words, a recognition that technological progress doesn't wait for us to figure out the answers. We push ahead, and only afterward do we begin to understand the full implications of what we've done.

As our conversation began to wind down, I couldn't help but feel that we had only scratched the surface of these profound questions. Warwick's experiences as a cyborg were both deeply personal and deeply philosophical, raising questions about identity, mortality, and what it means to be human in a world where technology can extend and transform our bodies and minds.

Before we ended the call, I asked him one final question: "Do you think we'll see more people following in your footsteps, embracing cybernetic enhancements?".

Warwick paused, his expression thoughtful. "I thought we would," he said, a note of disappointment in his voice. "When I first did these experiments, I fully expected that others would pick up where I left off, that this would become a field of serious academic and practical exploration. But it hasn't happened. Maybe people are afraid. Maybe the risks are too great. I don't have a clear answer."

His words left me with more questions than answers, but perhaps that was the point. Kevin Warwick's journey as a cyborg was not just about pushing the boundaries of technology. It was about exploring the very essence of what it means to be human in a world that is constantly changing.

As I ended the call and stepped out into the warm Barcelona afternoon, I couldn't shake the feeling that I had been given a glimpse into the future—a future where the lines between biology and technology are not just blurred, but erased entirely. And in that future, the question of what it means to live—and to die—takes on a whole new meaning.

**Funding** Jordi Vallverdú was supported by the ICREA 2019 and PAL-GAIT (PID2023-148336NB-100) grants.

MIX
Papier aus verantwortungsvollen Quellen
Paper from responsible sources
FSC® C105338

If you have any concerns about our products,
you can contact us on
**ProductSafety@springernature.com**

In case Publisher is established outside the EU,
the EU authorized representative is:
**Springer Nature Customer Service Center GmbH**
**Europaplatz 3, 69115 Heidelberg, Germany**

Printed by Libri Plureos GmbH
in Hamburg, Germany